Satellite Remote Sensing in Hydrological Data Assimilation

Mehdi Khaki

Satellite Remote Sensing in Hydrological Data Assimilation

Springer

Mehdi Khaki
Civil, Surveying, Environmental
Engineering
University of Newcastle Australia
Newcastle, Australia

ISBN 978-3-030-37374-0 ISBN 978-3-030-37375-7 (eBook)
https://doi.org/10.1007/978-3-030-37375-7

This Springer imprint is published by the registered company Springer Nature Switzerland AG
The registered company address is: Gewerbestrasse 11, 6330 Cham, Switzerland

Preface

Development of satellite remote sensing data has provided a unique opportunity to improve our understanding of Earth and its hydrological processes. This is especially of interest due to diverse types of satellite platforms with various measurement sensors offering vast coverage with unprecedented spatial and temporal resolution data. Such valuable datasets allow us not only to observe hydrologic components directly from the space but also to improve hydrologic models, which are designed to study the Earth's hydrology. Whereas the former is limited only to the observed components, the latter can enhance multiple model outputs through advanced statistical and numerical methods, which are generally called data assimilation. The approach can effectively be used to address models' drawbacks due to several reasons such as uncertainties in data and parameters, conceptual simplifications of complex physical processes and errors in forcing datasets. Data assimilation constrains the models using new observations based on uncertainties associated with both model and data. This approach has gained a lot of interests in hydrology, especially with the growth of satellite remote sensing data in recent years. The aim of this book is to present the fundamental of data assimilation in hydrology and review the application of satellite remote sensing for hydrological data assimilation. It presents data assimilation background in hydrology, separably discusses remotely sensed data and assimilation methods, and finally describes some of data assimilation applications in hydrology.

The first chapter (Chap. 1) contains a comprehensive review of data assimilation in hydrology and new opportunities offered by satellite remote sensing in the field. This is investigated in more detail in Chap. 2, in which models and potential data products for hydrologic data assimilation are described. Popular data assimilation filtering methods in hydrology are discussed and compared in Chap. 3. Various data assimilation techniques from the sequential methods are tested to investigate their capabilities in hydrological data assimilation. Different aspects of these filtering techniques, e.g., state estimate improvement, forecasting, uncertainty quantification and stability during the process are examined. In Chap. 4, satellite data assimilation and specifically the Gravity Recovery And Climate Experiment (GRACE) Terrestrial Water Storage (TWS) assimilation are investigated due to its recent developments and popularity in hydrology. Satellite data assimilation can remarkably improve hydrological models by providing information that is not limited to the Earth's surface, specific time and locations. The impacts of tuning techniques are also analysed to maximize the effect of remote sensing observation during data assimilation. Most recent advances for this objective are also explained.

Chapter 5 presents one of the well-known challenges for implementing data assimilation in hydrology and more specifically water balance systems, which refers to the inconsistency between hydrological water fluxes as a result of the data assimilation. This chapter benefits readers by deeply exploring the issue on a basin scale using two newly proposed data assimilation methods. These constrained filtering methods are introduced due to their ability to preserve fundamental rules underlying the models while assimilating new observations. Chapter 6 explores the development of data-driven approaches in hydrology for integrating satellite products with model outputs. This is especially of interest due to some recent progress with the techniques as well as their capability for simplifying the system dynamics, which can lead to more efficient data integration. In particular, data-driven approaches can be helpful when models are subject to large errors. Lastly, in Chap. 7, the application of (multi) satellite data assimilation for improving hydrologic models is explored over different globally distributed regions. The performance of the process for improving our information regarding the surface and sub-surface water resources is reviewed. These are usually difficult to be captured by models due to extreme climate-induced and anthropogenic impacts, which are extensively explained in this chapter.

Newcastle, Australia Mehdi Khaki

Summary of Chapters

Part I—Hydrological Data Assimilation

Hydrological models are important for monitoring water resources around the globe, which is essential for sustainable water management in different socioeconomic sectors. The models can be used to study water changes in different forms such as surface water, groundwater and soil moisture. Nevertheless, due to various factors, their performance can be degraded. Satellite remote sensing data assimilation is an effective approach for improving the models' simulations. This chapter starts by giving an introduction into the hydrological modelling and the existing challenges for an accurate simulation of the Earth hydrological processes. It then discusses the usefulness of satellite remote sensing for studying such processes, particularly through data assimilation. The importance of satellite data assimilation in the field of hydrology is then described by reviewing the opportunities that both advanced data assimilation techniques and satellite remotely sensed products have offered in the past few decades. The chapter also exemplifies the associated difficulties with the model-data integration in hydrology, which require special considerations when applying the method.

Part II—Model-Data

Ever-increasing computational power in computers during the last few decades has revolutionized our perception of Earth hydrology through models. These models are established to (partially) reflect the natural or human-induced hydrological processes, and further to predict system behaviours. Detailed explanations about hydrological models and their inputs are presented in this chapter. Specifically, the impact of satellite data development for enhanced hydrological modelling is discussed. Besides, independent remote sensing techniques that can be used to further improve models are described. Today, satellite platforms provide high spatial and temporal resolution datasets with global coverage that can be used for data assimilation objectives. Various datasets are currently available for this objective thanks to advanced satellite techniques. These include but not limited to the Gravity Recovery And Climate Experiment (GRACE) data, satellite soil moisture products from active and passive sensors, satellite-derived evaporation, Leaf Area Index (LAI), surface water extent observations, and snow and ice cover products. Assimilation of these products can improve the accuracy of hydrological models for not only better simulations but also for better forecasting.

Part III—Data Assimilation Filters

Various filtering techniques have been proposed and developed in various fields for data assimilation objectives. Among these, variational methods and sequential filtering have gained a lot of attention during the past few decades. While the former is more popular in the fields of atmospheric and oceanic sciences, sequential methods have largely been applied for hydrologic applications because of their reasonable computational requirements. This allows them to potentially be a better data assimilation method for high-dimensional systems as in hydrology. Besides, sequential characteristic of these methods eliminates the requirement for preserving historical states for data assimilation, which indeed could effectively increase the computational efficiency. Different types of sequential data assimilation filters and their characteristics for assimilating satellite observations into hydrological models are described in this chapter. Investigating these are important for a successful data assimilation to not only improve estimations of the system state but also to accurately quantify the system uncertainties and keep the model dynamically stable during the assimilation process.

Part IV—GRACE Data Assimilation

Assimilation of remotely sensed observations can be a challenging task. For efficient assimilation of these observations, in addition to choosing the right filtering method, it is important to effectively deal with their uncertainty during the process. For some satellite remote sensing such as the Gravity Recovery And Climate Experiment (GRACE), this is more important due to the unique characteristics of their data products, e.g., the coarser spatio-temporal resolution compared to most of the existing hydrological models. Correlations in observations can also cause the covariance matrices to be badly conditioned or not invertible leading to inefficiency of the filtering process. Therefore, a special focus should be invested to account for these problems during data assimilation. Rigorous modelling of the observation covariance matrices is required to maximize the contribution of observations into model state realistically. All these make remote sensing data assimilation difficult.

To address this, one not only requires an appropriate filtering method but also needs a robust assimilation strategy for dealing with different aspects of datasets and their inconsistency with models. This chapter describes an extensive framework for assimilating GRACE TWS data, which provides valuable information for assimilating satellite products in land surface modelling.

Part V—Water Budget Constraint

One of the important factors when assimilating new observations into land surface models, more specifically into water balance systems, is to preserve the balance between water fluxes after each assimilation cycle. Assimilation of observations, e.g., from satellite platforms can introduce artefacts in (non-assimilated) system states. This can break the underlying balance between water components in the model equations. To mitigate this, one should apply more sophisticated data assimilation approaches and involve additional observations such as water flux observations. For efficient data assimilation, uncertainties associated with these observations should also be included in the process. The modification to a standard data assimilation filter for retrieving the broken balance is usually imposed as a constraint to the process based on water balance equation. This constraint is applied at each assimilation cycle after a standard filtering step. In other words, the additional step appears as a second update step in a sequential filtering scheme. In this chapter, existing constrained data assimilation approaches are reviewed and details of two recently proposed approaches to deal with the water balance problem will be presented.

Part VI—Data-Driven Approach

Classical data assimilation methods, despite their success in hydrologic modelling, have their limitations. These include, but not limited to, their requirement of intensive computations for high-dimensional systems, as well as the availability of a physical model. These methods cannot be applied when the model is either unavailable or partially available. In such cases, data-driven (non-parametric) approaches can be used, which replace the actual model with a data-driven non-parametric reconstruction of the system's dynamic, usually based on the system trajectories. This method can sometimes reduce process complexity leading to more efficient filtering. The recently proposed data-driven approach for hydrologic applications is proposed in the chapter. The approach is based on the Kalman-Takens method to reconstruct the model for more complicated state observation transition systems in hydrology. Various aspects of the method such as the computational efficiency, its ability to effective sampling and stability during the process are discussed. All these factors are important to determine an effective assimilation method. Furthermore, detailed comparison between the data-driven approach and standard data assimilation techniques is presented in this chapter.

Part VII—Hydrologic Applications

The application of data assimilation can improve our understanding of hydrological processes over different parts of the world. This is particularly of interest over areas with limited datasets and especially those with water scarcity problems, which require accurate information of available water resources and their changes for future planning. Improving hydrological model simulations via data assimilation can provide this information. The approach allows for introducing missing trend and climate-related extreme events into hydrologic water compartments. It can also be used to downscale relatively coarse resolution data such as GRACE TWS by merging them into the models, and further to separate its aggregated water storage information into different water compartments based on the physical processes implemented in the model equations. This makes it possible to separately analyse various water compartments, e.g., soil moisture and groundwater, using the improved estimates. In this chapter, the application of data assimilation is investigated over various scales and different parts of the world. This is particularly of interest due to the fact that every region has unique characteristics in terms of hydrological processes, climate variabilities, human impacts on the environment and geographic properties, which make the data assimilation a challenging task. Its success, thus, depends on taking into an account various factors such as treating unmodelled (or poorly modelled) phenomena and considering hydroclimate connectivity while applying the process. This, generally, requires the use of additional information (e.g., for non-assimilated components) such as climate-related observations (e.g., precipitation and El Niño/Southern Oscillation) and data products related to anthropogenic impacts (e.g., irritation and human water use).

Contents

Part I
Hydrological Data Assimilation

Chapter 1
Introduction

1.1 Hydrologic Modelling, Challenges and Opportunities

Water resources are one of the most crucial essences of life on Earth. Due to its importance, studying water through hydrology has long been of attention. Any major changes in the availability, distribution, or interaction of water stored in the land can significantly affect the people locally as well as globally (Gain et al. 2012; Gosling and Arnell 2016). These changes can be imposed by various factors such as long-term climate change (e.g., Brekke et al. 2009; Cisneros et al. 2014), extreme climatic events (e.g., Beller-Simms et al. 2014; Nagy et al. 2018), and anthropogenic impacts (e.g., Sondergaard et al. 2007; Liyanage and Yamada 2017). Water storage can also be highly variable over different regions and through different seasons. Moreover, there is limited amount of fresh water available globally that can supply living and other socio-economic activities. Therefore, it is essential for sustainable water resource management to accurately monitor water storage variations and predict its changes. There are various tools that historically have been used to undertake these tasks including but not limited to deriving information for in-situ measurements, acquiring satellite remote sensing data, and applying hydrological models.

Monitoring water resources as in different forms such as surface water, groundwater, and soil moisture have been facilitated by developing hydrological models, which are used to simulate hydrological processes within the land (e.g., Devi et al. 2015; Sood and Smakhtin 2015; Gleason et al. 2018; Pérez-Sánchez et al. 2019). These models largely describe (simplified) physical processes underlying the Earth's hydrology using a set of mathematical formulations and parameters based on ingoing and outgoing energy, atmospheric components, and water fluxes. The models have gained a lot of attention in the past few decades due to recent developments in computational capacities, and also their increasing accuracy, high spatial and temporal coverage and resolutions, and their ability to represent multi-variate hydrological components in contrast to in-situ and satellite observations. The hydrological models have been used extensively to, for example, provide information of various water compartments on land, surface water flow, rainfall-runoff interactions, soil

M. Khaki, *Satellite Remote Sensing in Hydrological Data Assimilation*,
https://doi.org/10.1007/978-3-030-37375-7_1

moisture, water balance, and groundwater flow over a local region or globally. The use of hydrological models is particularly of interest over regions with limited to no ground-based measurements (e.g., Samaniego et al. 2011; Naschen et al. 2018; Khaki et al. 2018c). A majority of models are based on the water and energy balance to describe hydrological processes. The relationship between water fluxes and water storages are also reflected in most of the models.

Despite the unprecedented development of hydrological models and their ever-increasing complexity, they often pose to various source of errors that can affect their performance. The models largely rely on atmospheric forcing datasets as a driving force as well as various parameters to carry out simulation. Errors in inputs and forcing sets and any uncertainty associated with the parameters degrade the accuracy of the model outputs (see also Renard et al. 2010; Dong and Lu 2015; Teweldebrhan et al. 2018; Zhang and Al-Asadi 2019). Moreover, simplifications of physical processes and approximation within the models underlying mathematic formulas can introduce additional limitations and correspondingly negatively influence their simulations. Last but not least, limitations in the available input datasets, especially for global high-resolution modelling purposes have always been a major drawback factor for successful hydrological modelling. Many efforts have been invested to address these issues, e.g., through constant calibrating models' parameters, improving the quality, coverage, and resolution of inputs and atmospheric forcings, and making models even more complicated. Nevertheless, there are still uncertainties in the models' simulations that are difficult to deal with using these conventional approaches.

Alternatively, data assimilation approach can be used to improve the models' performance using new observations. The idea is to update model simulations when there is an observation of the same quantity available. This could push the simulations toward observations and potentially toward 'reality'. The approach has become more popular with the development of satellite remote sensing data, which provides valuable datasets of different hydrological components such as soil moisture, terrestrial water storage (TWS), canopy and snow cover (see, e.g., Moradkhani 2008; Reichle et al. 2008; Baguis and Roulin 2017; Helmert et al. 2018). The new information can be used to constrain the models' simulations and also their parameters through data assimilation. The approach can be applied to reduce the inconsistency between the observed variable in the model and its corresponding 'real' value by incorporating the incoming observations.

1.2 Data Assimilation

Data assimilation applies statistical/numerical processes to constrain the model simulations according to the available observations by taking into account uncertainty in both model and observations. This allows applying corrections at each assimilation step, i.e., whenever a new observation is available, realistically based on the accuracy of simulations and observations. In other world, these uncertainties determine the weight of observations and correspondingly the level of corrections to be applied to the models. This weighting process is done following the Bayesian approach (Hairer et al. 2005), which updates the probability distribution of system according to the observations. The process then uses the new (updated) probability distribution to derive following simulations until new observations become available for the next update assimilation cycle. In reality, however, these computations are complex because it is not possible to analytically derive the probability distribution function of the state anymore. This is even more difficult where a nonlinear or non-Gaussian system is present such those in hydrological models (see, e.g., Han and Li 2008; Anderson 2010; Vetra-Carvalho et al. 2018). In these cases, numerical solutions are used to solve for the Bayesian estimation problem.

Numerical solution of data assimilation, particularly with the advancement of high power computing systems, has helped the method to gain more popularity in various Earth science fields including oceanography, atmospheric sciences, and hydrology (e.g., Bouttier et al. 1993; Bauer et al. 1996; Houser et al. 1998; Reichle et al. 2001; Rodell et al. 2002; Lindskog et al. 2004; Balmaseda et al. 2007; Clark et al. 2008; Bormann et al. 2010; Edwards et al. 2015). A wide range of operational systems such as weather and water forecast have been using the approach not only to increase the accuracy of their modelling simulation but also to improve the reliability of their forecasts. In addition, data assimilation has been used in numerous researches for a better representation of historical simulations using a variety of observations from in-situ measurements, reanalysis data, and satellite remote sensing products. A number of works have also investigated different aspects of the numerical solutions and improved the method's efficiency and accuracy. High capability of data assimilation has made it an essential part of most of the modelling systems, especially in hydrology where there is a lot of new opportunities with the emerge of state-of-art satellite datasets being able to provide observations of various hydrological components.

1.3 Hydrological Data Assimilation

Data assimilation has been excessively used in the fields of ocean and atmospheric sciences (e.g., Bauer et al. 1996; Bennett 2002; Kalnay 2003; Lindskog et al. 2004; Balmaseda et al. 2007; Bormann et al. 2010; Hoteit et al. 2012; Edwards et al. 2015). In hydrological studies, however, this technique has been implemented more recently to increase the accuracy of hydrologic model simulations. The approach has been applied for different objectives and to enhance various hydrological components. Historically, data assimilation has been used in hydrology to improve streamflow simulation, especially in rainfall-runoff models and also to better monitor and predicts floods (e.g., Seo et al. 2003; Clark et al. 2008; Moradkhani and Sorooshian 2009; McMillan et al. 2013; Hutton et al. 2014; Ercolani and Castelli 2017). Different studies have used data assimilation to reduce errors in model simulations of soil moisture, especially over agricultural areas and some times to improve land-atmosphere interactions in a couple systems (e.g., Pauwels et al. 2002; Crow and

Wood 2003; Brocca et al. 2012; Alvarez-Garreton et al. 2015; Baguis and Roulin 2017; Maggioni and Houser 2017). Evapotranspiration data assimilation has been done mainly due to crucial impacts of evapotranspiration on terrestrial water cycles (e.g., Schuurmans et al. 2003; Xu et al. 2014; Lei et al. 2017). More recently, data assimilation has been applied in hydrology and land surface modelling to derive better simulations of surface water extent and its storage (e.g., Neal et al. 2009; Giustarini et al. 2011). With the availability of TWS data from the Gravity Recovery and Climate Experiment (GRACE), a number of studies have applied the dataset to update hydrological model estimates of various water compartments (e.g., Zaitchik et al. 2008; Schumacher et al. 2016; Kumar et al. 2016; Girotto et al. 2016, 2017; Khaki et al. 2017a; Shokri et al. 2019).

In addition to different datasets, different data assimilation techniques have also been implemented and tested in hydrological studies as well as hydrologic operational systems. These techniques range from a simple Nudging technique to more complicated data assimilation methods such as variational and sequential methods. Nudging (Newtonian relaxation) techniques add a term to the prognostic model equations resulting in the solution to be gradually relaxed towards the observations. A few other methods such as statistical and successive corrections have also been applied, which share common characteristics with Nudging technique. Variational methods including 3-D variational and 4-D variational (adding time to 3-D variational) have also been tested. Despite the popularity of these methods in the ocean and atmospheric sciences, they have not been used widely in hydrological applications mainly due to their reliance on the validity of the linearity assumption and executing a large adjoint model. The sequential techniques such as Ensemble Kalman Filter (EnKF) and Particle Filter (PF), on the other hand, have been frequently used for data assimilation objectives in hydrology. This could be explained by the fact that these filters are capable of dealing with high dimensional and non-linear models, as well as their efficient computational burdens.

Chapter 2
Data Assimilation and Remote Sensing Data

2.1 Satellite Remote Sensing, New Opportunities

Satellite remote sensing with a wide range of platforms and on-board sensors has changed our view of Earth and its hydrology remarkably. They offer various type of observations on large scales and now covering more than decades of measurements. The provided spatial and temporal resolutions have revolutionised the way we monitor hydrological processes. Satellite datasets have also provided valuable information over regions with no in-situ network or sparse ground-based measurements. The application of satellite datasets for studying hydrological components been successfully applied from regional to global scales with different temporal scales. For example these observations are used by scientists to monitor precipitation (e.g., Barrett et al. 1999; Alder et al. 2000; Huffman et al. 2007; Haile et al. 2010; Burdanowitz et al. 2015), evapotranspiration (e.g., Garatuza-Payan et al. 2003; Allen et al. 2007; Rafn et al. 2008; Glenn et al. 2011; Farg et al. 2012; Reyes-González et al. 2018), vegetation index (e.g., Bannari et al. 1995; Murray et al. 2009; Kamble et al. 2013; Roy et al. 2016), soil moisture (e.g., Njoku and Entekhabi 1996; Western et al. 2004; Schnur et al. 2010; Jonard et al. 2011; Li et al. 2013; Renzullo et al. 2014; Bartalis et al. 2018), water level height fluctuations (e.g., Frappart et al. 2005; Papa et al. 2006; Santos da Silva et al. 2010; Asadzadeh et al. 2013; Khaki et al. 2014), and more recently TWS changes (e.g., Landerer and Swenson 2012; Voss et al. 2013; Longuevergne et al. 2013; Awange et al. 2014; Forootan et al. 2014a; Yao et al. 2016).

Satellite remote sensing data has been extensively used for studying surface and even subsurface water storage changes. With a temporal resolution of one day to a few weeks and spatial resolution of meters to a few hundred kilometers, satellite products have been remarkably helpful for studying geophysical and hydrological variations. GRACE satellite mission, as an instance, has provided a great opportunity for studying water resources and more specifically TWS changes derived from variations of gravity fields. In addition to studying water compartments, GRACE-derived

M. Khaki, *Satellite Remote Sensing in Hydrological Data Assimilation*,
https://doi.org/10.1007/978-3-030-37375-7_2

measurements have been successfully applied to explore hydrological processes through water balance analyses, e.g., in the water balance models.

Based on these studies, one can clearly see great potential in satellite remote sensing data for improving hydrological models via data assimilation. They can be used to improve and calibrate models at regional, basin-scale, and global scales. They can also be applied separately or simultaneously to improve model simulations by providing measurements of real phenomena. This, however, comes with difficulties in reality as in many cases there are various discrepancies between model and observed data. Data assimilation provides a great tool with a variety of filtering and tuning methods that can increase the efficiency of the process while maximizing the impact of satellite observations on the model simulations.

2.2 Satellite Data Assimilation Challenges

The ability of satellite remote sensing for improving hydrological models has been proved by many studies. Nevertheless, successful assimilation of satellite remote sensing observations can be challenging due to a variety of factors. Models and satellite observations do not necessarily match in different aspects such as their different spatio-temporal resolutions. What's more, observations' error characteristics, and often non-linear model-observation transition system make data assimilation difficult. Furthermore, this process could become more difficult when satellite observations such as GRACE TWS data has a complicated error characteristic, which requires appropriate more sophisticated date merging algorithm. The success of the data assimilation process largely depends on accurate approximation of uncertainties associated with the model and particularly data. Any simplification or misinterpretation of these uncertainties can degrade the performance of data assimilation. Therefore, realistic assumption should be made to use full information of observation errors during the process to reach better outcomes. This, however, can introduce difficulties for data integration. In the case of GRACE TWS data, for example, the use of full error covariance information can be problematic as the matrix suffers from rank deficiency and thus makes the data assimilation complicated as it requires to deal with matrix inversions. These problems can be addressed by a robust data assimilation methodology.

In addition, assimilation of satellite data, while improving some components, can introduce artifacts to other hydrological components. For example, if not done carefully, satellite-derived soil moisture data assimilation may impose errors or unrealistic corrections to deep-zone soil moisture. Data assimilation can also break the existing water-energy balance within the models. Take GRACE TWS data assimilation as an instance. The process can negatively affect the balance between water fluxes, i.e., water storage changes, discharge, precipitation, and evaporation established by underlying equations in water balance models. GRACE TWS data assimilation can break this balance and therefore impose errors to the system. This will affect the models' simulations and the consistency between water storage changes on the one hand and water fluxes on the other hand. It is, therefore, essential to address this issue while assimilating observation in such systems. In other words, one should take any potential artefacts introduced by data assimilation into account for more efficient data integration. These issues have been seen or addressed in some hydrological data assimilation studies. Various schemes have been put forward to minimise artefacts through applying pre-processing filtering on satellite observations or introducing additional assimilation step to retrieve the imposed imbalance or unrealistic corrections by standard satellite data assimilation. These methods will be discussed in more details in Section V.

Part II
Model-Data

Chapter 3
Hydrologic Model

3.1 Background

Ever-increasing computational power in computers during the last few decades has revolutionized our perception of hydrology through models. These models are established to (partially) reflect the natural or human-induced hydrological processes (Dingman 2002), and further to predict system behaviours. There are different categorizations of hydrologic models in the science community. The main difference between these model is the way they either simulate or simplify a real system (e.g., surface water, soil water, wetland, and groundwater). This is mainly based on developed relationships between hydrological components (e.g., features, events, and processes) and applied various parameters that define the characteristics of the model. The more general classification of hydrologic models separates them into empirical, conceptual, statistical, and physically-based models. Analog models are non-mathematical approaches and they simulate processes that are used to represent a natural process. Conceptual models use entities (stores of water) and relationships between these entities (flows or fluxes between stores) to represent all of the hydrologic components. Statistical models are used in hydrology to describe data, as well as relationships between data, which are developed empirically by hydrologists using statistical methods (Wallis 1965).

According to Jajarmizadeh et al. (2012), physical models are a scaled-down form of reality and represent hydrologic systems using physical laws that have been obtained using scientific methods. The relationships between model components are specified using mathematical equations. These models then apply analytical or numerical procedures in logical steps to produce numerical outputs base on numerical inputs. The physical-based models are more popular these days due to advancements in numerical methods for representing hydrologic systems, computational capabilities and efficiencies, and satellite products for generating numerous model inputs. In a different definition, models are classified as deterministic and stochastic models. Deterministic models rely on constant mathematics that results in the same outputs for a single set of input values. This, however, is not the case in stochastic models as

M. Khaki, *Satellite Remote Sensing in Hydrological Data Assimilation*,
https://doi.org/10.1007/978-3-030-37375-7_3

a result of implemented random processes in one or more inputs over time. Various types of hydrologic models are available today such as land surface models, rainfall-runoff models, flood flow outlets and sewage systems, watershed models, streamflow routing models, and flood inundation models.

From all developed hydrologic models, land surface models have gained a lot of attention, particularly with the availability of unprecedented information about soil-water dynamics, plant physiology, micrometeorology and the hydrology that control biosphere-atmosphere relationships (Houser et al. 2012). These models mainly represent the interaction of mass, energy and momentum between a vegetated surface and the atmosphere (Sellers et al. 1986). They are often constructed in forms of lumped or distributed (or combined) as a function of space and time (Devi et al. 2015). Land surface models, developed in regional or global scales, also play a crucial role in atmospheric models in a coupled system providing valuable information on soil moisture as important inputs in forecast models. A majority of these models rely on water balance equation to describe landscape water stored in the land as the relationships between different water fluxes, i.e., changes in water storage, evaporation, precipitation, and water discharge (Sokolov and Chapman 1974). Simulating soil column as the fundamental hydrological unit is a common characteristic between various land surface models, which generally separate water storage into the topsoil, shallow soil and deep soil corresponding to tall, deep-rooted vegetation and short, shallow-rooted vegetation, leaf biomass, groundwater and surface water storages.

3.2 Forcing Observations

Meteorological forcing datasets are required to run physical-based hydrologic models, acting as a set of inputs. With the development of new satellite missions, remote sensing observations have been used broadly as forcing data. This, particularly, has become more popular more recently with more remote sensing platforms, specifically designed for hydrologic observations. Meteorological forcing datasets for land surface models vary but generally includes minimum and maximum temperature, surface radiation, and precipitation products. Land surface temperature measures the near-air temperature and is crucial to understand earth energy, water, and carbon cycles (Yao et al. 2013). This data can be provided from point-based measurements, retrievals, as well as satellite remote sensings such as Moderate Resolution Imaging Spectroradiometer (MODIS; Mostovoy et al. 2006) and National Oceanic and Atmospheric Administration (NOAA) Advanced Very-High-Resolution Radiometer (AVHRR) (Kilpatrick et al. 2001). Shortwave and longwave surface radiations are used to reflect the energy exchange between earth and atmosphere in the model by measuring total solar radiation, i.e., radiation that is reaching to the earth surface and backscattered to atmosphere. Similar to the air temperature, radiation data can be obtained from three main sources of ground observation, retrieval products, and satellites. There are multiple satellite platforms that can offer this dataset such as Nimbus-7 (Jacobowitz and Tighe 2004), Meteosat (Harries et al. 2005), and MODIS (Liang et al. 2006).

Perhaps the most important forcing component for land surface models is precipitation, which measures water content in any form that reaches the Earth's surface from clouds and is highly variable spatially and temporally (Michaelides et al. 2009; Yao et al. 2013). This necessitates the use of an accurate source of observations with the high spatiotemporal resolution, which can be offered by satellite remote sensing while available ground-based measurements are sparse in time and space. There are various sources to acquire satellite precipitation data from passive or microwave/IR sensors in, e.g., the Climate Prediction Center morphing method (CMORPH; Joyce et al. 2004), the Tropical Rainfall Measuring Mission (TRMM) Multi-satellite Precipitation Analysis (Huffman et al. 2007), the Global Satellite Mapping of Precipitation (GSMaP; Okamoto et al. 2005), and the GEO Operational Environmental Satellites (GOES) Precipitation Index (GPI; Arkin and Meisner 1987). In addition to these forcing sets, satellite remote sensing can further provide hydrologic observations that can be used for modelling, which include snow cover data, land cover, water vapour, and vegetation components, which are mostly being offered by recent satellite missions. The availability of such observations allow models to perform globally with even better resolutions.

Chapter 4
Remote Sensing for Assimilation

Various datasets are currently available to improve hydrologic models, and specifically land surface models thanks to advanced satellite developments for monitoring Earth's hydrology. Some of these have already been utilised for data assimilation objectives while there are more to be done in hydrologic modelling. One of the valuable data products is GRACE data, which can be provided from different sources such as the Centre for Space Research at University of Texas (CSR), the Jet Propulsion Laboratory (JPL) and the German Research Centre for Geosciences (GFZ). The data can be derived in a form of TWS, e.g., from Mass Concentration blocks (mascons) or spherical harmonics, which should be converted into TWS. In terms of spherical harmonics data, traditionally, low degree coefficients are replaced with more accurate estimates, e.g., degree 1 coefficients from Swenson et al. (2008) to account for the movement of the Earth's centre of mass and degree 2 and order 0 (C20) coefficients from Satellite Laser Ranging solutions (e.g., Chen et al. 2007). Furthermore, the post-glacial rebound should be considered, e.g., based on the estimated corrections by Geruo et al. (2013). There are also necessary filtering steps to account for the coloured/correlated noise in GRACE products. Furthermore, it is important to consider the spatiotemporal resolution of the data, which is usually coarser than models and make data assimilation challenging. TWS data can then be used to correct for the models' state estimates, i.e., the summation of various water compartments such as soil moisture at different layers, snow, vegetation, groundwater, and surface water storage as GRACE TWS measures total storage change.

Another important product for hydrologic data assimilation is satellite soil moisture, which usually measures top layer soil moisture (e.g., 0–5 cm) using passive or active sensors. The soil moisture measurements can be derived from different missions such as the Advanced Microwave Scanning Radiometer for EOS (AMSR-E), European Space Agency (ESA) Soil Moisture Ocean Salinity (SMOS), Tropical Rainfall Measuring Mission (TRMM), and Soil Moisture Active Passive (SMAP). There are, however, cautions that need to be taken into account when using these data products for data assimilation. Different platforms deliver soil

M. Khaki, *Satellite Remote Sensing in Hydrological Data Assimilation*,
https://doi.org/10.1007/978-3-030-37375-7_4

water content at different depths, and likely different than those of models so they can be used to update a fraction of model soil moisture via data assimilation. Soil moisture observations are mainly used to update model state variabilities rather than its absolute values. Moreover, there are sometimes bias and scale between model and satellite soil moisture observation, which should be considered, e.g., using cumulative distribution function (CDF) matching (Reichle and Koster 2004; Drusch et al. 2005) prior to data assimilation.

In addition, leaf area index (LAI) observations are now provided globally with very high resolutions that can be used to improve land surface models. For this purpose, NOAA Climate Data Record (CDR) of LAI from Advanced Very High Resolution Radiometer (AVHRR) and MODIS can be applied. LAI as well as Fraction of Absorbed Photosynthetically Active Radiation (FAPAR, an indicator of the presence and productivity of live vegetation) provided by NOAA, are important for hydrologic modelling as essential climate-related variables, which help to better understand the role of vegetation in the models (Claverie et al. 2016). LAI data can further be derived indirectly from the Normalized difference vegetation index (NDVI) datasets. Droogers and Kite (2002) showed that the use of LAI measurements improves the hydrologic modelling at different spatial scales. LAI variable plays a critical role in estimating the amount of precipitation interception and evapotranspiration (ET) and accordingly can be effectively used for data assimilation (Claverie et al. 2014). Furthermore, vegetation cover (also vegetation index) and canopy data from satellite remote sensing primary by AVHRR and MODIS can also be used to improve land surface models. Vegetation data based on remote sensing is useful for monitoring land surface conditions and analysing the influence of human activities and global or regional climate changes (Guo 2003).

Other land cover data such as snow and ice cover, inundation, and surface water extent can further be extremely valuable for hydrological modelling. There are various satellite missions such as Landsat Missions, Environmental Satellite (Envisat), MODIS, IceSat, AMSR, and AVHRR that provide land cover data using optical and microwave instruments with different spatial and temporal scales. Surface water extent data can particularly be applied to overcome one of the land surface models drawbacks with surface water compartments, mainly due to a lack of data with enough coverage. This dataset along with water level data from satellite radar altimetry, e.g., European Remote Sensing (ERS) satellites, ERS-1 and -2, TOPEX/POSEIDON, Jason missions, and Cryostat can be used to estimate surface water storage and consequently update the models' surface storage. The dataset can also be used for improving streamflow and hydrologic routing. There are, however, difficulties in dealing with satellite altimetry data due to their spatial resolution, which is limited to along-track observations, as well as errors for monitoring inland water. Satellite radar altimetry, originally designed to monitor sea level changes, nowadays are successfully used to monitor inland water bodies (see, e.g., Birkett 1995) and rivers (see e.g., Birkett et al. 2002; Berry et al. 2005; Tseng et al. 2013). Retracking algorithms can be used to improve the accuracy of measured ranges (or sea surface height; SSH) over inland waters (Brown 1977; Davis 1997; Khaki et al. 2014). Besides, with the launch of Surface Water and Ocean Topography (SWOT), jointly developed by NASA and Centre National D'Etudes Spatiales (CNES) with contributions from the Canadian Space Agency (CSA) and United Kingdom Space Agency, there will be new opportunities to use its terrestrial surface waters for hydrologic modelling.

Part III
Data Assimilation Filters

Various filtering techniques have been proposed and developed in various fields for data assimilation objectives. Among these, variational methods and sequential filtering have gained a lot of attention during the past few decades. While the former is more popular in the fields of atmospheric and oceanic sciences, sequential methods have largely been applied for hydrologic applications. This can be attributed to the fact that these methods do not require an adjoint and are becoming increasingly popular because of their reasonable computational requirements (Hoteit et al. 2002; Bertino et al. 2003; Robert et al. 2006). This allows them to be a better option for high dimensional systems as in hydrology. Besides, sequential characteristic of these methods eliminates the requirement for preserving historical states for data assimilation, which indeed could effectively increase the computational efficiency. Two variants of the sequential filtering techniques including the Particle filter (PF) and Ensemble Kalman filters (EnKF) have excessively been used in literature (see, e.g., Reichle et al. 2002; Moradkhani et al. 2005; Weerts et al. 2006; Zaitchik et al. 2008; Houborg et al. 2012; Moradkhani et al. 2012; Eicker et al. 2014; Renzullo et al. 2014; Tangdamrongsub et al. 2015).

All these efforts show the capability of different forms of sequential methods for assimilating hydrological components. Nevertheless, to determine the capability of assimilation filters, there are various aspects that should be considered. This can determine the right method for a specific satellite data assimilation objective. Data assimilation filters, in addition to improving the estimation of the system state and quantifying the associated uncertainties, should be able to keep the model dynamically stable during the assimilation process while incorporating additional data. More specifically for sequential filtering methods, effective sampling from the observation covariance is crucial and sometimes difficult (Whitaker and Hamill 2002). There are different resampling techniques, designed for PF [e.g., Multinomial and Systematic Resampling; Arulampalam et al. 2002), as well as EnKF (i.e., deterministic EnKFs; Sun et al. 2009; Hoteit et al. 2015) to address this issue. These indicate that how important and the same time challenging is to choose a proper data assimilation filtering technique. Therefore, a comprehensive analysis of the different sequential filtering techniques for assimilating satellite observations into hydrological models is presented in this chapter. This is done to assess the performance of the commonly used sequential filtering techniques for data assimilation in hydrological models.

Chapter 5
Sequential Data Assimilation Techniques for Data Assimilation

There is evidence that the different filtering schemes are more suited to different applications. This study considered the implementation of several data assimilation filtering techniques based on the two most commonly applied algorithms, ensemble Kalman filter, and Particle Filter, to assess their performances for hydrologic assimilation. The most promising results were obtained using SQRA and EnSRF both in terms of ensemble representation as well as an assessment against in-situ measurements, demonstrating significantly better performance than the other implemented filters.

—M. Khaki (This chapter is presented following Khaki et al. (2017b), "Assessing sequential data assimilation techniques for integrating GRACE data into a hydrological model")

5.1 Summary

The time-variable terrestrial water storage (TWS) products from the Gravity Recovery And Climate Experiment (GRACE) have been increasingly used in recent years to improve the simulation of hydrological models by applying data assimilation techniques. In this study, we assess the performance of established sequential data assimilation techniques for integrating GRACE TWS into the World-Wide Water Resources Assessment (W3RA) model. We implement six versions of the classical ensemble Kalman filters (EnKF), as well as Particle filters. These choices provide various opportunities for weighting observations and model simulations during the assimilation and, also for investigating error distributions. Monthly GRACE TWS data, covering March 2002 to December 2012, are assimilated into W3RA for the entire Australian continent. To evaluate the performance of the filters and analyze their impact on the model behaviours, their estimates are validated with independent in-situ measurements of groundwater and soil moisture. Our results indicate that the implemented filters improve the estimation of water storage differently. In particular, two extensions of the deterministic EnKF, the Square Root Analysis (SQRA)

M. Khaki, *Satellite Remote Sensing in Hydrological Data Assimilation*,
https://doi.org/10.1007/978-3-030-37375-7_5

scheme and the Ensemble Square Root Filter (EnSRF), provide the best estimations of water storage in terms of agreement with the groundwater observations, respectively decreasing the errors by 34 and 31%, compared to the model outputs without assimilation.

5.2 Introduction

Hydrological models offer important tools for simulating and predicting hydrological processes at global (e.g., Döll et al. 2003; Huntington 2006; Coumou and Rahmstorf 2012; van Dijk et al. 2013) and regional (e.g., Chiew et al. 1993; Wooldridge and Kalma 2001; Christiansen et al. 2007; Huang et al. 2016) scales. Models are still being developed to simulate all available hydrological processes (e.g., groundwater recharge) and the inclusion of all interactions between water cycle components (e.g., evapotranspiration, precipitation, and runoff). Currently, the most important deficiencies in hydrological models are caused by a high level of uncertainties in imperfect modelling of complex water cycle processes, data deficiencies on both temporal and spatial resolutions (e.g., limited ground-based observations), uncertainties in input and forcing data, and uncertainties of (unknown) empirical model parameters (Vrugt et al. 2013; van Dijk et al. 2011, 2014). Since making models more complex introduces ever increasing model parameters that cannot be well interpreted and makes computations more expensive, a logical step to address these limitations is the assimilation of observations into models (e.g., McLaughlin 2002; Zaitchik et al. 2008; van Dijk et al. 2014). Data assimilation techniques have found increasing interests with the availability of new data sources, such as those derived from satellite remote sensing observations. For example, time-variable gravity fields from the Gravity Recovery And Climate Experiment (GRACE) mission (Tapley et al. 2004) can be converted into terrestrial water storage (TWS) fields, a fundamental parameter of the water cycle that might be useful in reducing uncertainties in hydrological models.

Data assimilation is a procedure that constrains the dynamic of a model with available observations in order to improve its estimates (Bertino et al. 2003). The solution of the data assimilation problem is based on the Bayes' rule (Jazwinski 1970; van Leeuwen and Evensen 1996), which basically computes the Probability Density Function (PDF) of the state, i.e., the model variable of the system that should be estimated, given the data. The updated distribution is then propagated with the model to the time of the next available observation to obtain the prior PDF. In the case of a nonlinear or non-Gaussian system (as it is the case for hydrological models), it is not possible to analytically derive the posterior (analysis) PDF of the state (Hoteit et al. 2008; Vrugt et al. 2013). The Bayesian estimation problem, therefore, needs to be solved numerically, using either variational smoothing or sequential filtering methods (Subramanian et al. 2012).

Generally speaking, variational methods look for the model trajectory that best fits the data by minimizing a chosen cost function that measures the misfit between the model state and the observations (Talagrand and Courtier 1987). These methods require coding and executing an adjoint model, which is very demanding in terms of human and computational resources (Hoteit et al. 2005). Furthermore, the efficiency of variational methods might be limited to update the estimation statistics during the data assimilation process (Courtier et al. 1994; Kalnay 2003). In contrast, sequential techniques process the data as they become available following two steps; a forecast step to propagate the distribution forward in time and an analysis step to update the distribution with the newly available observation. Monte Carlo methods are commonly used in the forecast step (based on ensembles or particles) and Kalman (Ensemble Kalman filtering) or point-mass weight (Particle filtering) updates are applied in the analysis step (Evensen 2009; Hoteit et al. 2012). Sequential methods do not require an adjoint and are becoming increasingly popular because of their reasonable computational requirements (Hoteit et al. 2002; Bertino et al. 2003; Robert et al. 2006).

The Particle filter is based on a point-mass (particle) representation of the system state's PDF. It forecasts the PDF by propagating the particles forward in time. At the analysis time, the state PDF is updated by assigning new weights to the particles based on incoming observations (Doucet et al. 2001; Pham 2001; Hoteit et al. 2012). The fundamental problem of this technique is the degeneracy phenomenon of its particles, with only very few particles carrying most of the weights (Subramanian et al. 2012). Moreover, errors in the assimilated observations may propagate to the estimated distribution because the method was not designed to improve the structure of the model (Hoteit et al. 2008; Smith et al. 2008). This problem is addressed by the Ensemble Kalman filters (EnKFs), which assume a Gaussian forecast PDF at the analysis time, so a Kalman update-step is applied to the particles (Hoteit et al. 2015). This allows an efficient implementation of the Bayesian filtering approach for data assimilation into large systems using small ensembles (van Leeuwen and Evensen 1996; Hoteit et al. 2008).

EnKFs can be classified into stochastic and deterministic filters, depending on whether the observations are perturbed with noise before assimilation, or not (Tippett et al. 2003; Hoteit et al. 2015). In the stochastic EnKF, each ensemble member is updated with perturbed observations, readily providing an analysis ensemble for the next filtering cycle. In contrast, a deterministic EnKF updates only the mean and the covariance of the ensembles exactly as in the Kalman Filter, and thus require a resampling step to generate a new analysis ensemble. The resampling step is not unique, and as such several deterministic EnKFs have been proposed (Sun et al. 2009; Hoteit et al. 2015).

Sequential filtering methods have been extensively applied and compared in oceanic and atmospheric applications (Garner et al. 1999; Elbern and Schmidt 2001; Bennett 2002; Kalnay 2003; Schunk et al. 2004; Lahoz 2007; Zhang et al. 2012; Altaf et al. 2014). In hydrological studies, data assimilation has been used to estimate different water compartments, such as soil moisture (e.g., Reichle et al. 2002; Brocca et al. 2010; Renzullo et al. 2014) and surface water storage (e.g., Alsdorf et al. 2007; Neal et al. 2009; Giustarini et al. 2011). However, the efficiency of the various filtering methods in dealing with remotely sensed data in hydrology has not been fully investigated (McLaughlin 2002; Schumacher et al. 2016).

Global terrestrial water storage data derived from the GRACE satellite mission can now be employed to improve the behaviour of hydrological models (e.g., Zaitchik et al. 2008; Tangdamrongsub et al. 2015; Thomas et al. 2014; van Dijk et al. 2014; Eicker et al. 2014; Reager et al. 2015), providing unprecedented temporal and spatial coverage. For instance, Zaitchik et al. (2008) demonstrated the relevance of GRACE data in improving the estimation of groundwater variability over the four major sub-basins of the Mississippi through data assimilation into the Catchment Land Surface Model using an ensemble Kalman smoother. Houborg et al. (2012) investigated drought conditions in North America through GRACE data assimilation. The developed GRACE-based drought indicators in the USA led to an improved monitoring of soil moisture and groundwater conditions of deep layers. The impact of GRACE error correlation structure on the assimilation of GRACE data was very recently studied by Schumacher et al. (2016). Yet, to the best of our knowledge, a comparison of the different sequential filtering techniques for assimilating GRACE TWS into hydrological models has not been fully explored.

In this study, we investigate the performance of the most common sequential filtering techniques for data assimilation using the hydrological model of the World-Wide Water Resources Assessment (W3RA; van Dijk 2010) over Australia. The amount of rainfall in Australia is low in comparison to other inhabited continents on Earth, leading to prolonged drought in the interior regions (Forootan et al. 2016). Hence, accurate estimation of water storages (e.g., using hydrological models) is necessary to manage water resources in this region. Here, different filters are used to assimilate GRACE TWS into W3RA in order to improve its simulations and to explore the filters' performances in various aspects. Both stochastic and deterministic EnKFs are tested and their performances are compared against two standard Particle filters. In particular, the deterministic EnKF is tested to avoid perturbing observations before assimilation (that is the case in an ordinary EnKF). Gaussian-based random updates in the EnKF approaches likely do not fully represent the statistical properties of the model simulations and TWS observations. Therefore, the fully non-Gaussian Particle filter is also applied to estimate more realistic updates.

We apply the standard EnKF and its deterministic variants, including, the Square Root Analysis (SQRA) scheme following Evensen (2004) and Schumacher et al. (2016), the Ensemble Transform Kalman Filter (ETKF, Bishop et al. 2001), the Deterministic EnKF (DEnKF, Sakov and Oke 2008), and the Ensemble Square-Root Filter (EnSRF, Whitaker and Hamill 2002). We also implement the static-ensemble variant of the EnKF, the Ensemble Optimal Interpolation (EnOI, Evensen 2003), in an attempt to reduce the computational burden. To mitigate the deficiency that may arise from limited ensemble sizes and knowledge of model errors' statistics (Anderson et al. 2007; Oke et al. 2007), covariance inflation (e.g., Anderson et al. 1999, 2007; Ott et al. 2004) and localization techniques (e.g., Bergemann and Reich 2010; Hamill and Snyder 2002) are applied. The performance of these ensemble filters is assessed against two nonlinear Particle filters based on two different resampling strategies: (i) Multinomial Resampling and (ii) Systematic Resampling techniques (Arulampalam et al. 2002). A summary of the applied filters in this study is presented in Table 5.1. The results of assimilations are evaluated by comparing

Table 5.1 A summary of the applied filters for data assimilation

Filter	Acronym	Type	Reference
Ensemble Kalman filter	EnKF	Stochastic ensemble Kalman filter	Evensen (1994)
Square root analysis	SQRA	Deterministic ensemble Kalman filter	Evensen (2004)
Ensemble transform Kalman filter	ETKF	Deterministic ensemble Kalman filter	Bishop et al. (2001)
Ensemble square-root filter	EnSRF	Deterministic ensemble Kalman filter	Whitaker and Hamill (2002)
Ensemble optimal interpolation	EnOI	Deterministic ensemble Kalman filter	Evensen (2003)
Deterministic ensemble Kalman filter	DEnKF	Deterministic ensemble Kalman filter	Sakov and Oke (2008)
Particle filter, multinomial resampling	PFMR	Particle filter	Arulampalam et al. (2002)
Particle filter, systematic resampling	PFSR	Particle filter	Arulampalam et al. (2002)

their estimates against independent groundwater in-situ measurements within the Murray-Darling basin and measurements from the moisture-monitoring network in the Murrumbidgee catchment in New South Wales, Australia.

5.3 Model and Datasets

5.3.1 W3RA

The World-Wide Water Resources Assessment (W3RA), based on the Australian Water Resources Assessment system (AWRA) model (version 0.5) is used in this study (http://www.wenfo.org/wald/data-software/). The model was first developed in 2008 by the Commonwealth Scientific and Industrial Research Organisation (CSIRO) to monitor, represent and forecast Australian terrestrial water cycles. The W3RA is a grid-distributed biophysical model that simulates landscape water stores in the vegetation and soil systems (van Dijk 2010). The $1° \times 1°$ global daily fields of minimum and maximum temperature, downwelling short-wave radiation, and precipitation from Princeton University (http://hydrology.princeton.edu) are used for meteorological forcing data (Sheffield et al. 2006). This one-dimensional grid-based water balance model represents the water balance of the soil, groundwater and surface water stores in which each cell is modelled independently of its neighbours (van Dijk 2010; Renzullo et al. 2014). The model state is composed of the $1° \times 1°$ W3RA model storages of the top, shallow root and deep root soil layers, groundwater storage, and surface water storage in a one-dimensional system (vertical variability). In this study, we use W3RA providing daily model states for the period of March 2002 to December 2012. More detailed information on the W3RA model can be found in van Dijk (2010).

5.3.2 GRACE-Derived Terrestrial Water Storage

Here, we use monthly GRACE level 2 (L2) products along with their full error information between March 2002 to December 2012 as provided by the ITSG-Grace2014 gravity field model (Mayer-Gürr et al. 2014). The GRACE monthly Stokes' coefficients are truncated at spherical harmonic degree and order 90, resulting in approximately ~300 km spatial resolution at the equator.

Following Swenson et al. (2008), degree 1 coefficients are replaced to change the origin of the coordinate system to the Earth's center of mass. Degree 2 and order 0 (C20) coefficients from GRACE are not well determined and are replaced by more reliable estimations of the Satellite Laser Ranging solutions (Cheng and Tapley 2004). An anisotropic spatial sampling, instrumental noise (K-band ranging system and GPS), and temporal aliasing introduced by the incomplete reduction of short-term mass variations cause correlated noises in L2 products (Forootan et al. 2014b). These errors are reduced by smoothing based on a Gaussian averaging kernel with 300 km half radius and destriping following Swenson and Wahr (2006). However, the smoothing may cause signal attenuation (Klees et al. 2008) and can result in considerable spatial leakage, such as the apparent movement of masses from one region to another (Chen et al. 2007) especially over coastlines (see examples within Australia in e.g., Brown and Tregoning 2010; Forootan et al. 2012). In order to address this issue, following Swenson and Wahr (2002), we apply an isotropic kernel using a Lagrange multiplier filter to best balance signal and leakage errors over the basin of interest.

An additional post-processing step is applied to convert the filtered L2 gravity fields (after removing the mean field of the study period) to gridded TWS fields ($1° \times 1°$) following Wahr et al. (1998). The GRACE TWS data are gridded at the same spatial $1° \times 1°$ resolution of W3RA resulting in 794 grid points within Australia that covers an area of approximately 7.692 million km^2 located between 10 °S and 46 °S latitude, and 110 °E and 160 °E longitude. While GRACE data measures changes in TWS, W3RA provides absolute TWS. The offset between the two needs thus to be corrected. Accordingly, the mean TWS of the study period is taken from W3RA and added to the GRACE TWS changes time series in order to obtain absolute values, in accordance with the model outputs (Zaitchik et al. 2008). During assimilation, monthly variations of the model states are updated using GRACE TWS. In addition, the monthly full error information of the Stokes' coefficients is used to construct an observation error covariance matrix for the GRACE TWS fields (Eicker et al. 2014; Schumacher et al. 2016). The temporally averaged covariance matrix for the $1° \times 1°$ spatial resolution is shown in Fig. 5.1.

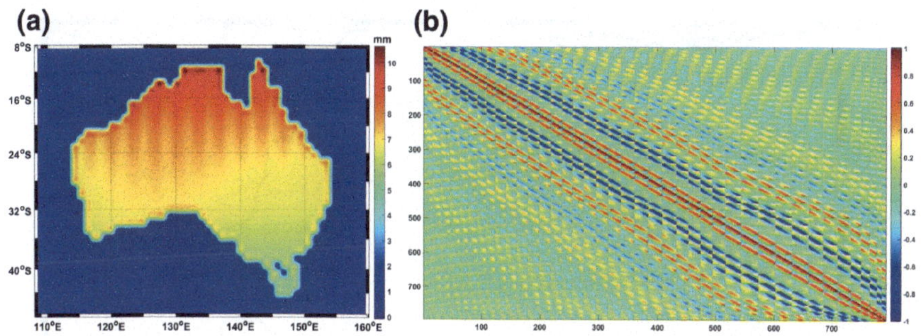

Fig. 5.1 Temporally averaged GRACE covariance matrix error (**a**) and the structure of its correlation matrix (**b**)

5.3.3 In-Situ Data

For validating the assimilation results, we use in-situ groundwater level data that are collected over the Murray-Darling River Basin. The independent in-situ measurements are provided by New South Wales Government (NSW) groundwater archive (http://waterinfo.nsw.gov.au/pinneena/gw.shtml). Monthly well measurements are acquired and time series of groundwater storage anomalies are generated by removing measurements with data gaps and those without showing seasonal variations that we assume belong to confined aquifers (Houborg et al. 2012; Tangdamrongsub et al. 2015). Selected bore-water levels are then converted to variations in groundwater (GW) storage. To this end, instead of using specific yield estimates (Rodell et al. 2007; Zaitchik et al. 2008) that are not available in the region, TWS variation from GRACE and Global Land Data Assimilation System (GLDAS, Rodell et al. 2004) soil moisture are used to scale the observed head following Tangdamrongsub et al. (2015). Through this approach, rather than assuming a constant specific yield everywhere (e.g., 0.1 by Tregoning et al. 2012), different yield values (e.g., between 0.1 and 0.2) can be derived leading to a more realistic representation of groundwater systems in different areas (Seoane et al. 2013). Tregoning et al. (2012) show that this approach can be used to find a scaling factor over the Canning and Murray basins in Australia. The scaled in-situ groundwater level fluctuations are then used to assess the assimilation results.

In addition, in-situ measurements of the moisture-monitoring network (http://www.oznet.org.au/) in Murrumbidgee catchment (Smith et al. 2012) are used to evaluate the results. These data are known as the OzNet network, which provides long-term records of measuring volumetric soil moisture at various soil depths at 57 locations across the Murrumbidgee catchment area. Following Renzullo et al. (2014), we averaged the measurements into a daily scale and use 0–8 cm to evaluate the estimated model top-layer soil moisture and the 0–30 and 0–90 cm measurements for the evaluation of the model shallow root-zone soil moisture estimation. The distribution of the in-situ moisture network, as well as in-situ groundwater stations, are shown in Fig. 5.2.

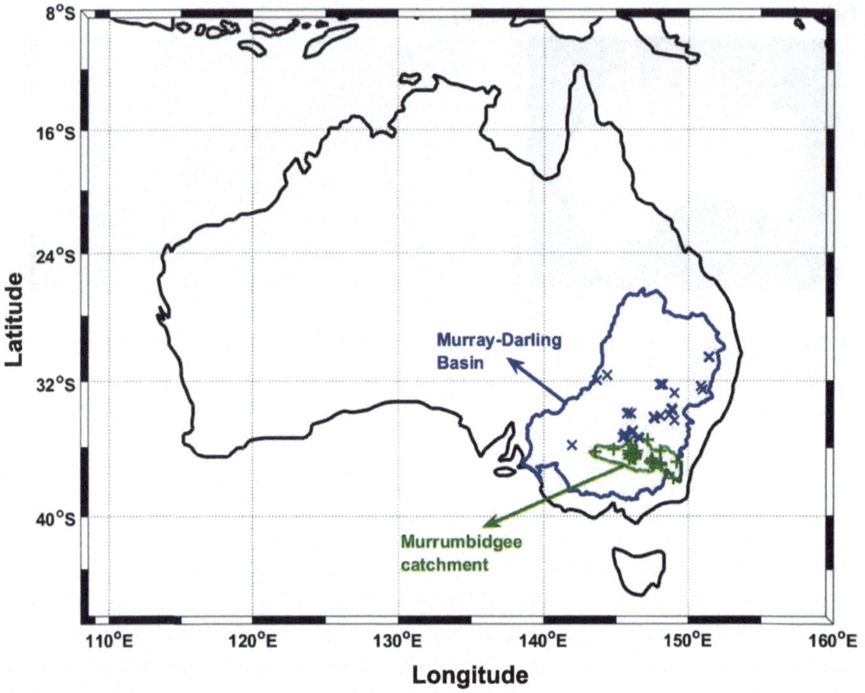

Fig. 5.2 The study area is represented by black solid line. The figure also contains the boundary of the Murray-Darling basin, the locations of the groundwater bore stations (blue), and the outline of the Murrumbidgee catchment with the OzNet soil moisture network (green), which are used for results assessment

5.4 Filtering Methods and Implementation

The selected Bayesian filtering procedures for data assimilation are presented here. The analytical process conditions a prior PDF of the state with available observations to compute the posterior PDF based on Bayes' rule (Koch 2007) in two steps; (1) forecasting the state PDF using a dynamical model and (2) updating the forecast PDF by assimilating observations using Bayes' rule. In the case of a linear system with Gaussian noise, the popular Kalman filter provides the Bayesian filtering solution by computing the first two moments of the state PDF, which remains always Gaussian (van Leeuwen and Evensen 1996). This two-step process is repeated whenever new observations become available. The basic Kalman filter equations are given by Kalman (1960) starting from an analysis of the state, X_t^a, and the associated error covariance, P_t^a, at a given time t. These can be summarized as:

(1) The forecast step consists of the evolution of the state estimate and its error covariance matrix with a linear dynamical model (M) to the assimilation step (the time of the next available observation),

$$X^f_{t+\Delta t} = MX^a_t + \eta, \tag{5.1}$$

$$P^f_{t+\Delta t} = MP^a_t M^T + Q, \tag{5.2}$$

where $X^f_{t+\Delta t}$ refers to the forecast state (X^f) at time $t + \Delta t$, with Δt representing the model time step, and T is the transpose index. In Eq. 5.1, η is the process noise, which is drawn from $N(0, Q)$ with covariance matrix Q, and $P^f_{t+\Delta t}$ (in Eq. 5.2) denotes the forecast error covariance (P^f) at time $t + \Delta t$.

(2) The analysis step updates the forecast state using new incoming observations Y that are related to the state vector by the linear observation operator (H) as,

$$Y = HX + \varepsilon, \tag{5.3}$$

where ε is the measurement noise. The analysis state (X^a) is then computed using

$$X^a = X^f + K(Y - HX^f), \tag{5.4}$$

$$P^a = (I - KH)P^f, \tag{5.5}$$

$$K = P^f H^T (HP^f H^T + R)^{-1}, \tag{5.6}$$

where K refers to the Kalman gain, R is the observation error covariance matrix, and I denotes the identity matrix.

The Kalman filter algorithm is not suited for high-dimensional or non-linear systems (Pham et al. 1998; Hoteit et al. 2002). The ensemble Kalman filter provides an efficient alternative for the implementation of the Kalman filter with these systems by representing the first two-moments of the state using a sample of state vectors, called ensemble. The forecast state and covariance matrix in Eqs. 5.1 and 5.2 are then estimated as the sample mean and covariance of the ensembles members $X^i, i = 1 \ldots N$:

$$\bar{X}^f = \frac{1}{N}\sum_{i=1}^{N} X^{f,i}, \tag{5.7}$$

$$P^f = \frac{1}{N-1}\sum_{i=1}^{N}(X^{f,i} - \bar{X}^f)(X^{f,i} - \bar{X}^f)^T = \frac{1}{N-1}A^f A^{fT}. \tag{5.8}$$

$\bar{X}^f$ is the forecast ensemble mean and A^f ($A^f = [A^{f,1} \ldots A^{f,N}]$) is the forecast ensemble of anomalies (perturbations; $A^{f,i} = X^{f,i} - \bar{X}$). When a new observation is available, the forecast ensemble is then updated with the data using Eq. 5.4, as in the Kalman filter. Several ensemble Kalman filters have been proposed, all sharing the same forecast step in which an available analysis ensemble ($[X^{a,1} \ldots X^{a,N}]$) is propagated forward with the model. The analysis step based on the Kalman filter, however, can be applied stochastically or deterministically.

5.4.1 Stochastic Ensemble Kalman Filter (EnKF)

The analysis step of the Stochastic EnKF updates each ensemble member with a perturbed observation written as,

$$X^{a,i} = X^{f,i} + K(Y^i - HX^{f,i}), \quad i = 1 \ldots N, \tag{5.9}$$

where $Y^i = Y + \varepsilon^i$, with ε^i a random error sampled from $N(0, R)$. The use of perturbed observations in the EnKF results in an analysis error covariance that matches that of the Kalman filter, in a statistical sense (Hoteit et al. 2012). The advantage of the stochastic update is that it readily provides a randomly sampled ensemble from the Gaussian-assumed state analysis PDF for the next forecast cycle (Hoteit et al. 2015). However, as illustrated by Whitaker and Hamill (2002), sampling error can be reflected in the EnKF background covariance matrix, especially for the small-size ensembles. This could be particularly pronounced when a large number of (independent) observations are assimilated (Nerger 2004), as the observation covariance cannot be properly sampled with a small ensemble (Hoteit et al. 2015).

5.4.2 Deterministic Ensemble Kalman Filters

Instead of updating each forecast member separately, deterministic EnKFs (DEnKFs) update the forecast ensemble in two steps, first the ensemble-mean and then the ensemble perturbations (Tippett et al. 2003) are calculated so that the sample mean and covariance of the updated ensemble exactly match those of the Kalman filter in Eqs. 5.4 and 5.5.

Various methods have been proposed in order to update the ensemble perturbations. SQRA resamples the new ensemble perturbations (A^a) from the forecast ensemble perturbations (A^f) as,

$$A^a = A^f V \sqrt{I - \Sigma^T \Sigma \Theta^T}, \tag{5.10}$$

where Σ is computed by applying the following singular value decomposition (SVD),

$$U \Sigma V^T = SVD(\Lambda^{-\frac{1}{2}} Z^T HA), \tag{5.11}$$

$$Z \Lambda Z^T = (HP^f (H)^T + R)^{-1}, \tag{5.12}$$

with Θ in Eq. 5.10 being a random orthogonal matrix for redistribution of the variance among the ensemble members (see Evensen 2004, 2007, for more details). This is very similar to the random rotation that has been introduced in the context of the Singular Evolutive Extended Kalman (SEEK) filter (Pham 2001; Hoteit et al. 2002).

ETKF introduces a transformation matrix to directly compute the analysis ensemble perturbations from their forecast counterparts,

$$A^a = A^f.T, \tag{5.13}$$

where $T = U(I + \Lambda)^{-1/2}$, with U and Λ, respectively being the orthogonal and diagonal matrices computed from an eigenvalue decomposition of $(HX^f)^T R^{-1}(HX^f)$.

DEnKF and EnSRF adopt a similar analysis step to the EnKF in the sense that they compute the analysis perturbations from the forecast perturbations by updating each ensemble perturbation with a Kalman-like update step. To match the Kalman filter covariance matrix by an ensemble of perturbations, DEnKF computes a first-order approximation of the Kalman gain (Sakov and Oke 2008) under the assumption that $||KH||$ is small compared to 1. The approximate gain $\tilde{K}$ is then used to compute the analysis perturbations as,

$$A^a = A^f - \frac{1}{2}KHA^f. \tag{5.14}$$

EnSRF exploits the serial formulation of the Kalman filter analysis step in which the observations are assimilated each at a time to compute the analysis perturbations that exactly match the Kalman filter covariance using the modified gain (αK) with,

$$\alpha = \left(1 + \sqrt{\frac{R}{HP^f H^T + R}}\right)^{-1}. \tag{5.15}$$

This requires the observations to be uncorrelated, which can always be satisfied by scaling the observations with the square-root inverse of the observational error covariance matrix (Hoteit et al. 2015).

Another form of ensemble Kalman filtering is the so-called Ensemble Optimal Interpolation (EnOI) scheme, which is basically the EnKF, but without update of the ensemble anomalies. More precisely, EnOI only updates the forecast state with a Kalman gain computed from a preselected static ensemble. The main advantage of not updating the ensemble is of course to reduce the computational load, but it can also be beneficial to retain the spread of the ensemble and to enforce climatological smoothness in the update step. EnOI can be stochastic or deterministic (Hoteit et al. 2002). Here we only test the more standard stochastic variant (Evensen 2003).

5.4.3 *Particle Filtering*

Particle filtering is also a sequential Monte Carlo method that was originally proposed by Gordon et al. (1993) and has since been applied in numerous studies (Doucet 1998; Arulampalam et al. 2002). The idea is to represent the state PDF by a set of weighted particles (Arulampalam et al. 2002), hence the name Particle Filter

(Gordon et al. 1993; Doucet 1998), which is similar to the ensemble members in the EnKF but with non-uniform weights. The state PDF is then decomposed as,

$$P(X_t|Y_{1:t}) \approx \sum_{i=1}^{N} \omega_t^i \delta(X_t - X_t^i), \tag{5.16}$$

where $\{X_t^i; i = 1 \ldots N\}$ are the particles at time t, observations between time 1 and t are denoted by $Y_{1:t}$, ω_t^i are the weights of the particles, and δ is the Dirac function. In the forecast step, the Particle filter just integrates the particles forward with the model, exactly as the EnKF, and their weights remain the same. In the analysis step, only the weights, and not the particles, are updated with the incoming observation as,

$$\omega_t^i = \frac{P(y_t|X_{t|t-1}^i)}{\sum_j P(y_t|X_{t|t-1}^j)}. \tag{5.17}$$

The Particle filter suffers from the degeneracy problem in which the weights of all particles become negligible except only for a very few, requiring a prohibitive number of particles to prevent particles collapse (Arulampalam et al. 2002). Degeneracy can be mitigated using the so-called resampling technique (Doucet et al. 2005), which resamples a new set of particles with uniform weights after every update step based on the analysis PDF. In this study, we consider two of the most common resampling techniques: the Particle filter with Multinomial Resampling (PFMR) and Particle filter with Systematic Resampling (PFSR), as proposed by Doucet et al. (2005). PFMR is the most straightforward resampling scheme, where N independent random numbers ($u \sim U(0, 1)$) are generated to select a particle from the old set. PFSR, which is also called universal sampling, draws only one random number $u_1 \sim U(0, 1/N)$ and the remaining $N - 1$ numbers are then calculated from u_1 (Doucet et al. 2005) as,

$$U_i = u_1 + \frac{(i-1)}{N}, \quad i = 2 \ldots N. \tag{5.18}$$

These are then used to select a new set of particles according to the multinomial distribution (Hol et al. 2006).

The Particle filter has been applied in few hydrological studies. Among them, Moradkhani et al. (2005) investigated the relevance of the Particle filter for estimating the joint posterior distribution of the parameters and state. In another effort, Moradkhani et al. (2012) proposed a modified version of the Particle filter, focusing on enhancing the sampling of the posterior with Markov chain Monte Carlo moves. Plaza et al. (2012) used the Sequential Importance Sampling for soil moisture assimilation and focused on the consequent effect on baseflow generation. Previous studies focused on the different implementations of the Particle filter using various resampling techniques. Very few studies have compared Particle filters performances with

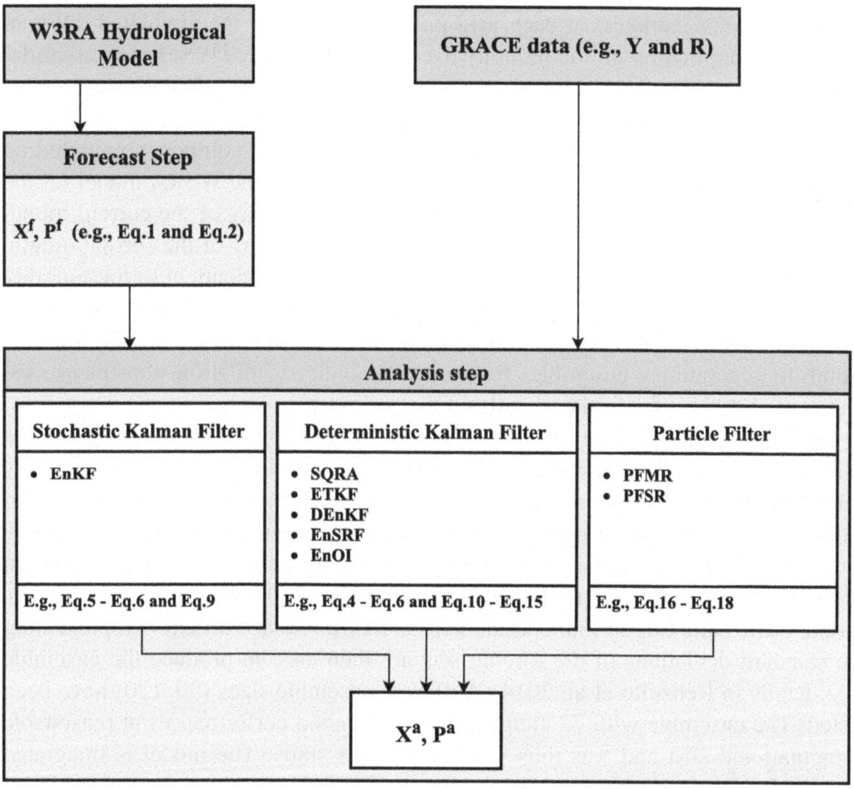

Fig. 5.3 A schematic illustration of the steps and filters applied for data assimilation in this study

diverse resampling techniques and EnKFs in hydrology (e.g., Weerts et al. 2006; Subramanian et al. 2012). Figure 5.3 outlines a summary of the filtering framework and the applied filters for data assimilation in this study.

5.4.4 Filter Implementation

An experimental framework is developed in order to assess the relevance and efficiency of the filtering techniques presented in the previous section for assimilating GRACE data into the W3RA model. All filters are implemented under identical conditions, using the same spatial scales ($1° \times 1°$) for both W3RA and GRACE TWS, and daily temporal scales for W3RA and monthly for the GRACE data. The state vectors for every grid point in our experiments are composed of the different water storages, including top soil, shallow soil, and deep soil water, canopy, snow, surface, and groundwater. The observations matrix (H) accumulates the state variables (the individual water storages) at each grid point to determine the simulated TWS in order to update them with the monthly averages of GRACE TWS during assimilation. In the ensemble filter update step, the monthly averaged update increment (i.e., the difference between the simulated monthly averaged TWS and averaged GRACE TWS) is vertically disaggregated into the individual water components, including soil moisture and groundwater storages. In order to restart the W3RA model for the next months, an update of the water variables of the last day of the current month is required. The monthly increment can be added to each day of the current month, which guarantees that the update of the monthly mean is identical to the monthly mean of the daily updates. In practice, the differences between the predictions and the updated states are added as offsets to the state vectors at the last day of each month to generate the ensembles for the next month assimilation step (please see Eicker et al. (2014) for further details on the approach).

Several steps need to be undertaken before assimilating GRACE TWS into the W3RA model. The initial ensemble members (or particles) are first generated by perturbing the three most important forcing fields; precipitation, temperature, and radiation. To generate the perturbations, we assume a multiplicative error of 30% for precipitation, an additive error of $50\,Wm^{-2}$ for the shortwave radiation, and an additive error of 2 °C for temperature (Jones et al. 2007; Renzullo et al. 2014). Monte Carlo sampling of multivariate normal distributions with errors representing the standard deviations of the forcing sets are then used to produce the ensemble (see details in Renzullo et al. 2014). Different ensemble sizes (30–120) have been tested. The ensemble with 72 members provided good performances at reasonable computational cost and was thus selected for this study. The model is integrated forward for two years (January 2001 to January 2003) using perturbed meteorological forcing datasets to obtain an ensemble of 72 states at the beginning of 2003 (study period), considered as the initial ensemble. All filters are initialized with this produced ensemble.

We use two tuning techniques of ensemble inflation and localization in order to enhance the assimilation performance of all EnKFs. Ensemble inflation uses a small coefficient to inflate the prior ensemble deviation from the ensemble-mean to increase its spread and alleviate the inbreeding problem (Anderson et al. 2007). Another auxiliary technique that has been proved to be helpful when using limited ensemble size is localization, initially proposed by Houtekamer and Mitchell (2001). We choose to use a Local Analysis scheme, which works by restricting the impact of a given measurement in the update step to the points located within a certain distance from the measurement location (Evensen 2003; Ott et al. 2004). We tested different inflation factors ($[1 - 1.8]$) and localization length scales ($[3^\circ - 10^\circ]$) and assessed their impacts on the ensemble Kalman filters. The ensemble spreads of these runs were monitored and their results evaluated against the in-situ measurements to determine the best values (i.e., 1.12 for inflations and 5° for localization length). Note that these two auxiliary techniques have not been applied in the Particle filters. In terms of computational requirement, all implemented filters require approximately the same processing time (when implemented with the same number of members/particles), with the forecast step of the ensemble being the most demanding.

5.5 Results

In this section, we review and analyze the performance of all the selected filtering techniques based on various factors. The implemented filters include (stochastic) EnKF, ETKF, SQRA, DEnKF, EnSRF, EnOI and Particle filter with Multinomial (PFMR) and Systematic (PFSR) resampling. In addition to improving the estimation of the system state and quantifying the associated uncertainties, a suitable data assimilation technique is expected to keep the model dynamically stable during the assimilation process while incorporating GRACE data. The assimilation is applied once a month due to the availability of the observations. Therefore, assimilating such a dataset is challenging, and requires a filtering scheme that is robust to the system error covariances, and also efficient in term of resampling representative ensembles after every assimilation step. Our analysis is organized in two parts; we first examine the filters performances by comparing their estimates (analysis and forecasts) against the assimilated GRACE data over the whole study area and then against the independent in-situ measurements over the Murray-Darling River Basin and Murrumbidgee catchment. We also compare the filters estimates with the outputs of a model-free run (open-loop) that is integrated with the filters initial condition without assimilation to evaluate the impact of assimilating GRACE data on the model behaviour. Next, the filters behaviours in terms of ensemble spread and the impact of assimilation on the forecast and analysis error covariances are investigated.

5.5.1 Assessment with GRACE and In-Situ Data

For every grid point, we compute the correlation between monthly averaged time series of the TWS from the model before and after data assimilation and the assimilated GRACE TWS observations (Fig. 5.4). The correlations are calculated at 95% confidence level and after removing the seasonal cycles from the time series. Spatial correlation maps with high correlations may suggest that the applied filtering method efficiently incorporates GRACE data into the model. The correlations between the model TWS outputs without assimilation and GRACE data presented in Fig. 5.4 range between 0.11 and 0.64, with the weaker correlations in the northern region. This can be due to the fact that GRACE measures vertically accumulated water storage including surface water discharge, which may not be accurately provided by the model. Moreover, TWS variations are generally higher in the northern part of the study area (Awange et al. 2009; Rieser et al. 2010; Seoane et al. 2013).

All filters improve the correlations between the estimates and the data, with some filters leading almost to a perfect correlation (e.g., EnSRF). The model is however not able to preserve the resulting high correlations during the forecast step for the following month with the correlations mainly decreasing in the center and southern regions. After monitoring the impact of the observations on the model states throughout the study period, it is found that the impact of the update step gradually decreases (more than 10% after approximately 3–5 days). Some of the information are

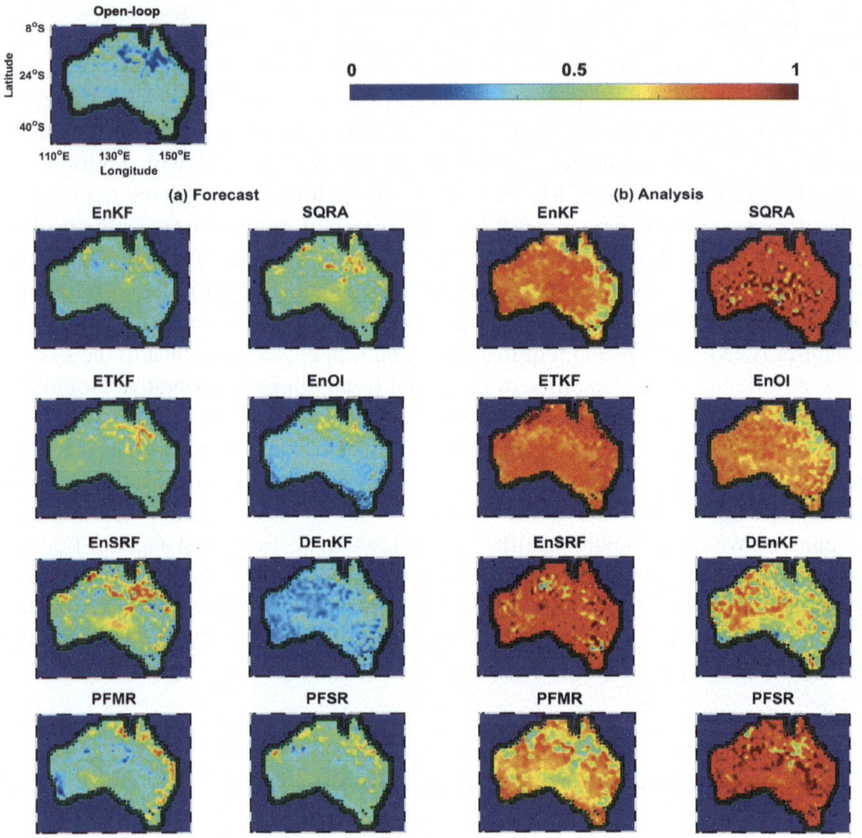

Fig. 5.4 Time average correlations between the assimilated GRACE TWS and different filters estimations at the (**a**) forecast and (**b**) analysis steps. Spatial correlation maps are generated at every assimilation step over the study period and their averages are presented

however retained, which can be seen from the comparison of the correlations between the filters estimates and the free-run. The daily effects of the forcing sets influence the model simulated water variables and gradually increase their uncertainties during the forecast month (between the assimilation steps). The level of improvement in the analysis step in terms of correlations is different for each filter. For instance, EnSRF, SQRA, and PFSR lead to higher correlations between the water variables and GRACE-derived TWS (0.95, 0.93, and 0.90, respectively), suggesting that these methods assimilate more efficiently the observations in the state estimates. Overall, the EnKFs perform better than the Particle filters, except for DEnKF which shows no remarkable impact on the model behaviour.

The Root-Mean-Squared Errors (RMSE) time series between the GRACE TWS data and the filters estimates are illustrated in Fig. 5.5). We use monthly averaged TWS time series from the filters forecast and analysis, and from the assimilated

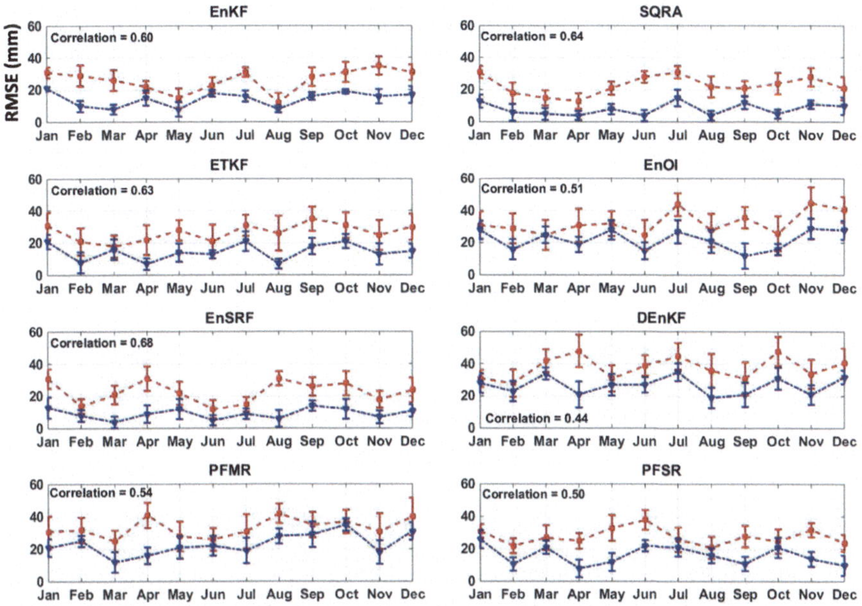

Fig. 5.5 RMSE time series between the assimilated GRACE TWS and the filters' forecasts and analyses which are calculated over all grid points at the forecast (red) and analysis (blue) steps and their averages at each month (with 95% confidence interval) are shown here

GRACE observations to compute the RMSE time series. Confidence intervals (0.05 significant level) for every month are also plotted in Fig. 5.5. In all cases, the analysis step decreases the RMSE with respect to the forecast. Nevertheless, the RMSE resulting from DEnKF, EnOI, and PFMR are significantly larger, indicating that these methods are not able to decrease the misfit between the model outcomes and GRACE data as efficiently as the other filters. Estimates by all filters exhibit the largest error in some periods (e.g., July and October), which may be caused by uncertainties in the forcing fields. The RMSEs from SQRA (average 8.12 in 0.05 significance level) and EnSRF (average 7.11 in 0.05 significance level) are smaller in comparison to the remaining filters. The smaller average errors during the study period are in agreement with the more robust performances of SQRA and EnSRF. The results in Figs. 5.4 and 5.5 suggest that the deterministic SQRA, ETKF, EnSRF filters, and to less extent PFSR, better incorporate GRACE data in the model estimates, in a statistical sense, considering the higher correlations and less errors between their estimates and the observations. A summary of the filters' performances, including correlations and RMSE in comparison with the assimilated observations, is presented in Table 5.2. We use the same method as for Fig. 5.5 to build time series at each grid cell over the study area and calculate statistics in Table 5.2. This table indicates higher correlations (84% average) and smaller RMSEs (35% average error reduction) after the analysis step for all the filters. The maximum RMSE decreases are achieved by EnSRF (58.88%) and SQRA (55.17%).

Table 5.2 Summary of statistical values derived from the implemented methods using the assimilated GRACE data. For each method the RMSE average and its range (±XX) at the 95% confidence interval is presented. The RMSE reductions are calculated using analysis states and the GRACE data in comparison to the forecast states and model-free run

Method	Forecast		Analysis		RMSE reduction (%)	
	RMSE (mm)	Correlation	RMSE (mm)	Correlation	Regarding forecast	Regarding model-free run
EnKF	26.51 ± 3.36	0.43	16.54 ± 3.58	0.90	30.61	39.59
SQRA	18.11 ± 3.65	0.48	8.12 ± 2.54	0.93	37.56	55.17
ETKF	21.84 ± 3.84	0.44	14.87 ± 3.41	0.91	31.92	41.92
EnOI	35.21 ± 4.76	0.39	22.93 ± 3.87	0.71	29.04	34.87
EnSRF	17.29 ± 3.67	0.49	7.11 ± 2.88	0.95	38.19	58.88
DEnKF	41.64 ± 4.00	0.36	36.74 ± 4.71	0.63	12.77	15.77
PFMR	37.60 ± 3.83	0.38	30.21 ± 3.91	0.81	15.63	19.63
PFSR	20.03 ± 3.87	0.47	13.87 ± 2.13	0.90	25.76	41.74

We investigate the performances of the filtering methods through comparison with the independent groundwater in-situ data over the Murray-Darling basin (cf. Sect. 5.3.3). We use the 54 in-situ measurements over the Murray-Darling basin for comparison with the estimated groundwater (Fig. 5.6). The filter estimates are spatially interpolated to the nearest observation bore. For each filter, the average RMSEs (overall 54 in-situ data) of the forecast state (red) and analysis state (blue) are determined. As for the assimilated GRACE data (cf. Fig. 5.5), all the filters decrease the RMSE with respect to the in-situ data, with the largest discrepancies resulting from DEnKF, EnOI, and PFMR. The average RMSEs are smaller in SQRA and EnSRF. The similar behaviour of the filters in the analysis steps can be found in both Figs. 5.6 and 5.5. For some months (e.g., March and July), the larger errors can be seen in Fig. 5.6 in comparison to Fig. 5.5. The reason for this can be due to an incompatibility between groundwater in-situ measurements and GRACE data. The absent of some water compartment terms, such as the surface water storage in the model and in-situ data in contrast to the GRACE data may also introduce a degree of error.

The agreement between the estimated state variables and both GRACE data and in-situ measurements (Fig. 5.7) demonstrates the filters capability to dynamically propagate the information extracted from GRACE data into system variables. In agreement with the previous results, the best performances are obtained using SQRA, EnSRF, and ETKF (Figs. 5.6 and 5.7).

The correlation coefficient and RMSE results are summarized in Table 5.3. For each filter, 54 time series are computed (i.e., for each individual well), and are used to calculate correlation and RMSE between filters' estimations and in-situ measurements. The results of all the filters summarized in Table 5.3 show improvements (by 35% average) in the analysis steps in all cases, similar to the assessment against GRACE data (cf. Table 5.2). SQRA and EnSRF lead to the highest correlations with

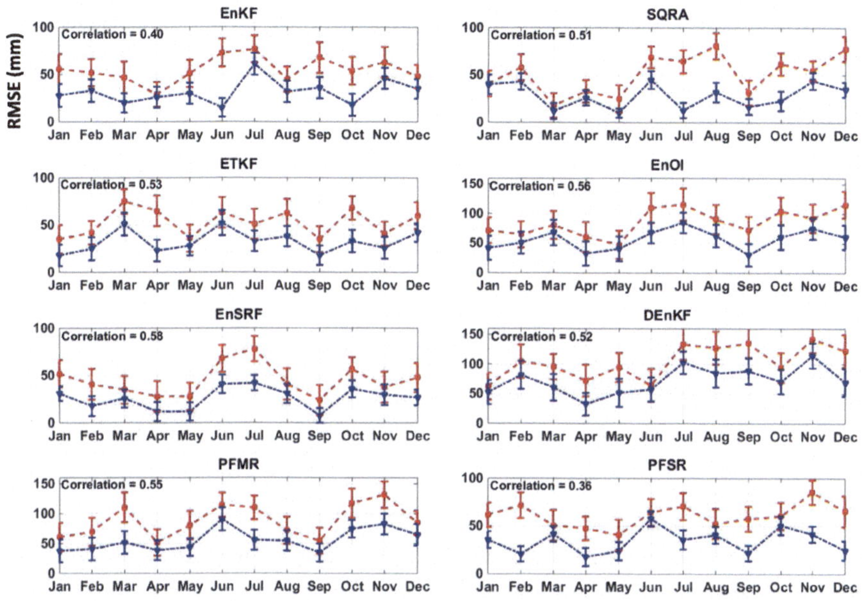

Fig. 5.6 Same as Fig. 5.5, but for the in-situ groundwater measurements and the filters estimates

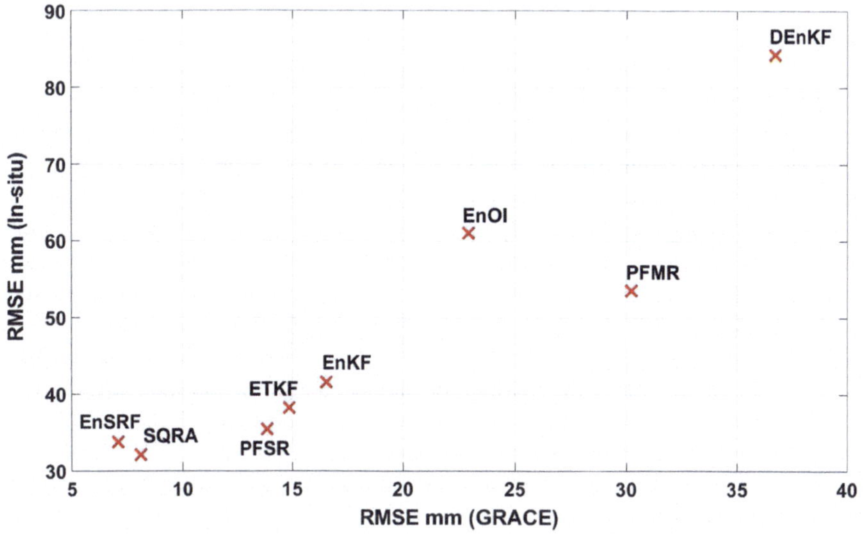

Fig. 5.7 Comparison between the computed average RMSEs of assimilation resulting from each filter applied using GRACE and the groundwater in-situ datasets. This figure presents the average of the best performances of the filters at the analysis steps from both assessments against GRACE and groundwater in-situ data

Table 5.3 Summary of statistical values derived from implemented methods using the groundwater in-situ measurements. For each method the RMSE average and its range (±XX) at the 95% confidence interval is presented. The improvements in the analysis state RMSE estimates are calculated using the in-situ measurements in comparison to the forecast states and model-free run

	Forecast		Analysis		Improvement (%)	
Method	RMSE (mm)	Correlation	RMSE (mm)	Correlation	Regarding forecast	Regarding model-free run
EnKF	62.65 ± 7.17	0.22	41.54 ± 5.98	0.64	24.65	31.68
SQRA	56.34 ± 6.18	0.28	32.13 ± 5.30	0.75	21.05	42.96
ETKF	60.77 ± 7.12	0.25	38.21 ± 5.21	0.67	15.54	33.12
EnOI	89.54 ± 8.87	0.17	61.05 ± 7.16	0.46	21.78	23.82
EnSRF	58.52 ± 6.85	0.23	33.74 ± 5.18	0.72	18.13	42.35
DEnKF	112.97 ± 10.84	0.14	84.31 ± 8.67	0.33	8.91	10.36
PFMR	75.37 ± 8.06	0.19	53.54 ± 6.85	0.55	10.55	14.96
PFSR	61.01 ± 7.05	0.22	35.45 ± 5.20	0.68	24.48	37.88

the in-situ measurements, i.e., 0.75 and 0.72, respectively. These filters also provide the best estimates in terms of estimated errors. DEnKF and to a lesser degree EnOI results in highest RMSEs. The Particle filters, especially the one using the systematic resampling technique, statistically exhibit reasonable results but not as good as the EnKFs. In terms of the assessment results against GRACE data, the deterministic ensemble filters provide the best performances (except for DEnKF). Overall, SQRA and EnSRF seem to be the most efficient for assimilating GRACE data into W3RA.

The correlations between the filters estimates and OzNet data also confirm the superiority of SQRA and EnSRF in the previous assessments (Table 5.4). Note that considering the difference between W3RA outputs (i.e., column water storage measured in mm) and the OzNet measurements (i.e., volumetric soil moisture) and the fact that converting the model outputs into volumetric units may introduce a bias (Renzullo et al. 2014), only a correlation analysis is considered here. For this purpose, we use different soil moisture layers from in-situ measurements including 0–8cm (compared to the model top soil moisture layer), 0–30cm (compared to the summation of the model top and shallow soil moisture layers), and 0–90cm (compared to the summation of the model top, shallow, and deep soil moisture layers). The correlations (at 95% confidence level) are computed between the OzNet data and model outputs after the forecast and analysis steps for each applied filter. The average improvements to the correlations (from forecast and analysis) with respect to open-loop outcomes are outlined in Table 5.4.

As expected, the highest correlations between the filters' estimates and the in-situ soil measurements are obtained in the analysis steps for the top, shallow, and root zones in comparison to the forecast steps. In general, the most significant improvements (19.88% average) are found for the root zone moisture. The GRACE observations show larger effects on the model in this soil layer for all the filters. The weakest correlations are found in the top layer soil moisture in both forecast and analysis steps with respectively 1.13 and 2.05% average improvement for all filters. The highest

Table 5.4 Summary of the improvements in correlations between state estimates derived from implemented methods and the soil moisture in-situ measurements at different layers. The improvements in the forecast and analysis state estimates are calculated using the in-situ measurements in comparison to the model-free run

	0–8 cm		0–30 cm		0–90 cm	
Method	Forecast	Analysis	Forecast	Analysis	Forecast	Analysis
EnKF	1.08	2.25	2.17	4.78	10.15	21.54
SQRA	2.42	3.45	3.34	8.36	15.84	26.17
ETKF	1.19	2.05	2.47	6.52	12.11	20.25
EnOI	–	0.78	0.27	2.77	8.78	13.00
EnSRF	2.35	3.34	2.99	8.91	16.73	28.94
DEnKF	–	0.66	–	1.26	4.18	12.24
PFMR	0.47	1.70	0.34	2.87	8.25	14.19
PFSR	1.56	2.19	2.72	7.40	13.36	22.73

correlations with the OzNet soil moisture measurements result from EnSRF, with 28.94% correlation improvement in the analysis step for the root zone comparison. SQRA also demonstrates a significant impact on the model predictions with 12.66% (average) correlation improvement for all the soil layers. The weakest performance, with respectively an average 1.39 and 4.72% correlation increases in the forecast and analysis step, is achieved by DEnKF. These results suggest the capability of EnSRF and SQRA in improving the non-observed model variables through data assimilation, performing significantly better than the other implemented filters against the in-situ measurements as outlined in both Tables 5.3 and 5.4. Their estimates are indeed more correlated (at 0.05 significant level) with the soil moisture and groundwater measurements and, in a statistical sense, exhibit the least RMSEs against the in-situ groundwater data.

5.5.2 *Error Analysis*

Analyzing the filters sampled error covariance, particularly the ensemble spread is important to understand the filters behaviors and performances. The performance of ensemble-based filters relies on their ability to represent and propagate the estimation error statistics, which of course depends on how the ensemble members are sampled and updated at the analysis steps (Sun et al. 2009). An efficient filtering method should be able to preserve the variation of its ensemble to properly span the error subspace while gradually reducing the uncertainties of the updates. The error covariance matrices are analyzed hereafter in terms of estimated errors and correlations.

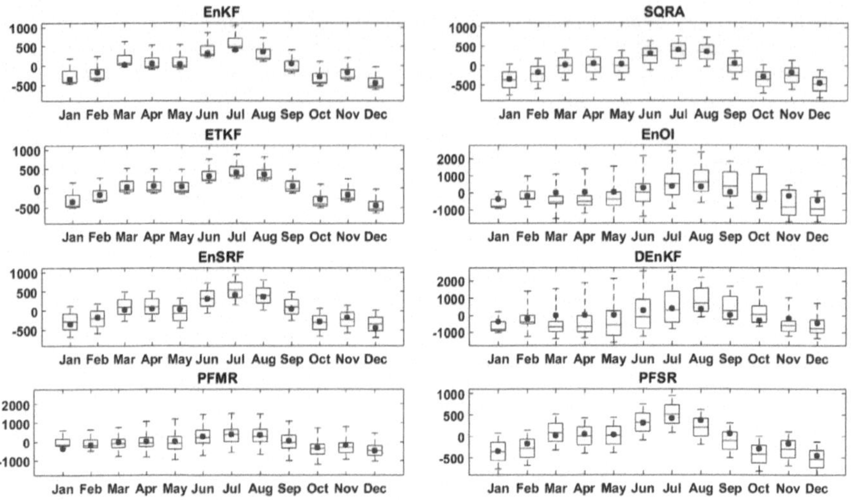

Fig. 5.8 TWS variations of ensembles (dashed black lines) at the assimilation steps represent by black boxes (with 0.05 significant level) for each filtering method (units are mm). The blue dotes are the observations and horizontal red lines show the mean ensembles at each analysis step

One important aspect of a filter performance refers to its ability to sample representative ensembles (or particles) at the analysis steps. Figure 5.8 outlines how the different filtering techniques adjust the ensemble members during the assimilation procedure. The average TWS variations time series over Australia and their ensembles at the analysis steps are calculated for all filters. For every month, we plot the range of TWS variations as predicted by the ensemble members for each filter, as well as the upper and lower margins of the ensemble results at the 95% significance interval (blue box), the ensemble mean (red lines), and observations (blue dots) to depict the average ensemble spread at each month by the applied filters.

The ensemble spread by each filter (Fig. 5.8) provide important insights about the filters behaviours. The filters should be able to preserve their ensemble spread to allow maximum exploitation of observations without under- nor over-fitting the data. As can be seen in Fig. 5.8 most of the filters generate ensemble means close to the assimilated observations, suggesting that the filters provide good estimates of the observed variables. However, one should also evaluate the distribution of the ensemble members. Those of EnKF, SQRA, ETKF, EnSRF, and PFSR are consistent over the assimilation window, which suggests the robustness of these techniques over time. The ensemble members, especially those of the EnSRF and SQRA, are uniformly distributed around the mean, suggesting a good coverage of the error sub-space. The ensemble distributions for DEnKF and EnOI, on the other hand, are different and exhibit an excessively large spread. In most of the cases, the range of the ensemble (with 95% confidence) in DEnKF and EnOI are either misplaced or overestimated. This would result in an overestimation of the forecast error and possibly lead to inaccurate assimilation results. In the case of the Particle filter, the

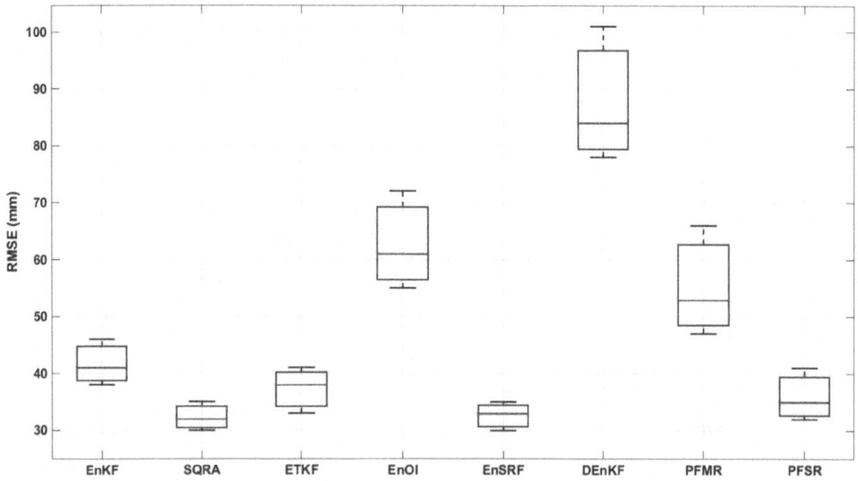

Fig. 5.9 The average groundwater RMSE of ensemble members for each method calculated using groundwater in-situ measurements

Systematic Resampling technique seems to be more robust; the PFMR ensembles and their variation (black dashed lines) span an unrealistically wide range space, even though the mean appears fairly close to the observations.

To evaluate the filters ability to preserve a good ensemble spread, we plot the RMSE variations of the filters' ensembles in Fig. 5.9. Groundwater time series of each ensemble member (for each filter) are spatially interpolated to the nearest in-situ bore and are assessed against in-situ groundwater measurements. We then calculate the average RMSE (of 54 bore stations) for each ensemble member to explore the proximity of the ensemble spread to the RMSE of the ensemble mean (indicated by a red line). The closer the ensemble members RMSE to that of the ensemble mean the better the performance is expected, suggesting that the estimated errors are reliable estimates of the true errors. It can be seen that this is best reflected by the EnSRF and SQRA. The ensemble RMSE variations (with 95% confidence interval) of these two filters (and to a lesser degree ETKF, PFSR, and EnKF) cover a smaller range, which suggests good estimations by all ensemble members. The lesser RMSEs reported by Fig. 5.9 for EnSRF (33.74 ± 5.18) and SQRA (32.13 ± 5.30) ensemble members, in a statistical sense, are more significant than other applied filters, which confirm the superiority of these two filters. DEnKF, EnOI, and PFMR, on the other hand, do not suggest good behaviours, exhibiting large ranges of ensemble errors. This shows that the ensemble members of these filters provide relatively different groundwater estimates.

More information can be inferred about the filter ensemble distributions by evaluating the ensembles skewness and kurtosis. These indicate the departure of the ensembles distributions from a Gaussian distribution (with a skewness 0 and a kurtosis 3). Kurtosis quantifies the distribution shape (i.e., heavy-tailed or light-tailed, in

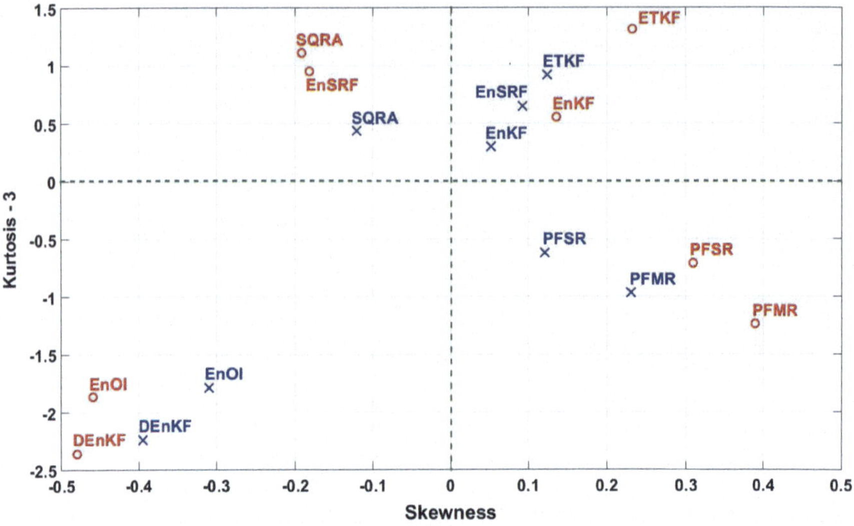

Fig. 5.10 Comparison between the average skewness and kurtosis of each filter for forecast (red circles) and analysis (blue crosses). Note that a normal distribution has a kurtosis of 3 and uses as a reference so the excess kurtosis is usually presented by kurtosis–3

comparison to a normal distribution) and skewness measures the distribution asymmetry (Joanes and Gill 1998). Ensembles are expected to span the model subspace and this might be better satisfied by small skewness and kurtosis (Lawson and Hansen 2004). The average ensembles skewness and kurtosis of all filters (Fig. 5.10) show skewness and kurtosis changes during assimilation (in forecast and analysis steps). Figure 5.10 depicts the reduction in skewness and kurtosis after the analysis steps for all filters, suggesting that the filters posteriors become closer to Gaussian as assimilation proceeds. This is, however, more pronounced for skewness than for kurtosis, suggesting the filters higher impact on the ensembles distribution asymmetry. The stochastic EnKF ensemble is closer to a Gaussian distribution, which is related to the random perturbations of the observation (Hoteit et al. 2015). In contrast, the DEnKF and EnOI ensembles are not uniformly distributed, showing an important departure from Gaussian distributions. Although a Gaussian distributed ensemble is not required for EnKFs, a non-Gaussian distributed ensemble might result in unrealistic approximations of the filters posteriors. The larger values for skewness and kurtosis for the DEnKF and EnOI could explain their weaker performances in comparison to the often ensemble Kalman-based filters.

We further perform an analysis to investigate the influence of the filtering methods on the forecast and analysis error statistics. To this end, we investigate the time-evolution of the forecast and analysis error covariances for each filter. The reductions in the all diagonal elements of the error covariance matrices before and after data assimilation are calculated at each assimilation step and their minimum, maximum, and average are calculated. The results show how the different filtering techniques

Table 5.5 Effects of filtering methods on the model state covariance matrix as a percentage improvement

Error reduction (%)		Method							
		EnKF	SQRA	ETKF	EnOI	EnSRF	DEnKF	PFMR	PFSR
	Minimum	29	35	22	15	34	6	18	28
	Maximum	47	52	44	38	55	20	35	48
	Average	35	**44**	33	21	**47**	8	27	**38**

decrease the errors using GRACE data (Table 5.5). The best performances are found for EnSRF (47% average), SQRA (44% average), and to a lesser degree, PFSR. Again, DEnKF and EnOI show the weakest effects on error covariance.

Only two filters of SQRA and EnSRF show good performances in both analyses. These filters not only better incorporate GRACE data into the model state and improve its estimates compared to the (groundwater level and soil moisture) in-situ measurements, but also efficiently spread the ensemble and decrease errors. The resulting estimates of groundwater storage by these filters also exhibit less RMSE against independent groundwater level in-situ data.

5.6 Summary and Conclusions

There is evidence that the different filtering schemes are more suited to different applications (Reichle et al. 2002). This study considered the implementation of several data assimilation filtering techniques based on the two most commonly applied algorithms, ensemble Kalman filter, and Particle Filter, to assess their performances for assimilating GRACE TWS data into the hydrological model W3RA. GRACE-derived TWS over Australia was assimilated into the W3RA hydrological model using the various filters. Among the ensemble Kalman filters, we tested the stochastic and the deterministic schemes including standard EnKF, the Square Root Analysis (SQRA), the Ensemble Transform Kalman Filter (ETKF), the Deterministic EnKF (DEnKF), the Ensemble Square-Root Filter (EnSRF), and the Ensemble Optimal Interpolation (EnOI) along with two different resampling approaches of Multinomial Resampling and Systematic Resampling techniques for Particle Filters (PFMR and PFSR, respectively). The effects of the filtering methods on the ensembles spread and the estimation error covariance matrices were investigated. The filter state estimates were further evaluated against in-situ groundwater measurements and in-situ soil moisture data. The most promising results were obtained using SQRA and EnSRF both in terms of ensemble representation as well as an assessment against in-situ measurements, demonstrating significantly better performance than the other implemented filters. The largest errors reduction with minimum error covariances was achieved by EnSRF (47% average) and SQRA (44% average). These two filters (along with EnKF) also showed a good ability to sample representative ensembles with enough spread. In a statistical sense (at 95% confidence interval), these two filters provided the least RMSEs (32.14 and 33.74 mm) and maximum correlations for both groundwater level (0.75 and 0.72) and soil moisture (0.78 and 0.82 for the 0–90 cm soil depth) in-situ measurements. In contrast, DEnKF was the least successful in the experiment and suggested a larger error in the state estimates. Considering the model-specific and application-specific characteristics of the present study, a general conclusion on the preference of ensemble filters might not be possible. Thus, further research should be undertaken to investigate the filters in different hydrological applications and also to explore other filters like new designed Particle filters that were not considered here.

Part IV
GRACE Data Assimilation

For efficient assimilation of satellite observations, in addition to choosing the right filtering method, it is important to effectively deal with their uncertainty during the process. For some satellite remote sensing such as GRACE, this is more important due to their unique characteristics. In order to efficiently assimilate GRACE data in a land surface model, a special focus should be undertaken to consider these characteristics, e.g., the coarser spatio-temporal resolution compared to most of the existing hydrological models. GRACE data has been mostly applied for data assimilation using GRACE level 2 solutions provided in terms of spherical harmonics potential coefficients or mass concentration (mascon) solutions, both with coarser spatial (e.g., 3°) and temporal (e.g., monthly) resolutions of models. These make GRACE data assimilation quite challenging compared to other satellite products. For example, applying GRACE data with higher spatial resolution such as 1° results in a highly correlated error covariance, which can later be problematic during the assimilation process. GRACE correlated errors cause the covariance matrices to be badly conditioned or not invertible leading to inefficiency of the filtering process during data assimilation. A successful assimilation filtering method should be able to account for these limitations in GRACE products (see, e.g., Schumacher et al. 2016).

Different studies have assimilated GRACE data in either basin scales (e.g., Zaitchik et al. 2008; Houborg et al. 2012; Li et al. 2012) or grid element scales (e.g., Eicker et al. 2014; Tangdamrongsub et al. 2015; Schumacher et al. 2016). Moreover, an assumption of Gaussian error for observations is often made to deal with the badly conditioned error covariance problem (Zaitchik et al. 2008; Reichle et al. 2013; Tangdamrongsub et al. 2015; Kumar et al. 2016; Tian et al. 2017) mainly either for simplification or lack of information about GRCAE full error covariance. Some of these issues have been investigated, e.g., Eicker et al. (2014) studied the relationship of different GRACE spatial resolutions on the data assimilation process and Girotto et al. (2016) considered using GRACE error covariances with 3° spatial resolution to avoid error correlation (see aso Girotto et al. 2017). Accordingly, rigorous modelling of the full error covariance matrix of the GRACE TWS is required to maximize the contribution of observations into model state realistically. To achieve this, one not only requires an appropriate filtering method but also needs a robust assimilation strategy. This chapter describes an extensive framework for assimilating GRACE TWS data, which provides a valuable opportunity for land surface modelling.

Chapter 6
Efficient Assimilation of GRACE TWS into Hydrological Models

Data assimilation as an inverse problem uses the covariance information of model simulations and observations. Significantly correlated errors yield covariance matrices that are bad conditioned or not invertible leading to inefficiency in filtering process during data assimilation. Due to the lack of information (or to enhance computations), the decision of uncorrelated data (Gaussian error for observations) is often made to deal with this problem, which cannot be realistic.

—M. Khaki (This chapter is presented following Khaki et al. (2017a), "Accounting for spatial correlation errors in the assimilation of GRACE into hydrological models through localization")

6.1 Summary

Assimilation of terrestrial water storage (TWS) information from the Gravity Recovery And Climate Experiment (GRACE) satellite mission can provide significant improvements in hydrological modeling. However, the rather coarse spatial resolution of GRACE TWS and its spatially correlated errors pose considerable challenges for achieving realistic assimilation results. Consequently, successful data assimilation depends on rigorous modelling of the full error covariance matrix of the GRACE TWS estimates, as well as realistic error behavior for hydrological model simulations. In this study, we assess the application of local analysis (LA) to maximize the contribution of GRACE TWS in hydrological data assimilation. For this, we assimilate GRACE TWS into the World-Wide Water Resources Assessment system (W3RA) over the Australian continent while applying LA and accounting for existing spatial correlations using the full error covariance matrix. GRACE TWS data is applied with different spatial resolutions including 1° to 5° grids, as well as basin averages. The ensemble-based sequential filtering technique of the Square Root Analysis (SQRA) is applied to assimilate TWS data into W3RA. For each spatial scale, the performance of the data assimilation is assessed through comparison with independent

M. Khaki, *Satellite Remote Sensing in Hydrological Data Assimilation*,
https://doi.org/10.1007/978-3-030-37375-7_6

in-situ ground water and soil moisture observations. Overall, the results demonstrate that LA is able to stabilize the inversion process (within the implementation of the SQRA filter) leading to less errors for all spatial scales considered with an average RMSE improvement of 54% (e.g., 52.23 mm down to 26.80 mm) for all the cases with respect to groundwater in-situ measurements. Validating the assimilated results with groundwater observations indicates that LA leads to 13% better (in terms of RMSE) assimilation results compared to the cases with Gaussian errors assumptions. This highlights the great potential of LA and the use of the full error covariance matrix of GRACE TWS estimates for improved data assimilation results.

6.2 Introduction

The Gravity Recovery And Climate Experiment (GRACE) satellite mission provides global time-variable gravity field solutions that have been used to obtain global terrestrial water storage (TWS) changes (Tapley et al. 2004). Several studies indicate that GRACE TWS can play an important role in better understanding surface and sub-surface physical processes related to water redistribution within the Earth system (e.g., Huntington 2006; Chen et al. 2007; Kusche et al. 2012; Forootan et al. 2014b; van Dijk et al. 2014; Wouters et al. 2014). A growing number of studies has also been applying GRACE TWS to constrain the mass balance of hydrological models (e.g., Zaitchik et al. 2008; Thomas et al. 2014; van Dijk et al. 2014; Eicker et al. 2014; Tangdamrongsub et al. 2015; Reager et al. 2015; Khaki et al. 2017b). This combination is motivated by the fact that hydrological models use conceptual or physical knowledge (or both) to simulate hydrological processes at global (e.g., Huntington 2006; Coumou and Rahmstorf 2012) and regional (e.g., Zaitchik et al. 2008; Chen et al. 2013; Munier et al. 2014) scales. The accuracy of simulations might be limited due to imperfect models (i.e., lack of knowledge about the processes or simplified model equations) and uncertainties in input and forcing data (Vrugt et al. 2013). Data limitation (both on temporal and spatial scales) also plays a substantial role in land hydrological modeling, especially for closing the water balance that requires reliable information about all storage compartments from which that of groundwater is very challenging. In this regard, GRACE TWS estimates are of great importance since they can be used through data assimilation to constrain the vertical summation of water storages (including groundwater) in the models.

Data assimilation is a technique to incorporate observations into a dynamic model in order to improve its state estimation (Bertino et al. 2003; Hoteit et al. 2012). It has been widely applied in the fields of ocean and climate science (Garner et al. 1999; Elbern and Schmidt 2001; Bennett 2002; Kalnay 2003; Schunk et al. 2004; Lahoz 2007; Zhang et al. 2012). In hydrological studies, different in-situ measurements (e.g., river discharge and soil moisture) have been assimilated into models (Liu et al. 2012) to improve their estimates of different hydrological quantities (see, e.g., Crow and Wood 2003; Seo et al. 2003; Vrugt et al. 2005; Weerts et al. 2006; Reichle et al. 2010).

The application of remotely sensed data in data assimilation for hydrological purposes has gathered interests in the past few years. This is especially due to the increased development and availability of satellite remote sensing systems such as Sentinel, Soil Moisture Active Passive (SMAP), GRACE, and satellite radar altimetry (e.g., Moradkhani et al. 2006; Clark et al. 2008; Houborg et al. 2012; van Dijk et al. 2014; Renzullo et al. 2014; Reager et al. 2015; Kumar et al. 2016). Data assimilation can improve various water compartments of hydrological models such as soil (e.g., Reichle et al. 2002, 2008; Brocca et al. 2010; Kumar et al. 2014; Renzullo et al. 2014), surface water (e.g., Alsdorf et al. 2007; Neal et al. 2009; Giustarini et al. 2011), and snow (e.g., Liu et al. 2013; Kumar et al. 2015) storages. A number of studies has also investigated the possibility of using GRACE data to improve hydrological models (e.g., Zaitchik et al. 2008; Houborg et al. 2012; Li et al. 2012; Eicker et al. 2014; van Dijk et al. 2014; Tangdamrongsub et al. 2015; Kumar et al. 2016; Schumacher et al. 2016).

GRACE data with a suitable coverage, both temporally and spatially, provide a unique opportunity to study water storages in lands on global and regional scales. The mission now provides 15 years of data with a global coverage, which provides the chance to study seasonal to decadal changes in TWS. Before using GRACE TWS in any assimilation framework, however, there are some important aspects which should be considered such as the temporal and spatial resolution mismatch between GRACE observations and model simulations, as well as existing spatial and temporal correlations in the time series of GRACE TWS and model simulations. Its spatial resolution is limited to a few hundred kilometers depending on the signal strength and the inversion technique applied to recover time-variable gravity fields (Schmidt et al. 2008). This coarse spatial resolution exists in both GRACE level 2 solutions provided in terms of spherical harmonics potential coefficients or mass concentration (mascon) solutions. Although mascon is provided on a finer spatial scale (e.g., 0.5°), the native resolution of the data is smaller (e.g., 3°; Watkins et al. 2015; Wiese 2015). Different studies have tried to assimilate GRACE data in either basin scales (e.g., Zaitchik et al. 2008; Houborg et al. 2012; Li et al. 2012) or grid element scales (e.g., Eicker et al. 2014; Tangdamrongsub et al. 2015; Schumacher et al. 2016). GRACE level 2 products have been truncated (e.g., at degree and order 60–120). They also have been filtered (e.g., Swenson and Wahr 2006; Kusche 2007) resulting in low spatial resolutions. Upscaling of the original established TWS with a limited spatial resolution to create a high spatial resolution data (e.g., 1°) with grid points that are not independent of each other increases spatial correlation significantly (see e.g., Schumacher et al. 2016). Accounting for these correlations is important especially in the context of data assimilation, where complete knowledge of the data error structure including uncertainties and existing correlations is necessary.

Data assimilation as an inverse problem uses the covariance information of model simulations and observations. Significantly correlated errors yield covariance matrices that are bad conditioned or not invertible leading to inefficiency in filtering process during data assimilation. Due to the lack of information (or to enhance computations), the decision of uncorrelated data (Gaussian error for observations) is often made to deal with this problem, which can be realistic when observations are denser than models' grid, e.g., independent grid points of neighbours (Berger and Forsythe 2004; Stewart et al. 2008). In contrast, when the spatial resolution of models is finer than the assimilated observations, it can lead to no improvement in the accuracy of final assimilation results (e.g., Liu and Rabier 2003; Dando et al. 2007; Stewart et al. 2008). In this regard, it is necessary to precisely consider the full GRACE error covariance for different spatial resolutions in data assimilation applications especially where the model spatial scale is finer than GRACE TWS, and the existing correlations in the observations are problematic (see e.g., Schumacher et al. 2016).

Most of the previous studies assimilated GRACE TWS (e.g., grid-based or basin averaged) into models while assuming white noise (i.e., uncorrelated observations). This, for basin averaged applications, might be justified to some extent as the spatial averaging of TWS observations adds up the non-Gaussian noise distributions and generates a mixture that is closer to Gaussian distribution according to the central limit theorem (Stone 2004, Chap. 5). In this regard, for example, Zaitchik et al. (2008) applied GRACE TWS on a sub-basin scale (sub-basins of the Mississippi River) and assumed a Gaussian error (with zero correlation) for GRACE TWS measurements. Reichle et al. (2013) investigated the effects of coarse-scale satellite observations (e.g., GRACE) and vertically integrated measurements (such as TWS) on model variables within the assimilation system. For a grid-based assimilation of GRACE-TWS in models, Eicker et al. (2014) studied the relationship of different GRACE spatial resolutions on the data assimilation process and reported that there is always a trade-off between employing GRACE data in a higher spatial resolution while keeping the GRACE error covariance matrices reasonably well conditioned. Girotto et al. (2016, 2017) have considered the fact that 1° GRACE error covariances are spatially highly correlated and to address this issue, they have used a spatial correlation length of 3° for the observation errors (see also Kumar et al. 2016; Khaki et al. 2017b). Schumacher et al. (2016) indicated that both the characteristics of GRACE error correlation and spatial discretization of TWS observations are important on the performance of the data assimilation process. In another effort, van Dijk et al. (2014) proposed an alternative approach for estimating GRACE TWS errors in data assimilation. The triple collocation technique was used to merge model-derived storage in (sub-) surface compartments with TWS estimates from GRACE measurements (van Dijk et al. 2014). In the studies discussed above, GRACE error covariance for different spatial resolutions is hardly treated. For example, Eicker et al. (2014) considered error covariance of various spatial resolutions that were rescaled (e.g., rescaling 0.5° to 5°) rather than solving for distinct spatial resolution individually (e.g., 0.5°, 1°, and 5°).

In the present study, we extend the works above by employing a Local Analysis (LA) technique. LA allows utilization of different GRACE TWS spatial resolutions by addressing instability in data assimilation that arises from the GRACE covariance matrices of the corresponding spatial resolutions. The contribution of this study is, therefore, twofold: (i) we mathematically assess the efficiency of the localization technique to use GRACE TWS with its full error information and with high spatial resolution in an assimilation framework; and (ii), we compare the performance of a localization technique to in-situ data in a real case study covering the entire Australian continent. These will assess the ability of local data assimilation in maximizing the contribution of GRACE TWS into a hydrological model by considering its full error covariance matrix. Here, we use the full variance-covariance of GRACE to establish the observation error covariance matrices for the grid resolutions of 1°, 2°, 3°, 4°, 5°, and a basin scale, and examine their effects on data assimilation. More importantly, for the first time, we offer a solution to increase the performance of data assimilation in using GRACE data. A localization technique is applied to account for correlations in high spatial resolution observations, which can lead to a rank deficiency problem and correspondingly an instability in the data assimilation procedure. In terms of localization technique, Local Analysis (LA) of the filter (Evensen 2003; Ott et al. 2004) is considered mainly due to its ability in dealing with correlations by spatially limiting the use of ensemble-based covariance information of high-dimensional systems to the limited local region (Ott et al. 2004). LA effects on each data assimilation scenario (i.e., using different spatial resolutions) are assessed to explore its ability for improving the results. In addition, the application of LA has the potential to minimize a large part of error sources in the ensemble filtering methods when a small number of ensembles is used (Mitchell and Houtekamer 2000; Houtekamer and Mitchell 2001).

GRACE TWS data is assimilated into the World-Wide Water Resources Assessment (W3RA, van Dijk 2010) over Australia. The ensemble-based sequential technique of the Square Root Analysis (SQRA) filtering scheme represented by Evensen (2004) is used to assimilate GRACE TWS into W3RA. SQRA, which is a deterministic form of ensemble Kalman filtering, has considerable advantages in comparison to some existing methods in terms of the computational speed, simplicity, and its independency to an observation perturbation unlike traditional Kalman filtering methods (see detail in Sect. 6.4.1 and Khaki et al. 2017b). In addition to implementing the LA, in order to further address possible problems that arise from ensemble size, sampling errors, and insufficient ensemble variance in ensemble-based techniques (Anderson et al. 2007; Oke et al. 2007), ensemble inflation is applied. This technique, which has frequently been used in previous works (e.g., Anderson et al. 1999, 2007; Ott et al. 2004), tries to increase the variance of ensembles around the ensemble mean by inflating prior ensembles (Anderson et al. 2007).

6.3 Datasets

6.3.1 Grace

Monthly GRACE level 2 (L2) potential coefficients products along with their full error covariance information are obtained from the ITSG-Grace2014 gravity field model (Mayer-Gürr et al. 2014). The solution is computed up to degree and order (d/o) 90 resulting in approximately ∼300 km spatial resolution at the equator. The study period (February 2003 to December 2012) is limited by the availability of the climate data (see Sect. 6.3.2) to force the hydrological model.

Following Swenson et al. (2008), degree 1 coefficients (http://grace.jpl.nasa.gov/data/get-data/geocenter/) are replaced to account for the movement of the Earth's centre of mass. Degree 2 and order 0 (C_{20}) coefficients (http://grace.jpl.nasa.gov/data/get-data/oblateness/) are not well determined and are replaced by those from Cheng and Tapley (2004). Correlated noise in the TWS data products is reduced by applying de-striping and smoothing using a Gaussian averaging kernel with 300 km half radius following Swenson and Wahr (2006). This causes some degree of signal attenuation (Klees et al. 2008) and moving anomalies from one region to another (Chen et al. 2007). This leakage effect can lead to some degree of signal inference especially at the land-ocean boundary. In order to address this issue, following Swenson and Wahr (2002), we apply an isotropic kernel using a Lagrange multiplier filter to best balance signal and leakage errors over the entire Australia. This filter uses a basin averaging kernel method expanded in spherical harmonic coefficients and subsequently combined with L2 potential coefficients to improve the GRACE estimates (see details in Swenson and Wahr 2002).

The filtered gravity fields, are then converted to TWS changes (following Wahr et al. 1998) over the entire Australia in both grid and basin scales. The amount of rainfall over Australia, especially over its northeast, western, and central parts, is low in comparison to other inhabited continents on Earth leading to prolonged drought in the interior regions (Forootan et al. 2016). This effect can be seen from the average precipitation (between February 2003 and December 2012) in Fig. 6.1. This map shows small amount of rainfall over most parts of Australia (e.g., the western and eastern parts). Therefore, an accurate estimation of water storages (e.g., using hydrological models) is necessary to manage water resources in this region. TWS changes from GRACE are gridded into the spatial grid resolutions of 1°, 2°, 3°, 4°, 5°, and also a basin scale for 12 major Australian drainage divisions and river basin (cf. Fig. 6.1). As a number of studies have used basin averaged GRACE TWS for data assimilation (e.g., Zaitchik et al. 2008; Houborg et al. 2012), we test the LA in both grid and basin scales. Accordingly, for each grid size as well as basin scale error covariance matrices are calculated using the full error information of the L2 potential coefficients for each month. Note that the errors in lower degree potential coefficients provided along with degree 1 coefficients and C_{20} are substituted into the GRACE covariance matrix. No correlation is considered between the GRACE covariance matrix and errors in the lower degree potential coefficients. This error information is then used to reach observation errors for data assimilation. To this end, following Schumacher et al. (2016), an error propagation is implemented to convert the full error information of the GRACE coefficients to TWS errors.

6.3.2 W3RA

In this study, we use the World-Wide Water Resources Assessment system (W3RA), which was developed in 2008 by the Commonwealth Scientific and Industrial Research Organisation (CSIRO) to monitor, represent and forecast Australia's

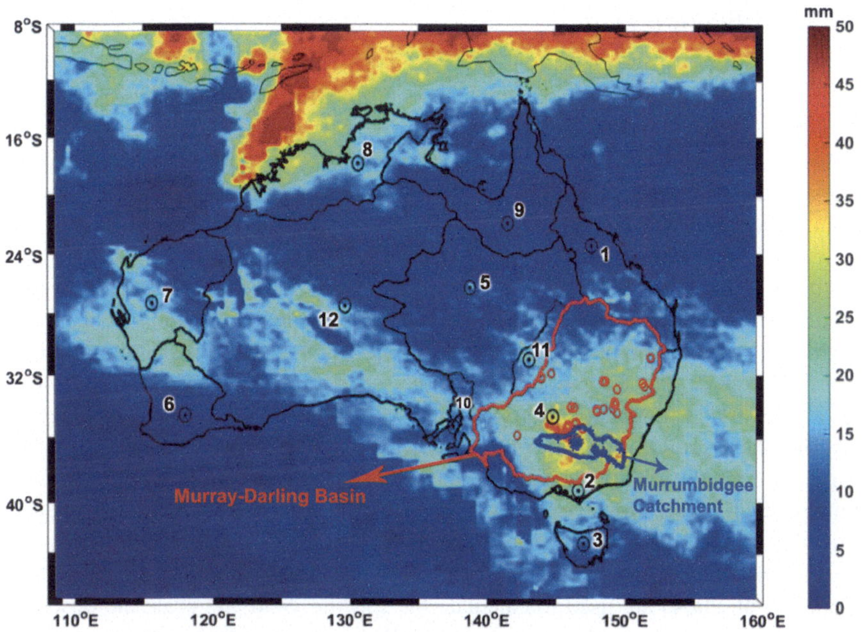

Fig. 6.1 Overview of the study area. The black polygons indicate the twelve river basins that are considered for spatial aggregation of GRACE data to basin scale. The red and blue polygons indicate the Murray-Darling Basin and Murrumbidgee Catchment, respectively. Data from in-situ groundwater stations (red circles) and data from the OzNet soil moisture network (blue circles) are used in these regions for independent validation of the data assimilation results. The underlaying map shows temporally averaged precipitation between 2003–2013 from TRMM-3B42 products (Huffman et al. 2012) on a $1° \times 1°$ grid

terrestrial water cycles (http://www.wenfo.org/wald/data-software/). W3RA is a grid distributed biophysical model that simulates water stores and flows with significant information of water storages over Australia (van Dijk 2010; Renzullo et al. 2014). Globally distributed $1°\times1°$ minimum and maximum temperature, downwelling short-wave radiation, and precipitation from Princeton University (http://hydrology.princeton.edu) are used as meteorological forcing data (see detail in Sheffield et al. 2006). The model parameters include effective soil parameters, water holding capacity and soil evaporation, relating greenness and groundwater recession, and saturated area to catchment characteristics (van Dijk et al. 2013). Model state in this study includes the W3RA water storages in the top, shallow, and deep root soil layers, groundwater storage, and surface water storage in a one-dimensional system (vertical variability). Here, we use W3RA (with a daily scale) for the same temporal coverage of GRACE (e.g., February 2003 to December 2012) and the spatial resolution of $1°\times1°$. More detailed information on W3RA can be found in van Dijk et al. (2013).

6.3.3 Validation Data

We use groundwater in-situ measurements over the Murray-Darling basin extracted from the New South Wales Government (NSW) groundwater archive (http://waterinfo.nsw.gov.au/pinneena/gw.shtml) to evaluate the performance of applied data assimilation. Although data assimilation is done over entire Australia, due to limited availability of in-situ stations, the existing in-situ measurements over the Murray-Darling basin are used for result assessment. Measurements with data gaps and those that did not exhibit seasonal variations are flagged as belonging to confined aquifers and are excluded (Houborg et al. 2012; Tangdamrongsub et al. 2015). Therefore, daily and monthly well measurements of 54 spatially distributed stations over the basin (cf. Fig. 6.1) are used and time series of groundwater storage anomalies are generated for each station. Selected well-water levels need to be converted to variations in groundwater (GW) storage in terms of equivalent water heights. This is usually done through the specification of yield estimates (e.g., Rodell et al. 2007; Zaitchik et al. 2008). However, such information does not exist in this study. Hence, following Tangdamrongsub et al. (2015), TWS variations from GRACE and soil moisture products from Global Land Data Assimilation System (GLDAS) NOAH (Rodell et al. 2004) are used to calculate the specific yield and scale the observed head water by modifying the magnitude of GW time series (Tregoning et al. 2012). As Tregoning et al. (2012) showed, the GW component can be extracted over Australia by removing the soil moisture component from GRACE TWS data. Other water compartments including biomass and surface water variations can be excluded due to their small contribution to regional scale mass variations within Australia. Through this approach, rather than assuming a constant specific yield everywhere (e.g., 0.1 by Tregoning et al. 2012), different yield values can be derived leading to a more realistic representation of groundwater systems in different areas.The calculated specific yields range between 0.08 and 0.16, falling within the 0.05–0.2 range suggested by the Australian Bureau of Meteorology (BOM) and Seoane et al. (2013), hence justifying the application of the method. The extracted yield factor is used at each in-situ location to scale the observed in-situ head time series (see also Rodell et al. 2007; Longuevergne et al. 2013). After removing temporal averages of in-situ groundwater time series, the anomaly time series are used in this study to assess W3RA estimates after the assimilation process.

Further result assessment is done using in-situ soil moisture measurements. These datasets are obtained from the moisture-monitoring network (http://www.oznet.org.au/) known as OzNet network and spotted in the Murrumbidgee catchment (Smith et al. 2012). OzNet network provides long-term records of measured volumetric soil moisture at various soil depths at 57 locations across the Murrumbidgee catchment area (cf. Fig. 6.1). The anomalies of in-situ soil moisture measurements are calculated and then averaged into daily scale. Following Renzullo et al. (2014), 0–8 cm data is used to evaluate the estimated model top-layer soil moisture and the 0–30 cm and 0–90 cm measurements are applied for the evaluation of the model shallow root-zone soil moisture estimation.

6.4 Data Assimilation

6.4.1 Methods

6.4.1.1 Square Root Analysis (SQRA)

The solution of the data assimilation problem is based on Bayes' theorem (Jazwinski 1970; van Leeuwen and Evensen 1996), which tries to improve the model state by updating the prior Probability Density Function (PDF) whenever new observations are introduced. The sequential data assimilation technique solves the Bayesian estimation problem numerically by providing a probabilistic framework and sequentially estimates the whole system using propagated information (ensembles) only forward in time (Jardak et al. 2007). There are various filtering methods in this framework, however, one of the mostly applied techniques is ensemble-based Kalman filter. In this study, we use the square root analysis (SQRA) scheme for the Ensemble Kalman Filter (EnKF), represented by Evensen (2004) as a data assimilation filtering method. SQRA is a deterministic form of ensemble-based Kalman filters and uses a statistical sample of state estimates (Sakov and Oke 2008). The model state contains N different vectors (N is the number of ensembles), each with the same size of the model state variables. The forecast model state is represented by $X^f = [X_1{}^f \ldots X_N{}^f]$, where $X_i{}^f$ $(i = 1 \ldots N)$ is the ith ensemble (hereafter 'f' stands for forecast and 'a' stands for analysis). The model state forecast error covariance of P^f is defined by:

$$P^f = \frac{1}{N-1}\sum_{i=1}^{N}(X_i{}^f - \bar{X}^f)(X_i{}^f - \bar{X}^f)^T = \frac{1}{N-1}A^f A^{f^T}, \tag{6.1}$$

where $\bar{X}^f$ is the ensemble mean and can be calculated using,

$$\bar{X}^f = \frac{1}{N}\sum_{i=1}^{N}(X_i). \tag{6.2}$$

Forecast ensemble of anomalies, $A^f = [A_1{}^f \ldots A_N{}^f]$, is the deviation of model state ensembles from the ensemble mean,

$$A_i{}^f = X_i{}^f - \bar{X}^f. \tag{6.3}$$

SQRA eliminates the need for the perturbation of measurements, which is essential in traditional EnKF (Burgers et al. 1998). Instead, SQRA uses unperturbed observations without imposing any additional approximations like uncorrelated measurement errors (Evensen 2004) by introducing a new sampling scheme. Rather than updating each sample separately in the analysis step, SQRA updates all of them in two stages; firstly by updating the ensemble-mean using $\bar{X}^f$ (cf. Eq. 6.1) as,

$$\bar{X}^a = \bar{X}^f + K(y - H\bar{X}^f), \quad i = 1 \ldots N, \tag{6.4}$$
$$K = P^f(H)^T(HP^f(H)^T + R)^{-1}, \tag{6.5}$$

where $\bar{X}^a$ is the mean analysis state, K represent the Kalman gain, y and R are the observation vector and associated covariance matrix. The transition matrix from the state vector space to the observation space is shown by H. Next, SQRA computes the ensemble anomalies. In this regard, one needs to first calculate the ensemble version of the analysis error covariance matrix, which can be done using Eq. 6.6. Afterward, by inserting the forecast (P^f from Eq. 6.1) and analysis (P^a from Eq. 6.6) error covariances in Eq. 6.7 and solving for A^a, analysis ensemble of anomaly can be computed.

$$P^a = \frac{A^a(A^a)^T}{N-1} \tag{6.6}$$
$$P^a = (I - KH)P^f \tag{6.7}$$

After a few simplification steps (cf. Evensen 2004), A^a can be obtained by,

$$A^a = A^f V \sqrt{I - \Sigma^T \Sigma \Theta^T}, \tag{6.8}$$

where Σ and V are calculated using singular value decomposition of A^f ($A^f = U \Sigma V^T$). Γ refers to the singular value decomposition and Θ is a random orthogonal matrix (e.g., the right singular vectors from a singular value decomposition of a random $N \times N$ matrix) for ensemble redistribution of the variance reduction (cf. Evensen 2004, 2007; Khaki et al. 2017b).

6.4.1.2 Filter Tuning

Many studies have previously investigated the sensitivity of ensemble-based schemes on ensemble size (e.g., Houtekamer 1995; Houtekamer and Mitchell 1998; Keppenne 2000; Mitchell et al. 2002; Keppenne and Rienecker 2002). It has been proven that a large number of ensemble members in ensemble data assimilation systems causes computation time to significantly increase while using a small ensemble size can also be problematic, as it can lead to filter divergent or inaccurate estimation (Tippett et al. 2003). A successful ensemble-based filter needs to adequately span the model sub-space for a better approximation of probability distribution of the background errors (Ott et al. 2004). This, however, can be very challenging once a small ensemble number (considerably less than the model dimension) is used. To tackle this problem, we apply ensemble inflation, which uses a small coefficient to separately inflate prior ensemble deviation from the ensemble-mean and increases their variations (Anderson et al. 2007). Here, we use a constant factor ($S = 1.12$; Anderson 2001) to inflate the ensemble perturbations as,

$$X'^{f} = S(X^{f} - \bar{X}^{f}) + \bar{X}^{f}, \tag{6.9}$$

with X'^{f} representing the new forecast state, which contains the inflated ensemble perturbation.

A further solution when dealing with a limited ensemble number is the application of localization techniques initially proposed by Houtekamer and Mitchell (2001). In this study, we use the Local Analysis (LA) scheme not only to address the issue of the small ensemble number, but also to investigate its effects in dealing with the GRACE error covariance for different spatial resolutions. LA works by restricting the information used for the covariance matrix computation to a spatially limited area and uses only measurements located within a certain distance from a grid point (Evensen 2003; Ott et al. 2004; Khaki et al. 2017b).

In using LA, at each horizontal grid point (m, n), with m and n representing geographic latitude and longitude directions, respectively, the selected measurements close to the grid point contribute to the SQRA filtering process. This means that only particular state variables close to the point (m, n) within an assumed distance and corresponding observations at the same locations are used in the assimilation process. To do this, a local system state vector, observations, and their covariance matrix need to be chosen at each grid point separately. Following Ott et al. (2004), a model state vector $X(r)$ (r is a two-dimensional vector with r_{mn}) is used to achieve the local forecast state vector $X^{f}{}_{mn}$ in Eq. 6.9 using a linear operator M_{mn} by,

$$X'^{f}{}_{mn} = M_{mn} X'^{f}(r). \tag{6.10}$$

At the specific grid point of (m_0, n_0), $X'^{f}{}_{mn}$ contains the information of $X'^{f}(r_{m+m_0,n+n_0})$ with $-l \leq m - m_0, n - n_0 \leq l$ (l localization length) and limited to grid points close to (m_0, n_0) within a $(2l + 1)$ by $(2l + 1)$ patch (Ott et al. 2004).

Local state vectors and observations within the local region (y_{mn}) with covariance matrix R_{mn} can then be used in SQRA to locally estimate the model state for each grid point. In case of using a gridded GRACE TWS dataset in a finer spatial resolution (e.g., 1° and 2°), the calculated error covariances have rank deficiency mainly due to correlation errors (see more details in Sect. 6.5.1). This problem can cause instability in the data assimilation procedure. Applying LA, therefore, can be helpful since it numerically resolves the possible singularity in the filtering process during data assimilation. Ott et al. (2004) proved that LA yields a good approximate representation of the background covariance matrix using a small ensemble number with a rank much lower than the state dimension. LA localization can also be used in the vertical direction, where different water compartments (e.g., shallow and deep soil moisture, groundwater storage, and surface water storage) exist. This can be helpful to vertically decrease the influence of the layers on each other by limiting the filtering process to specific layers, especially when there is a high correlation between the observed components at different layers. Here, however, LA is applied only horizontally because the GRACE TWS observation at each grid point is assimilated to an aggregate of water compartments at the same point. Therefore, a vertical variability in system states is not reflected in the observation error covariance. Furthermore, we are more interested in monitoring the performance of the localization scheme on the GRACE covariance matrix rather than a state covariance matrix. Different trial localisation lengths (2° to 10° for gridded TWSs) are applied in this study and their results are assessed against independent groundwater in-situ measurements (cf. Sect. 6.3.3) to find the best case (see details in Sect. 6.5.2).

6.4.2 Assimilating GRACE Data

In order to address the rather low temporal resolution of GRACE (approximately 30 days), its monthly data and errors are interpolated to 5-day data following Tangdamrongsub et al. (2015), the spline interpolation between consecutive months is used to generate these time series, which allows the ensemble to gradually change between updates. Next, the mean water storage over the study area between 2003 and 2013 is calculated from the W3RA and is added to GRACE TWS changes time series in order to achieve the absolute values. The provided observations are assimilated into W3RA for the 5 different grid resolutions of 1°, 2°, 3°, 4°, and 5° and also in basin scales.

The 1° × 1° spatial resolution of the model leads to a model state vector (X^f) with 794 elements within the Australian continent. Each of these elements contains different water compartments. This means that the state vectors for every grid point in our experiments are composed of the different water storages, including top soil, shallow soil, and deep soil water, canopy, snow, surface, and groundwater. The observations matrix (H) accumulates the state variables (the individual water storages) at each grid point to determine the simulated TWS in order to update them with the GRACE TWS during assimilation. In the update steps, the 5-day temporal average update increment (i.e., the difference between the simulated TWS and GRACE TWS) is applied.

Initial ensemble members are generated by perturbing the meteorological forcing fields following Renzullo et al. (2014). In this regard, the three most important forcing variables including precipitation, temperature, and radiation and their reported error characteristics (Sheffield et al. 2006) are used. To generate the perturbations, we assume a multiplicative error of 30% for precipitation, an additive error of 50 Wm^{-2} for the shortwave radiation, and an additive error of 2 °C for temperature (Jones et al. 2007; Renzullo et al. 2014). Monte Carlo sampling of multivariate normal distributions with the errors representing the standard deviations without considering correlations (spatial and/or temporal) are used to produce an ensemble (according to Renzullo et al. 2014). Different ensemble sizes (30–120) and their spread are tested. The selected number of 72 members agrees with the suggestion by Oke et al. (2008) and shows promising performance and is used in this study. The perturbed meteorological forcing datasets, then, are integrated forward with the model for two years (January 2001 to January 2003). This provided a set of state vectors at the beginning of the study period, considered as the initial ensemble. A schematic illustration of the assimilation process steps is provided in Fig. 6.2.

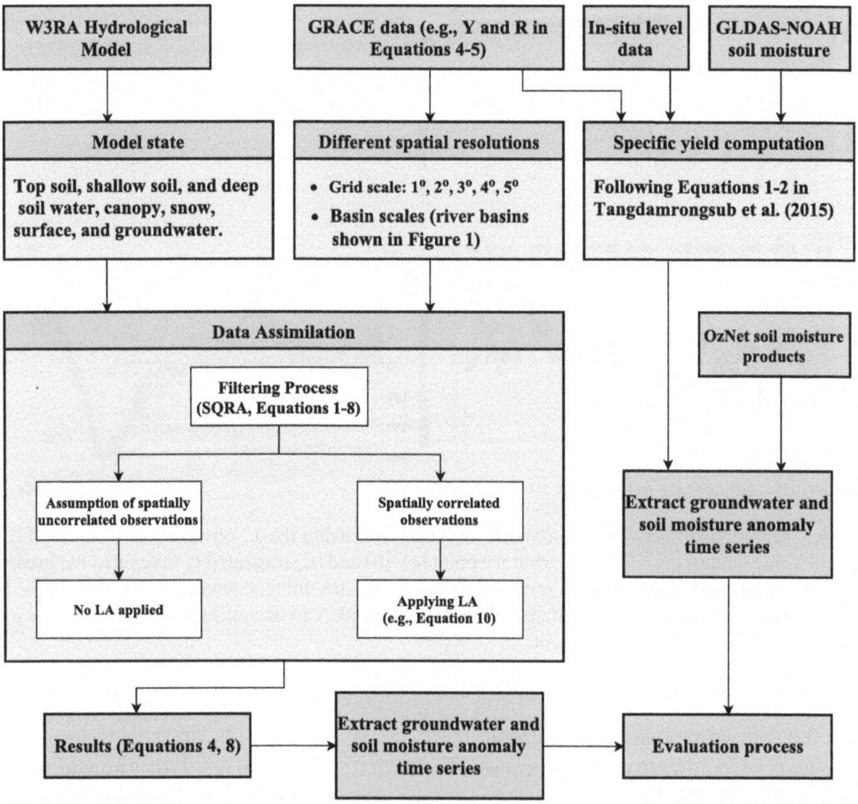

Fig. 6.2 A schematic illustration of the data assimilation approach implemented for this study and of the considered data sets

6.5 Results

In the following, we first analyze the effects of GRACE TWS spatial scaling on the error covariance matrix. Then, LA behavior in dealing with GRACE error covariance with different spatial resolutions is addressed. Afterwards, we evaluate the results of data assimilation using LA with respect to different resolutions against the in-situ groundwater and soil moisture products. These results are also compared with the data assimilation process without applying LA (with a consideration of zero correlation in GRACE data) to be able to better investigate its effects on the model estimations.

6.5.1 Scaling Effect

In this section, we review the behavior of assimilating GRACE TWS data for different spatial resolutions into the W3RA model. To this end, GRACE TWS is assimilated

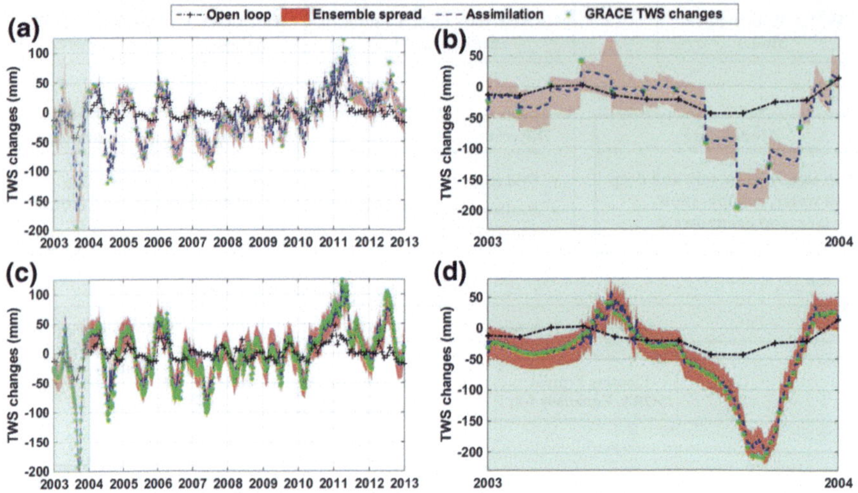

Fig. 6.3 Comparison between the assimilated time series using the 1° observations in a monthly (**a**) and 5-day temporal scale for an arbitrary point (**c**). (**b**) and (**d**) respectively magnified the green areas of (**a**) and (**c**) representing a zoom-in for one year. Ensemble spread represents the spread of the ensemble of updated TWS states. Note that we use LA to account for correlated errors in GRACE error covariance for this figure

with the following spatial resolutions, 1°, 2°, 3°, 4°, 5°, and a basin scale to monitor the effects of localization on the process. For each of the spatial resolution considered, 5-day GRACE TWS data (cf. Sect. 6.4.2) are assimilated into the model to address the coarser GRACE temporal scale in comparison to the model. As an example, in Fig. 6.3, we compare the assimilated time series using the $1° \times 1°$ observations for a monthly (Fig. 6.3a) and 5-day temporal scale over an arbitrary point (Fig. 6.3c) to show the effect of temporal rescaling. The denser temporal resolution in Fig. 6.3c eventuates in a much smoother time series. This is more obvious in Fig. 6.3b, d, which show only one year of the time series, respectively presented in Fig. 6.3a, c. Given daily time steps of W3RA, assimilating GRACE TWS data once a month (e.g., in the middle of the month) causes unnatural jumps at the assimilation steps (cf. Fig. 6.3b). Such a jump is much smaller in magnitude in Fig. 6.3d where a 5-day sampling interval is used. This leads to keeping the ensemble spread smoother without significant artifacts or temporal discontinuities. It should be mentioned that another solution for keeping ensemble spread smooth is the application of ensemble Kalman smoother (EnKS), which redistributes analysis increments evenly over all days of the month with the expense of more computational cost (see, e.g., Zaitchik et al. 2008; Houborg et al. 2012).

We can now assess the behaviour of LA on data assimilation when the full error covariance of GRACE is used for the different applied spatial scales. Figure 6.4 shows the estimated correlation matrices for each grid resolution (following Eicker et al. 2014). This figure helps in understanding how different grid resolutions affect the

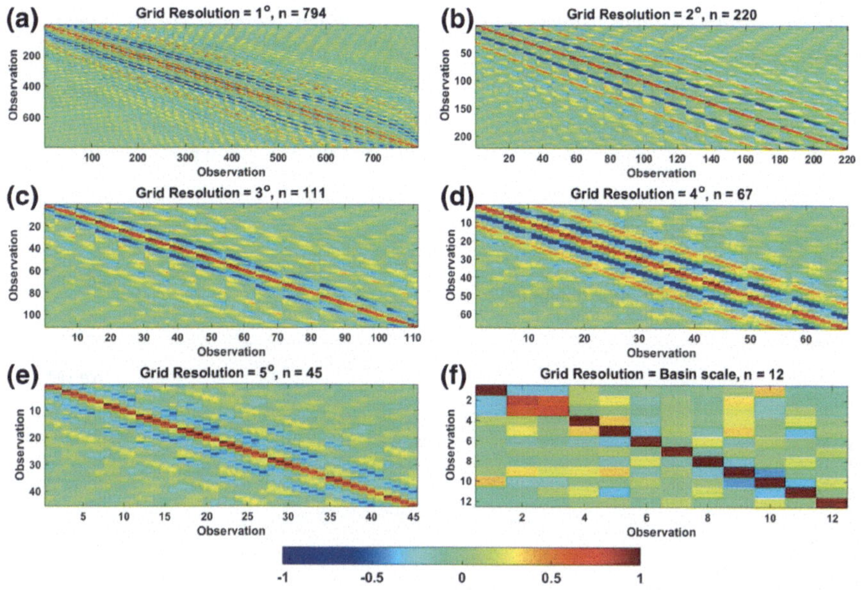

Fig. 6.4 Correlation matrices of the GRACE observations corresponding to various spatial aggregations. Here, no localization is applied. The variable n refers to the number of assimilated observations

corresponding observation covariance matrix. It can be seen that the spatial scaling influences the correlation between points. The correlation (off-diagonal elements) between grid points decreases for larger grid resolutions, with the least for the 5° gridded TWSs, which is significantly less than that of 1° grid resolution. This correlation is even smaller when the basin scale GRACE data is considered. To clarify how this affects the data assimilation procedure, Table 6.1 indicates the number of gridded observations in various grid resolutions and the estimated ranks of covariance matrices. We find that there is a close relationship between the grid resolution and covariance matrix rank (cf. Table 6.1). As mentioned earlier, rank deficiency problem in covariance matrices causes instability in the data assimilation procedure and inaccurate estimations. The application of LA, however, numerically addresses this issue. It can be seen that LA affects the estimated covariance matrix rank for each grid resolution. Details on the number of observations and the rank of the respective covariance matrices (cf. Table 6.1) demonstrates the LA effect on improving the process by solving the mathematical problem related to the rank deficiency especially in the cases of 1° and 2°.

Rank deficiency likely happens for error covariance matrices of GRACE TWS with grid resolutions that GRACE can resolve (e.g., 3° or coarser). However, when using smaller grid resolutions, the matrix does not have a full rank leading to instabilities in the data assimilation procedure. Although applying GRACE data at lower spatial resolutions might be helpful in dealing with the covariance matrix, this will reduce the number of observations during data assimilation process (cf. Table 6.1)

Table 6.1 The details of GRACE observations used in each grid resolution

Spatial scale	Observation Number	Rank	LA Rank[a]
1°	794	268	794
2°	220	211	220
3°	111	111	111
4°	67	67	67
5°	45	45	45
Basin scale	12	12	12

[a]Computed rank after the implementation of LA

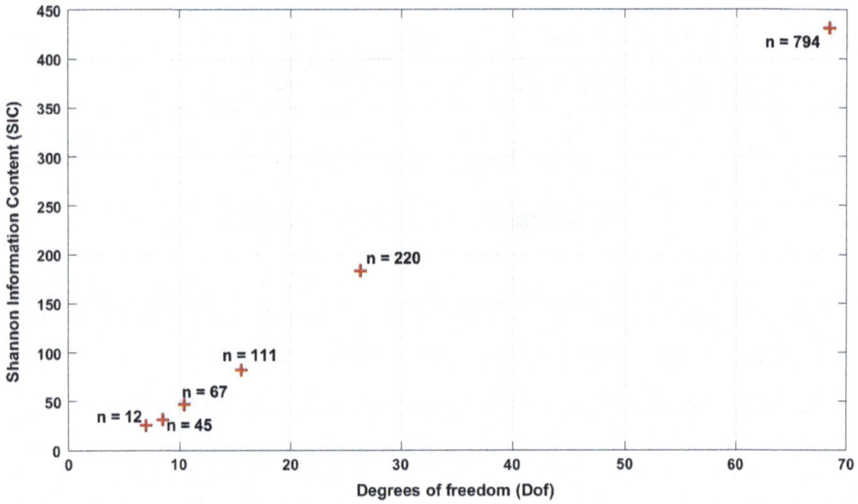

Fig. 6.5 Shannon Information Content (SIC) and Degrees of freedom (Dof) with respect to the number of assimilated GRACE observations (n)

leading to some loss of signal in the observations. This might not be obvious considering the spatial correlation between grid points for higher resolution GRACE TWS. However, we show that using more observations and considering their full error covariance information in the assimilation process allows more information to be transferred with a higher number of observations into the system states. In this regard, we use the frequently employed indexes of Shannon Information Content (SIC or entropy reduction) and degrees of freedom (Dof) to measure information, which is transferred from observations into the system states (Rodgers 2000) at the assimilation steps. SIC ($\frac{1}{2}ln(P^f/P^a)$) uses the information in the state probability density function (pdf) before and after assimilation to reflect a real-valued functional (Shannon and Weaver 1949). Dof ($n - trace(P^a/P^f)$, with n number of observations) on the other hand, is a measure of the amount of information from observations that is used (Stewart et al. 2008). For each grid resolution, the indexes of SIC and Dof are measured (Fig. 6.5).

It can be seen in Fig. 6.5 that by decreasing the spatial resolution, some information contained within the observations is lost. Therefore, although increasing the scale size (reducing the resolution) might be helpful in dealing with GRACE error covariance, it is at the cost of losing part of the signal. This justifies the application of LA, which allows us to use information with a higher spatial resolution in datasets.

As outlined in Sect. 6.4.1.2, one important effect of LA is underestimating the influences of spatially distant grid points on each other. The distance in localization preserves the information in observations close to each other while at the same time making it possible to use full error covariance information. To demonstrate this, we consider the correlation coefficient of the arbitrary point (at a location 136.6854 °E and 23.9015 °S) to the other grid points in Fig. 6.6. This point is chosen to be approximately in the middle of the study area for a better visual representation while similar results are achieved for all other grid points. We integrate the model and performed data assimilation using the 1° GRACE TWS (as the worst case among different applied resolutions) during the study period. The average correlation coefficients between the arbitrary point and the other grid points before and after assimilation using LA are then measured. Figure 6.6b shows how LA successfully reduces the correlation coefficients for more distant grid elements but maintains the correlations in the close vicinity.

The important point to consider when using LA is the removal of some information from the data, which is not desirable. Thus, attention needs to be taken when choosing the localization length to preserve the adequate continuity of analysis on adjacent points (Zeng 2014). LA length depends on the observation density and can be chosen arbitrarily. After testing different localization lengths, it is found that a small length (e.g., less than 5° for $1° \times 1°$ GRACE TWS) can result in large errors even though there would be no inverse problem in assimilation filter. We use groundwater in-situ measurements to assess the results of applying different localization lengths (2° to 10° for gridded TWSs). For every scenario (different grid resolutions), we interpolate assimilation time series at the location of the groundwater in-situ and calculate the root-mean-square error (RMSE). The average computed RMSE of each grid resolution for the applied lengths (Fig. 6.7) show that better results are obtained using the 5° localization halfwidth length compared to the other applied localization lengths.

A similar experiment is implemented to find efficient localization length for a basin scale spatial resolution. For each basin, we test different lengths mostly larger than those for grid scales (e.g., 5° to 15° with the best performance of 10° radii in average) and estimate TWS errors using the GRACE TWS data where in-situ measurements are not available for all basins. The localization length with the least error for each basin (Fig. 6.8) is used to assess the LA effects at the basin scale and also to compare corresponding results with grid scale resolutions.

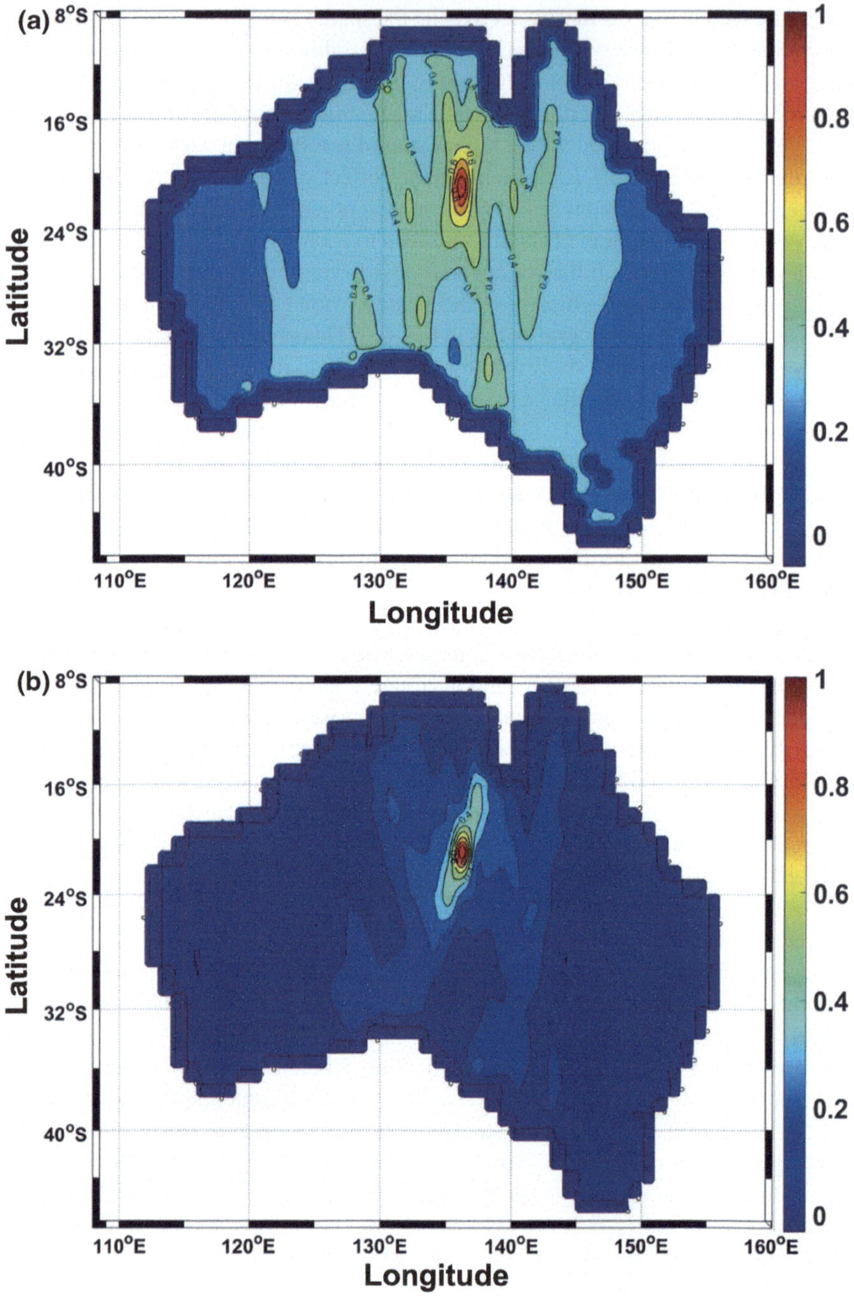

Fig. 6.6 2-D representation of correlation coefficients between the TWS anomalies of the arbitrary point (136.6854°E and 23.9015°S) and the rest of the grid points. The temporal average of the correlation coefficients before and after assimilation using LA are shown in (**a**) and (**b**) respectively

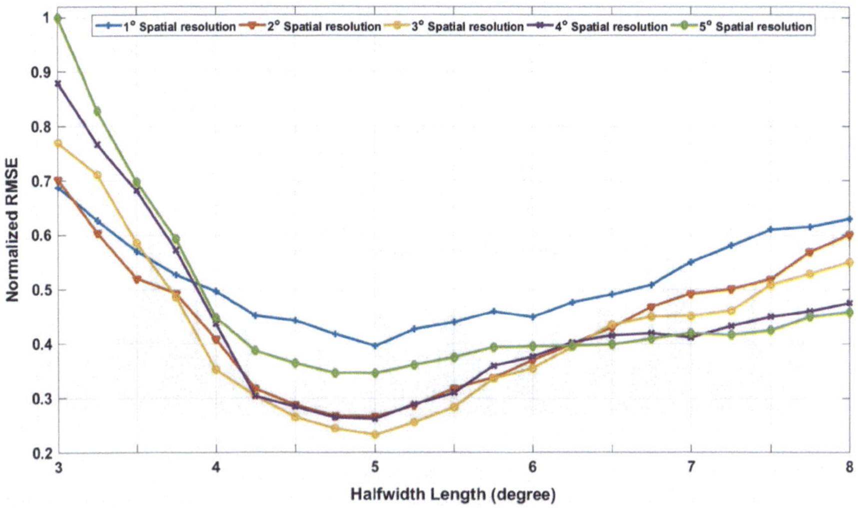

Fig. 6.7 Comparison between normalized RMSE of TWS anomalies for different localisation radii (degree) applied for each case of GRACE TWS spatial resolution used for assimilation. RMSEs are calculated in mm, however, for a better visual presentation, normalized values are presented

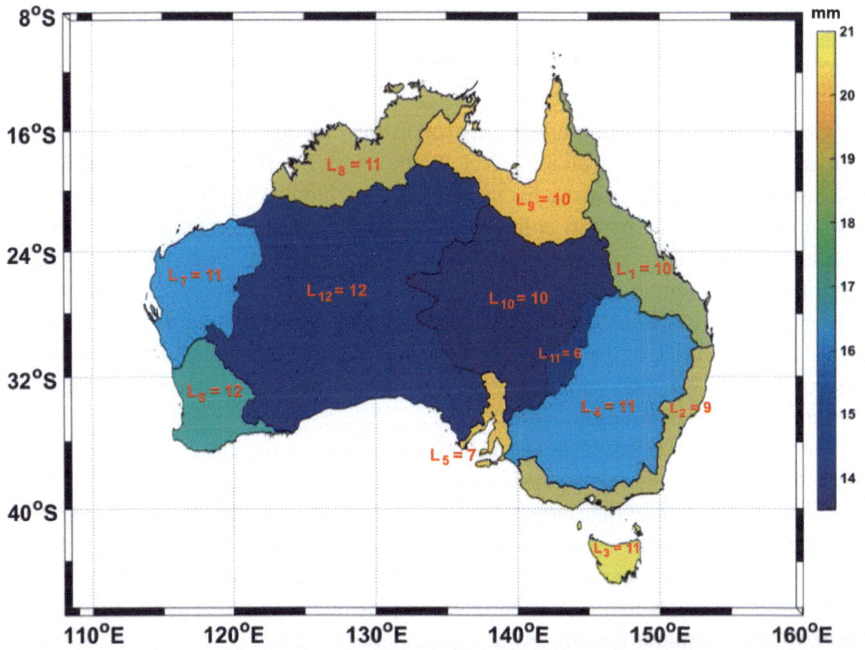

Fig. 6.8 The estimated optimized localisation radii (in degree and presented by L) and corresponding TWS errors with respect to the GRACE data for each basin

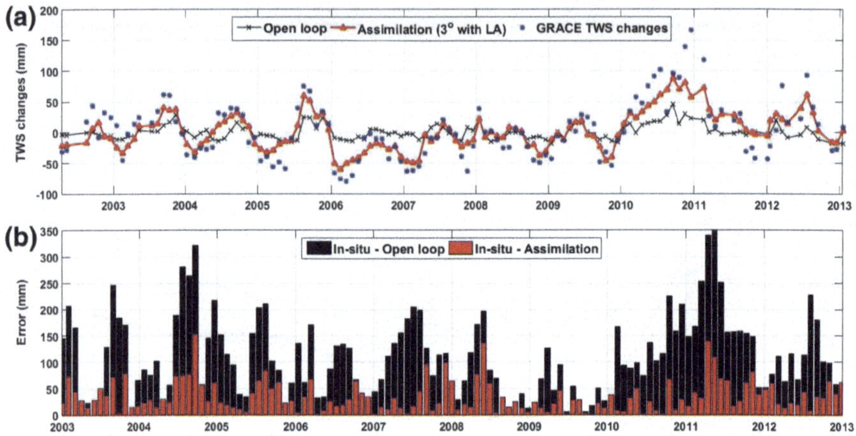

Fig. 6.9 **a** Comparison between the TWS time series of the assimilation process for the case of 3° spatial resolution (red), the GRACE observation (blue), with the open loop referring to the model estimation without applying data assimilation (black). **b** Absolute groundwater (GW) error bars before (black) and after (red) data assimilation process in comparison to the GW in-situ measurements. The time series shown in (**a**) and (**b**) are spatially averaged over the Murray-Darling Basin

6.5.2 Assessment with In-Situ Data

Post processed in-situ measurements of groundwater changes (cf. Sect. 6.3.3) over the Murray-Darling basin as well as OzNet soil moisture network in the Murrumbidgee catchment (see Fig. 6.1 for the location of the catchment) are used to evaluate the assimilation results. First, to compare the time series obtained from assimilation results with those of in-situ measurements, the GW results for each spatial resolution considered are spatially interpolated using the nearest neighbor (the closest four data values) to the location of the in-situ measurements. Afterward, the error time series are computed as the difference between the estimated GW and in-situ GW measurements. We then estimate average errors using these time series for each scenario of data assimilation.

The TWS time series of the assimilation process for the case of 3° is shown in Fig. 6.9a. Data assimilation with this spatial resolution results in a minimum GW error compared to the in-situ measurements. This figure also contains the open loop time series which refers to the estimations without implementation of any data assimilation and the assimilated observations. The absolute errors, i.e., the difference between the in-situ measurements and either the open loop or the assimilated estimates (for the best case of 3° spatial resolution) are presented in Fig. 6.9b. The assimilated time series fits well with the groundwater in-situ measurements (cf. Fig. 6.9a) and results in a higher correlation than the open loop time series (85% average). Note that in terms of representing the hydrology, sometimes the estimates do not really depict the signal of the in-situ measurements. In some instances, the error (for no assimilation)

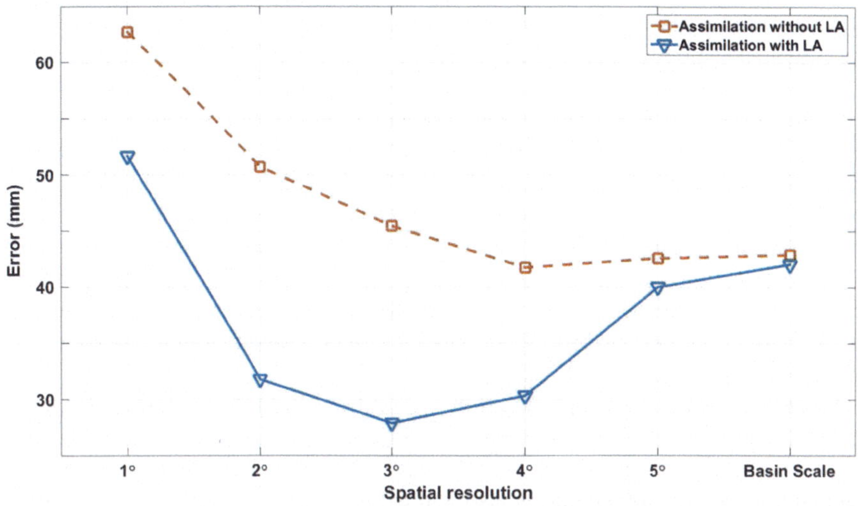

Fig. 6.10 Average estimated error in groundwater anomalies from assimilating GRACE data for different spatial scales with (blue) and without (red) implementation of LA. The results are spatial averages over all groundwater data points within the Murray-Darling Basin

is as large as the signal itself. This could be due to the fact that W3RA only simulates the dynamics of unconfined aquifers, that is, groundwater that receives soil drainage and discharges into streams. In some cases, a deeper (confined) aquifer underneath can also affect groundwater measurements. Nevertheless, data assimilation causes the updated time series to reflect better the real fluctuations in groundwater storage in most of the cases as given by the in-situ measurements.

The average estimated error of all GW in-situ stations during the study period for each scenario illustrates the LA performances for the different spatial resolutions (Fig. 6.10). The least error is obtained from the 3° spatial resolution by comparing assimilation results of all scenarios. In addition, to be able to monitor the effectiveness of LA, data assimilation is also applied using GRACE-derived TWS and only diagonal elements of its error covariance matrix. Results without applying LA (represented in Fig. 6.10) refers to this case where correlations between grid points are neglected. This comparison is of interest because many of the previously presented studies in using GRACE for hydrological data assimilation have neglected the existing correlation in observations (see e.g., Zaitchik et al. 2008; Houborg et al. 2012; Li et al. 2012; Tangdamrongsub et al. 2015; Sun et al. 2015; Kumar et al. 2016).

It can be seen that locally applying the GRACE observations effectively reduces errors for every grid resolution considered in comparison to the uncorrelated observation assumption. This, however, is more obvious for higher spatial resolution (e.g., 3° and higher) where a large difference between the assimilation results with and without the application of LA can be found. Although LA mathematically solves the inverse problem for using 1° gridded GRACE TWS data and associated error covariance (cf. Table 6.1) in the filtering process, this spatial resolution results in a larger error in comparison to the other scenarios. From Sect. 6.3.1, we know that truncating and smoothing procedures cause losing a part of GRACE data, especially in higher frequencies. Rescaling such a data into 1° spatial resolution results in an error in gridded GRACE TWS and correspondingly in the assimilation result (cf. Fig. 6.10). Figure 6.10 shows that increasing the spatial resolution results in a better estimation when LA is not applied. This error reduction by using a higher spatial resolution is also true when LA is applied but only to the point of 3°. After this point, errors start increasing, which can be explained by fewer observations used leading to less information content to be transferred to model states. The application of LA, however, reduces the error for all spatial resolutions while in an absolute sense, the smallest errors are obtained for 3°. Interestingly, this spatial resolution is about the spatial resolution that GRACE can resolve.

More detailed results are proposed in Fig. 6.11 and Table 6.2 in terms of RMSE and correlation analysis. As mentioned before, first, assimilation time series are interpolated at the location of the groundwater in-situs and then, their anomalies are calculated. A similar procedure is also applied to achieve assimilation time series over the soil moisture in-situ stations. Then for all stations, RMSE and correlation factor between assimilation results (for various scenarios) and in-situ measurements are calculated and their averages are used for assessment. Note that considering the difference between W3RA estimations (column water storage) and the OzNet measurements (volumetric soil moisture), only correlation analysis is assumed for

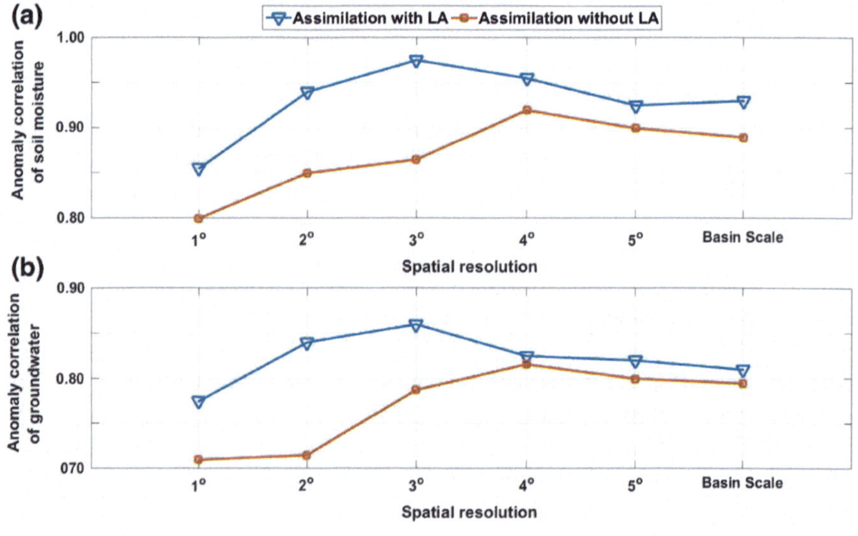

Fig. 6.11 Comparison between the correlation of assimilation results (using different spatial resolutions) with OzNet soil moisture anomalies spatially averaged over the Murrumbidgee Catchment (**a**) and with anomalies of groundwater in-situ level measurements spatially averaged over the Murray-Darling Basin (**b**). The correlation results in both cases of data assimilation using LA (blue) and without using LA (red) are shown

Table 6.2 A summary of the results belonging to each scenario of data assimilation. Improvements in groundwater are calculated using the estimated RMSE with and without applying data assimilation (open loop) in relation to groundwater in-situ measurements

	Assimilation without LA		Assimilation with LA	
Spatial scale	RMSE (mm)	Improvement (%)	RMSE (mm)	Improvement (%)
1°	68.54	17.76	52.23	37.33
2°	51.09	38.70	35.11	57.87
3°	47.41	43.11	26.80	67.84
4°	43.18	48.19	32.35	61.18
5°	44.37	46.76	41.19	50.58
Basin scale	43.84	47.40	41.93	49.69

assessing results against soil moisture in-situ data. The reason for this refers to the fact that converting model outputs (with unit 'mm') into volumetric units may introduce a bias (Renzullo et al. 2014). Estimated correlations between assimilation results and OzNet soil moisture (an average correlation for the total soil column; Fig. 6.11a) as well as groundwater in-situ level data (Fig. 6.11b) demonstrate the ability of LA in dealing with GRACE data. Also, both correlation analyses show that applying GRACE TWS with 3° leads to closer results to the in-situ measurements.

Based on the results in Table 6.2, all the results successfully improved the model estimation of water storage variation. Applying LA in data assimilation leads up to 24.73% (13% average) better estimations in comparison to the non-correlated assumption. This proves the importance of using local data assimilation for incorporating GRACE data into the hydrological model. We know from Eicker et al. (2014) that spatial upscaling of GRACE data to coarser resolutions (e.g., 5°) can significantly stabilize the assimilation process leading to more reliable results, however, LA can improve the results not only for these resolutions but also for smaller grid sizes (cf. Table 6.2).

It can be seen from Table 6.2 that using gridded TWS observation with 3° shows the best performance in terms of RMSE. Although there is no rank deficiency in using the full error covariance matrix for this grid resolution, local implementation of the assimilation process helps to improve the agreement with the in-situ measurements. The reason why LA does not have a similar impact on finer spatial resolutions, especially for a 1° resolution in comparison to 3°, could be due to the characteristic of GRACE L2 product as a degree limited data, e.g., truncated spherical harmonics sets. An interesting observation from Table 6.2 refers to the results of using GRACE TWS for a 2° spatial resolution. Considering Table 6.1, employing the 2° grid resolution causes a rank deficiency in covariance matrix leading to the unstable data assimilation. LA successfully solves this problem and significantly improves the results with a better performance (57.87% improvement). Fewer observations incorporated in the assimilation on a basin scale and for 5° resolution in comparison to the other spatial scales (e.g., 3°) leads to a weaker performance for these two cases.

6.6 Conclusion

The global time variable terrestrial water storage (TWS) data from the Gravity Recovery And Climate Experiment (GRACE) has provide an important opportunity for a hydrological model adjustment. In this study, we assessed the performance of local analysis (LA) method in accounting for the existing correlation in GRACE data and improving its effect on model states. To this end, we assimilated the GRACE-derived TWS changes into the World-Wide Water Resources Assessment system (W3RA) during 2003 to 2012 using Square Root Analysis (SQRA) filtering technique. LA was applied to (i) solve the mathematical problem of using correlated data for assimilation especially when the observation spatial resolution is high (e.g., 1° gridded TWS), and (ii) improve the assimilation results using GRACE TWS data for different spatial resolutions (1° to 5° and a basin scale). The observations were applied for a 5-day temporal scale and for 5 different grid resolutions to monitor the impact of LA on each scenario. The results showed that implementing LA successfully reduced data assimilation errors for all the cases (54.08% on average). This improvement is larger for the cases with smaller grid sizes along with the higher error correlations. LA addressed the rank deficiency problem in using the full information from the error covariance matrix for a higher spatial resolution of GRACE TWS data (e.g., 1°). This, to the best of our knowledge, for the first time, allowed us to be able to apply GRACE TWS considering spatial error correlation information at finer spatial resolutions (e.g., 1° and 2°) for the hydrological data assimilation. LA also improved the assimilation results at all grid resolutions and basin scale especially in comparison to using non-correlated observations (13.76% average). This highlights the great potential of LA in different scenarios for improved data assimilation. The best performance with 67.84% improvement was found with the application of GRACE data in assimilation with 3° spatial resolution. Overall, the importance of the application of LA in hydrological data assimilation is: (1) stabilising the assimilation of GRACE TWS observation using its full error covariance for finer spatial resolutions (e.g., 1° and 2°), and (2) improving the results for all the spatial grid sizes without the assumption of white noise. This study offered a method to deal with the GRACE error covariance matrix during data assimilation, however, further assessment needs to be undertaken to examine other potential methods like inflation of the observation error variances and circulant approximation.

Acknowledgements We would like to thank professor Harrie-Jan Hendricks-Franssen, the associate editor of Advances in Water Resources, and Dr. Manuela Girotto for their useful comments, which contributed to the improvement of this study. M. Khaki is grateful for the research grant of Curtin International Postgraduate Research Scholarships (CIPRS)/ORD Scholarship provided by Curtin University (Australia). This work is a TIGeR publication.

Part V
Water Budget Constraint

One of the important factors when assimilating new observations into land surface models, more specifically into water balance systems, is to preserve the balance between water fluxes after each assimilation cycle. In order to enhance the estimation of model water storages, one should also involve flux observations and their associated errors in the assimilation procedure. This is necessary because assimilating water budget components such as GRACE TWS and/or soil moisture measurements can break the balance between changes in water storage ($\Delta\mathbf{s}$), evaporation ($\mathbf{e}$), precipitation ($\mathbf{p}$) and discharge ($\mathbf{q}$), i.e. $\Delta\mathbf{s} = \mathbf{p} - \mathbf{e} - \mathbf{q}$ (Sokolov and Chapman 1974). Thus, advance data assimilation filtering is required to account for this issue, specifically for water balance models to retrieve the broken balance while including new observations into the system. These methods usually work by imposing a constraint to the process based on the water budget equation. This constraint is considered at each assimilation cycle after a standard filtering step. In other words, the additional step appears as a second update step in a sequential filtering scheme.

In this chapter, two recently proposed approaches will be presented to deal with the water balance problem. Both these methods are based on Ensemble Kalman Filter (EnKF) for updating system state, i.e. comprising two interactive EnKF-based updates to (1) update the system state based on incoming observations and (2) to enforce water budget constraint. The main objective of introducing the new method is to tackle the limitation in existing data assimilation studies, which refers to the inconsistency between hydrological water fluxes with assimilation observations. More specifically, contrary to existing methods, which do not include water balance constraint error, these two-step algorithms incorporate the uncertainty associated with the flux observations into the data assimilation process and optimize the water budget closure. The major difference between the two approaches is that contrary to the first one, so-called Weak Constrained Ensemble Kalman Filter (WCEnKF), the second method, unsupervised WCEnKF (UWCEnKF), incorporates water budget error terms into the process and estimates them along with the state. WCEnKF, however, relies on an assumption that considers these error terms to be known. UWCEnKF, on the other hand, develops the WCEnKF formulation to a more general (unsupervised) framework, in which the covariance associated with the water balance model is estimated online, together with the model states.

Chapter 7
Constrained Data Assimilation Filtering

Models water storage states are in balance since model structure, e.g., its equations, governs variations in the water state changes due to the incoming and outgoing hydrological water fluxes. An assimilation of water storage states (e.g., GRACE data) does not constraint the assimilated state to be balanced.

—M. Khaki (This chapter is presented following Khaki et al. (2017c), "A two-update ensemble Kalman filter for land hydrological data assimilation with an uncertain constraint")

7.1 Summary

Assimilating Gravity Recovery And Climate Experiment (GRACE) data into land hydrological models provides a valuable opportunity to improve the models' forecasts and increases our knowledge of terrestrial water storages (TWS). The assimilation, however, may harm the consistency between hydrological water fluxes, namely precipitation, evaporation, discharge, and water storage changes. To address this issue, we propose a weak constrained ensemble Kalman filter (WCEnKF) that maintains estimated water budgets in balance with other water fluxes. Therefore, in this study, GRACE terrestrial water storages data are assimilated into the World-Wide Water Resources Assessment (W3RA) hydrological model over the Earth's land areas covering 2002–2012. Multi-mission remotely sensed precipitation measurements from the Tropical Rainfall Measuring Mission (TRMM) and evaporation products from the Moderate Resolution Imaging Spectroradiometer (MODIS), as well as ground-based water discharge measurements are applied to close the water balance equation. The proposed WCEnKF contains two update steps; first, it incorporates observations from GRACE to improve model simulations of water storages, and second, uses the additional observations of precipitation, evaporation, and water discharge to establish the water budget closure. These steps are designed to account for error information associated with the included observation sets during the assimilation process. In order to evaluate the assimilation results, in addition to monitoring

M. Khaki, *Satellite Remote Sensing in Hydrological Data Assimilation*,
https://doi.org/10.1007/978-3-030-37375-7_7

the water budget closure errors, in-situ groundwater measurements over the Mississippi River Basin in the US and the Murray-Darling Basin in Australia are used. Our results indicate approximately 24% improvement in the WCEnKF groundwater estimates over both basins compared to the use of (constraint-free) EnKF. WCEnKF also further reduces imbalance errors by approximately 82.53% (on average) and at the same time increases the correlations between the assimilation solutions and the water fluxes.

7.2 Introduction

Terrestrial water storage plays an important role in both human life and environment all around the world. Quantifying this major water resource is, therefore, essential and can be done using different tools including ground-based in-situ measurements, satellite remote sensing data, and hydrological models. In the last few decades, hydrological models have extensively been used to determine and monitor stored water and fluxes in different forms within landscapes such as ice and snow, glaciers, aquifers, soils, and surface waters (e.g., Chiew et al. 1993; Wooldridge and Kalma 2001; Döll et al. 2003; Huntington 2006; van Dijk 2010). The models have been designed to reflect the behavior of a system of interest while satisfying known physical properties reliably (Smith et al. 2011). However, various sources of uncertainty, due for example, imperfect modeling, data limitations on both temporal and spatial resolutions, errors in forcing fields, as well as empirical model parameters, limit the accuracy of hydrological models (Vrugt et al. 2013; van Dijk et al. 2011, 2014). Assimilating accurate observations into models is an effective approach to overcome these limitations (e.g., McLaughlin 2002; Zaitchik et al. 2008; van Dijk et al. 2014; Gharamti et al. 2016).

Data assimilation is a procedure for incorporating observations of one or more variables (according to their uncertainties) into a numerical (physical) model to increase consistency of model simulations of a certain variable with its changes in the 'real world' (Bertino et al. 2003; Hoteit et al. 2012). Therefore it has been widely applied in hydrological studies to improve different water compartments, such as soil moisture (e.g., Reichle et al. 2002; Brocca et al. 2010; Renzullo et al. 2014), surface water (e.g., Alsdorf et al. 2007; Neal et al. 2009; Giustarini et al. 2011), and snow storages (e.g., Liu et al. 2013; Kumar et al. 2015). During past few years, some studies have assessed the capability of Gravity Recovery And Climate Experiment (GRACE) data, available since March 2002, to improve terrestrial water storages (TWS) (e.g., Zaitchik et al. 2008; Eicker et al. 2014; Tangdamrongsub et al. 2015; Schumacher et al. 2016; Tangdamrongsub et al. 2017; Khaki et al. 2017a, b; Tian et al. 2017) simulated by land (surface) hydrological models.

The water balance equation is applied in land hydrological models to describe the relationships between changes in water storage ($\Delta\mathbf{s}$), evaporation ($\mathbf{e}$), precipitation ($\mathbf{p}$), and discharge ($\mathbf{q}$), i.e., $\Delta\mathbf{s} = \mathbf{p} - \mathbf{e} - \mathbf{q}$ (Sokolov and Chapman 1974). However, the application of data assimilation may destroy the dynamical balances between water fluxes and water storage changes (Pan and Wood 2006). In another words, models water storage states are in balance since model structure, e.g., its equations, governs variations in the water state changes due to the incoming and outgoing hydrological water fluxes. An assimilation of water storage states (e.g., GRACE data) does not constraint the assimilated state to be balanced. Eicker et al. (2016) found distinct changes in the linear rates and seasonality of water storage from GRACE and the flux deficit ($\mathbf{p} - \mathbf{e} - \mathbf{q}$) even over large-scale river basins. Therefore, after assimilation, one can expect mismatches between the model estimation of $\Delta\mathbf{s}$ and the flux deficit after each assimilation cycle. This issue must be mitigated to better interpret model derived water storage changes after implementing data assimilation (see, e.g., Roads et al. 2003; Pan and Wood 2006; Sahoo et al. 2011).

In order to enhance the estimation of model water storages (e.g., for $\Delta\mathbf{s}$), it is important that the water variables satisfy the water closure equation. One way to do this is to impose a balance constraint based on the water budget equation after each assimilation cycle (Pan et al. 2012). Few assimilation schemes have been proposed in this context. Pan and Wood (2006) developed a constrained ensemble Kalman filter (CEnKF) based on the ensemble Kalman filter (EnKF; Evensen 1994) to solve the disclosure of the water balance equation after implementing a data assimilation over the southern Great Plains region of the United States. In addition to using CEnKF, Sahoo et al. (2011) and Pan et al. (2012) applied a data merging algorithm to prepare the datasets for data assimilation and to check for imbalance over various major river basins. They merged data from different sources (e.g., in situ observations, remote sensing retrievals, land surface model simulations, and global reanalyses) so that their errors can be used to achieve optimal weights leading to the best estimates for each terrestrial water cycle. These data were then used to resolve water balance errors by applying CEnKF (see also Zhang et al. 2016). In these studies, information about the uncertainties associated with water balance observations, however, have not been incorporated during data assimilation. The strong constraint imposed by assuming observation to be perfect is unrealistic and can cause estimation errors such as over-fitting issues (Tangdamrongsub et al. 2017). This motivates the new filtering technique, which is proposed in this study to involve observation errors in the assimilation procedure.

In this study, a new constrained ensemble Kalman filter, which we refer to as weak constrained ensemble Kalman filter (WCEnKF), is introduced that satisfies the closure of the water balance equation while taking the uncertainties in datasets into the account. WCEnKF is formulated based on the EnKF and imposes the closure constraint as a second update step, where the EnKF analysis members are updated to remain in balance with other variables (hereafter called pseudo-observation, and includes $\mathbf{p}$, $\mathbf{e}$, and $\mathbf{q}$ through the water balance equation). Water storages are therefore first updated using GRACE observations as in the EnKF in the first step, and the broken water balance is then mitigated using the pseudo-observations in the second EnKF update step. The novelty of the proposed scheme is that it accounts for the uncertainties in the pseudo-observations so that the budget closure is not strongly imposed. Moreover, in contrast to existing schemes, the filter does not seek to redistribute the imbalance between all compartments (i.e., Δs, $\mathbf{p}$, $\mathbf{e}$, and $\mathbf{q}$) and only adjusts the already estimated water storage ($\Delta\mathbf{s}$). WCEnKF treats **p**, **e**, and **q** and their uncertainties as a new set of observations, similar to any other observation in a standard EnKF. The imbalance problem requires a particular formulation of the state-space system, for which the process does not only depend on the state at the filtering time but also on the previous time.

The proposed WCEnKF with the dual update steps is used to assimilate GRACE TWS data into the World-Wide Water Resources Assessment (W3RA) hydrological model globally during 2002–2012. In addition to GRACE TWS data, remotely sensed measurements of **p** and **e** are also used to constrain the water balance in the filter estimates. For this purpose, we use the Tropical Rainfall Measuring Mission (TRMM-3B43; Huffman et al. 2007) precipitation products for **p**, the Moderate Resolution Imaging Spectroradiometer (MODIS) evaporation data (MOD16; Mu et al. 2007) for **e**, and the water discharge measurements from various ground stations for **q**. Although the imbalance constraint is spatially limited to locations, where ground-based discharge data are available, the Kalman-like second update step of WCEnKF spreads the imbalance adjustments to all model grid points. For a better presentation of results, we choose eight globally distributed major basins with a dense network of water discharge measurements and analyze the assimilation solution separately over each basin. Among these basins, the Mississippi River Basin and the Murray-Darling Basin are selected subject to the availability of ground-based data to evaluate the performance of WCEnKF against in-situ groundwater measurements.

7.3 Model and Data

7.3.1 W3RA Hydrological Model

We use a grid distributed biophysical model of W3RA from the Commonwealth Scientific and Industrial Research Organisation (CSIRO). The model is designed to simulate landscape water stored in the vegetation and soil systems (van Dijk 2010). The $1^\circ \times 1^\circ$ version of W3RA is applied to represent the water balance of the soil, groundwater and surface water stores, in which each cell is modeled independently from its neighbors (van Dijk 2010; Renzullo et al. 2014). The model parameters include effective soil parameters, water holding capacity and soil evaporation, relating greenness and groundwater recession, and saturated area to catchment characteristics (van Dijk et al. 2013). Forcing datasets consist of the daily meteorological fields of minimum and maximum temperature, downwelling short-wave radiation, and precipitation by Princeton University (Sheffield et al. 2006). The model state is composed of storages of the top, shallow root and deep root soil layers, groundwater storage, and surface water storage. The simulation covers the period from April 2002 to December 2012.

W3RA represents the storage of water in small river channels and consequently surface water storage changes in reservoir and lakes are not simulated by the model.

Therefore, it is necessary to remove surface water storages from GRACE TWS data before assimilation even though it has much lesser effects than other water storages such as groundwater and soil moisture. For this purpose, we use the Water-GAP Global Hydrology Model (WGHM; more details on Döll et al. 2003) surface storage estimations. WGHM models the vertical and horizontal water fluxes on a $0.5^\circ \times 0.5^\circ$ grid resolution and describes the major hydrological components, such as snow accumulation, runoff, and the lateral transport of water within the river networks (Forootan et al. 2014b). The surface water storages from WGHM are removed from the GRACE TWS before assimilation. Note that after updating the model states using the adjusted GRACE data (first update step in WCEnKF), the removed surface water storages are added to the filtered TWS estimates before applying the water budget closure step (second update step).

7.3.2 Terrestrial Water Storage (TWS) Data

Monthly TWS derived from GRACE level 2 (L2) gravity field data are used in the first step of the proposed filtering scheme to update the summation of the model derived water storage simulations including top soil, shallow soil, deep soil water, snow, vegetation, and groundwater. GRACE data are provided in terms of the gravity potential Stokes' coefficients, truncated at spherical harmonic degree and order 90, together with their full error information from the ITSG-Grace2016 gravity field model (Mayer-Gürr et al. 2014). Some post-processing steps are applied on the coefficients before converting them into TWSs. Degree 1 and degree 2 (C20) coefficients are replaced by more accurate coefficients that are calculated by Swenson et al. (2008) and the Satellite Laser Ranging solutions (Cheng and Tapley 2004), respectively. We also apply DDK2 (Kusche et al. 2009) to mitigate colored/correlated noise in the coefficients. The L2 gravity fields are then converted to $1^\circ \times 1^\circ$ TWS fields following Wahr et al. (1998). The mean TWS is taken from the model for the study period and is added to the GRACE TWS change time series to obtain absolute values in accordance with W3RA (Zaitchik et al. 2008). We further exploit the provided full error information of the Stokes' coefficients to construct an observation error covariance matrix for data assimilation. This is done by converting GRACE spherical harmonic error coefficients to error covariances associated with TWS data as suggested by Eicker et al. (2014) and Schumacher et al. (2016). Eicker et al. (2014) showed that applying GRACE TWS data on a $1^\circ \times 1^\circ$ grid resolution results in a rank deficiency problem during data assimilation (see also Khaki et al. 2017b). However, as shown by Khaki et al. (2017b), the application of local analysis (LA) successfully mitigates this problem by spatially limiting the use of ensemble-based covariance information in high-dimensional systems. Therefore, here, we follow Khaki et al. (2017b) and apply LA to cope with rank deficiency problem (see details in Sect. 7.4.3).

7.3.3 Water Fluxes

Precipitation data of TRMM-3B43 products (TRMM 2011; Huffman et al. 2007) is used. This dataset is limited spatially between 50° × 50°S in latitude, and −180° to +180° in longitude. The data is re-sampled from 0.25° × 0.25° to a monthly 1° × 1° spatial resolution. We also use the relative error available for each gridpoint and different times (Huffman et al. 1997).

We also acquire MOD16 evaporation data from the University of Montana's Numerical Terradynamic Simulation group with eight days temporal resolution and one km spatial resolution (Mu et al. 2011). The gridded data is converted to a monthly temporal scale and 1° × 1° spatial resolution. Following Aires (2014) and Munier et al. (2014), 10 mm uncertainty is considered for the evaporation data.

Different data sources are used to provide water discharge data with a maximum global coverage. In this regard, the largest part of runoff products (1970 globally distributed stations) is acquired from the Global Runoff Data Centre (GRDC). Over Africa, 83 stations are obtained from SIEREM (Systeme d'Informations Environnementales sur les Ressources en Eau et leur Modelisation), an environmental information system for water resources (Boyer et al. 2006). In additions, two dense networks of discharge stations over the United State (3800 stations), Southeast Asia (1700 stations), and Australia (1250 stations) are provided from the United States Geological Survey (USGS), China Hydrology Data Project (Henck et al. 2010; Schmidt et al. 2011), and the Australian Bureau of Meteorology under the Water Regulations (2008). In addition, a number of discharge stations are also obtained from the National River Flow Archive (NRFA), Department of Hydrology and Meteorology of Nepal, the Hydrology and Geochemistry of the Amazon basin (HYBAM) for the Amazon, Orinoco, and Congo basins. Figure 7.1 shows the locations of discharge stations distributed globally.

As mentioned, the water budget closure relies on $\mathbf{p}$, $\mathbf{e}$, and $\mathbf{q}$. Wherever a discharge station is located, it is possible to impose water budget closure adjustment. At each 1° × 1° grid point we use the nearest discharge stations to spatially interpolate the observations $\mathbf{q}$. To this end, an average of data from discharge stations located within 0.5° radius of each grid point is assigned to this grid point. Since no straight information on the data uncertainty is available, two approaches are applied here to specify errors on the data. Sheffield et al. (2009) suggested that the standard errors in the gauge-based data are 5–10% of the discharge values and Pan et al. (2012) proposed a formula to estimate the discharge error for a basin within a given area A as,

$$\mathit{Relative\ Error}\ (\%) = 5\frac{(A_1 - A)}{(A_1 - A_2)} + 5, \tag{7.1}$$

where A_1 and A_2 are the areas of Amazon Basin ($4.62 \times 10^6\,\text{km}^2$) and Ural Basin ($0.19 \times 10^6\,\text{km}^2$), respectively. Here we use Eq. (7.1) to assign errors to discharge stations located in the major basins of Amazon, Indus, Mississippi, Orange, Danube,

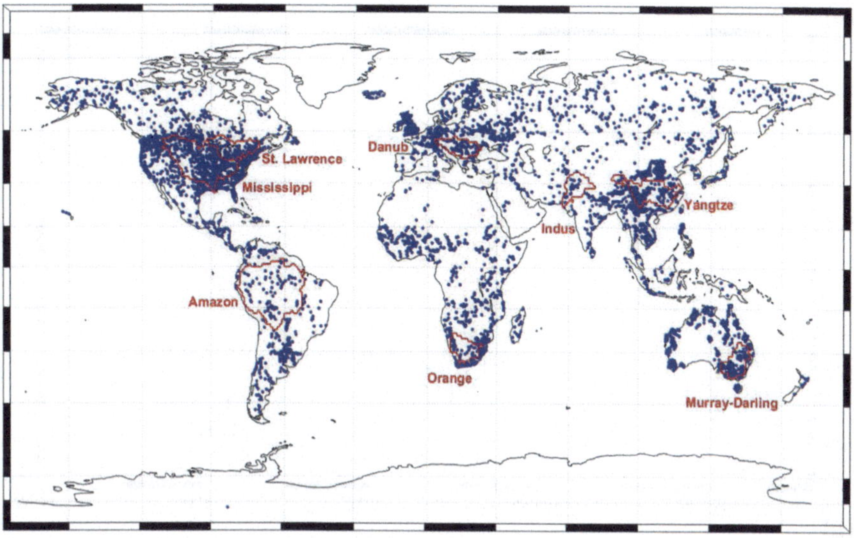

Fig. 7.1 Distribution of water discharge stations used in this study

St. Lawrence, Murray-Darling, and Yangtze, and 10% of discharge value for any station outside of these areas as suggested by literature (e.g., Pan et al. 2012; Aires 2014; Munier et al. 2014).

7.3.4 In-Situ Measurements

In addition to monitoring water budget closure errors using the water fluxes observations, we use in-situ groundwater measurements over the Mississippi Basin and Murray-Darling Basin to evaluate the performance of the proposed filter. The distribution of groundwater well stations is presented in Fig. 7.2. In the Mississippi Basin, independent data are collected from USGS. Additional measurements are provided for the Murray-Darling Basin by the New South Wales Government (NSW) groundwater archive. Monthly well measurements are acquired and time series of groundwater storage anomalies are generated. Generally, a specific yield is required to convert well-water levels to variations in groundwater storage regarding equivalent water heights (Rodell et al. 2007; Zaitchik et al. 2008). This information, however, is not available in our case, so TWS variation from GRACE and Global Land Data Assimilation System (GLDAS) soil moisture are used to calculate the specific yield and scale the observed headwater by modifying the magnitude of groundwater time series (Tregoning et al. 2012; Tangdamrongsub et al. 2015). As Tregoning et al. (2012) showed, the GW component can be extracted by removing the soil moisture component from GRACE TWS data while other compartments like biomass and

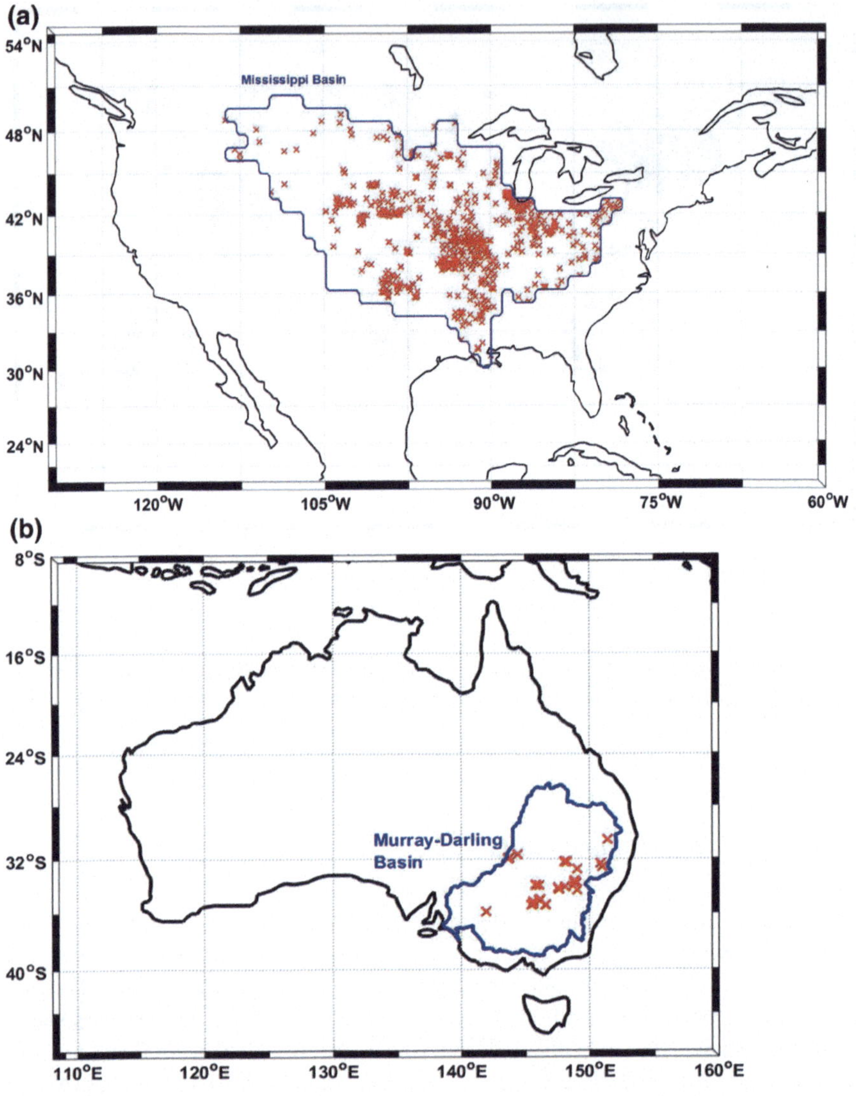

Fig. 7.2 Locations of groundwater stations within (**a**) the Mississippi Basin in the US (**a**) and (**b**) the Murray-Darling Basin in Australia

Table 7.1 A summary of the datasets used in this study

Description	Platform	Data access
Terrestrial water storage (TWS)	GRACE	https://www.tugraz.at/institute/ifg/downloads/gravity-field-models/itsg-grace2014/
Daily accumulated precipitation (**p**)	TRMM-3B42	http://disc2.gesdisc.eosdis.nasa.gov/data/TRMM_L3/TRMM_3B42_Daily.7
MODIS global evapotranspiration (**e**)	MOD16	http://www.ntsg.umt.edu/project/mod16
Water discharge (**q**)	GRDC	http://www.bafg.de/GRDC/EN/Home/homepage_node.html
q		http://www.hydrosciences.fr/sierem/consultation/choixaccess.asp?lang=en
q	USGS	https://waterdata.usgs.gov/nwis/sw
q		http://www.bom.gov.au/waterdata/
q	NRFA	http://nrfa.ceh.ac.uk/data/
q		http://www.ore-hybam.org/
q		http://www.hydrology.gov.np/new/bull3/index.php/hydrology/home/main
Hydrological model	W3RA	http://www.wenfo.org/wald/data-software/
Groundwater in-situ measurements	USGS	https://water.usgs.gov/ogw/data.html
	NSW	http://waterinfo.nsw.gov.au/pinneena/gw.shtml

surface water variations can be excluded due to their small contribution to regional scale mass variations. The calculated specific yields range between 0.08 and 0.16 over the Murray-Darling Basin, falling within the 0.05–0.2 range suggested by the Australian Bureau of Meteorology (BOM) and Seoane et al. (2013), and range between 0.15 and 0.22 over the Mississippi Basin along with those suggested by Gutentag et al. (1984) (i.e., 0.1 to 0.3), thereby justifying the application of the method. Using extracted yield factors, one can extract the groundwater components from the measured well-water levels. The scaled groundwater time series are then used to evaluate the data assimilation results over each basin. To this end, we compare groundwater estimates after data assimilation with ground-based groundwater measurements. Details of the datasets used in this study are outlined in Table 7.1.

7.4 The Weak Constrained Ensemble Kalman Filter (WCEnKF)

7.4.1 Problem Formulation

Let $\{\mathbf{x}_t\}_{t=0}^{T} \in \mathbb{R}^{n_x}$ denote the (unknown) system state process formed by top soil, shallow soil, deep soil water, snow, vegetation, and groundwater. Note that except for groundwater, all the other components are simulated with two hydrological response units (HRU) of tall, e.g., deep-rooted vegetation and short, e.g., shallow-rooted vegetation, which leads to 11 state variables ($5 \times 2 + 1$) of W3RA at each grid cell (24,509 cells in total). Although in general, t refers to model time steps, for the sake of simplicity, we assume that the model time step is equal to the assimilation time step (monthly scale). $\{\mathbf{y}_t\}_{t=0}^{T} \in \mathbb{R}^{n_y}$ represents the GRACE TWS observed process. The state and observed processes are related through a dynamical state-space system of the form,

$$\begin{cases} \mathbf{x}_t = \mathcal{M}_{t-1}(\mathbf{x}_{t-1}) + \nu_t, \\ \mathbf{y}_t = \quad \mathbf{H}_t\mathbf{x}_t + \mathbf{w}_t, \end{cases} \tag{7.2}$$

for which the state transition operator, $\mathcal{M}(.)$, is nonlinear. $\mathbf{H}$ is the (observation) design matrix containing 11 ones in each of the 24,509 rows, representing the sum of the individual compartments to TWS at each grid cell with all the other elements of the rows being zero (total 269,599 columns). The proposed scheme can be easily extended to the case of nonlinear observation operator (i.e., in which $\mathbf{H}_t\mathbf{x}_t$ is replaced by $h_t(\mathbf{x_t})$), as for example discussed in Liu and Xue (2002). The state transition noise process, $\nu = \{\nu_t\}_t$, and the observation noise process, $\mathbf{w} = \{\mathbf{w}_t\}_t$, are assumed to be independent, jointly independent, and independent of the initial state, $\mathbf{x}_0$. Furthermore, $\mathbf{x}_0$, ν_t, and $\mathbf{w}_t$ are assumed to be Gaussian; ν_t and $\mathbf{w}_t$ with zero mean and covariances $\mathbf{Q}_t$ and $\mathbf{R}_t$, respectively.

Data assimilation can destroy the balance between water fluxes. It is therefore essential to incorporate the water balance equation by imposing an equality constraint to restore the balance problem. Changes in monthly mean water storage at two different time steps (e.g., t and $t-1$) should be equal, up to uncertainties in the involved data, to the difference between the monthly mean input ($\mathbf{p}$) and output ($\mathbf{e}$ and $\mathbf{q}$) water storages. This can be formulated as:

$$\mathbf{d}_t = -\mathbf{x}_t + \mathbf{x}_{t-1} + \mathbf{p}_t - \mathbf{e}_t - \mathbf{q}_t + \boldsymbol{\xi}_t, \tag{7.3}$$

where $\{\boldsymbol{\xi}_t\}_t$ is the noise process accounting for errors associated with the different water fluxes data. Here we assume $\boldsymbol{\xi}_t$ Gaussian white noise with zero mean and covariance $\boldsymbol{\Sigma}_t$, and independent of $\mathbf{x}_0$ and $\{\mathbf{w}_t\}_t$. Defining $\mathbf{z}_t = \mathbf{d}_t - \mathbf{p}_t + \mathbf{e}_t + \mathbf{q}_t$, the constraint Eq. (7.3) is rewritten as,

$$\mathbf{z}_t = \mathbf{G}\mathbf{x}_t + \mathbf{L}\mathbf{x}_{t-1} + \boldsymbol{\xi}_t, \tag{7.4}$$

where $\mathbf{G}$, in general, is the $n_x \times n_x$ (with n_x being the length of x) identity matrix while in this study $\mathbf{G} = \mathbf{H}$ to aggregate different water compartments at each grid point and $\mathbf{L} = -\mathbf{G}$.

In the constrained state-space system Eqs. (7.2)–(7.4), we focus on the filtering problem, say, on the estimation, at each time t, of the system state, $\mathbf{x}_t$, conditional on both GRACE TWS observations, $\mathbf{y}_{0:t} \stackrel{\text{def}}{=} \{\mathbf{y}_0, \mathbf{y}_1, \ldots, \mathbf{y}_t\}$ and "pseudo-observations" $\mathbf{z}_{0:t}$. Let $\mathbf{r}_t = [\mathbf{y}_t^T, \mathbf{z}_t^T]^T$. As known in the Bayesian estimation theory, the computation of any estimator of $\mathbf{x}_t$ from $\mathbf{r}_{0:t}$ is based on the so-called posterior (filtering or analysis) probability density function (pdf), $p(\mathbf{x}_t|\mathbf{r}_{0:t})$. For instance, the posterior mean (PM) estimator, $\hat{\mathbf{x}}_{t|t}$, which minimizes the mean squared error, is given by

$$\begin{aligned} \hat{\mathbf{x}}_{t|t} &= \mathbb{E}[\mathbf{x}_t|\mathbf{r}_{0:t}], \\ &= \int \mathbf{x}_t p(\mathbf{x}_t|\mathbf{r}_{0:t}) d\mathbf{x}_t. \end{aligned} \tag{7.5}$$

The conditional independence property of the system Eqs. (7.2)–(7.4) enables for efficient *recursive* computation of this analysis pdf. Indeed, starting at time $t-1$ from $p(\mathbf{x}_{t-1}|\mathbf{r}_{0:t-1})$, one can compute $p(\mathbf{x}_t|\mathbf{r}_{0:t})$ following forecast and update steps as follows:

- *Forecast step.* The state transition pdf, $p(\mathbf{x}_t|\mathbf{x}_{t-1})$, is first used to compute the forecast pdf as (e.g., Ait-El-Fquih et al. 2016),

$$p(\mathbf{x}_t|\mathbf{r}_{0:t-1}) = \int p(\mathbf{x}_t|\mathbf{x}_{t-1}) p(\mathbf{x}_{t-1}|\mathbf{r}_{0:t-1}) d\mathbf{x}_{t-1}. \tag{7.6}$$

- *Update step with the GRACE TWS data.* Once available, the observation $\mathbf{y}_t$ is first used to update forecast pdf, $p(\mathbf{x}_t|\mathbf{r}_{0:t-1})$ as,

$$p(\mathbf{x}_t|\mathbf{r}_{0:t-1}, \mathbf{y}_t) \propto p(\mathbf{y}_t|\mathbf{x}_t) p(\mathbf{x}_t|\mathbf{r}_{0:t-1}), \tag{7.7}$$

  and

$$p(\mathbf{x}_{t-1}|\mathbf{r}_{0:t-1}, \mathbf{y}_t) \propto p(\mathbf{y}_t|\mathbf{x}_{t-1}) p(\mathbf{x}_{t-1}|\mathbf{r}_{0:t-1}). \tag{7.8}$$

  While the likelihood $p(\mathbf{y}_t|\mathbf{x}_t)$ in the update (7.7) is given through the observation model, $p(\mathbf{y}_t|\mathbf{x}_{t-1})$ in (7.8) is not known and needs to be computed beforehand as,

$$p(\mathbf{y}_t|\mathbf{x}_{t-1}) = \int p(\mathbf{y}_t|\mathbf{x}_t) p(\mathbf{x}_t|\mathbf{x}_{t-1}) d\mathbf{x}_t. \tag{7.9}$$

  By ignoring the pseudo-observations, $\mathbf{z}_{0:t-1}$, in Eqs. (7.7)–(7.8), these equations translate as a one-step-ahead (OSA) smoothing process, which computes the OSA smoothing pdf, $p(\mathbf{x}_{t-1}|\mathbf{y}_{0:t})$, from the previous analysis pdf $p(\mathbf{x}_{t-1}|\mathbf{y}_{0:t-1})$ (Ait-El-Fquih et al. 2016). For simplicity, we refer to pdf $p(\mathbf{x}_{t-1}|\mathbf{r}_{0:t-1}, \mathbf{y}_t)$ as the

OSA smoothing pdf (note that the actual OSA smoothing pdfs are $p(\mathbf{x}_{t-1}|\mathbf{r}_{0:t})$, $p(\mathbf{x}_{t-1}|\mathbf{y}_{0:t})$ or $p(\mathbf{x}_{t-1}|\mathbf{z}_{0:t})$).

- *Update step with* $\mathbf{z}_t$. The pdf $p(\mathbf{x}_t|\mathbf{r}_{0:t-1}, \mathbf{y}_t)$ that stems from the update of the forecast pdf with $\mathbf{y}_t$ (Eq. (7.7)) is in turn updated with $\mathbf{z}_t$ based on the Bayes' rule, leading to the analysis pdf of interest:

$$p(\mathbf{x}_t|\mathbf{r}_{0:t}) \propto p(\mathbf{z}_t|\mathbf{x}_t, \mathbf{y}_t, \mathbf{r}_{0:t-1})p(\mathbf{x}_t|\mathbf{r}_{0:t-1}, \mathbf{y}_t). \tag{7.10}$$

The unknown likelihood $p(\mathbf{z}_t|\mathbf{x}_t, \mathbf{y}_t, \mathbf{r}_{0:t-1})$ is computed beforehand as,

$$p(\mathbf{z}_t|\mathbf{x}_t, \mathbf{y}_t, \mathbf{r}_{0:t-1}) \approx \int p(\mathbf{z}_t|\mathbf{x}_t, \mathbf{x}_{t-1})p(\mathbf{x}_{t-1}|\mathbf{r}_{0:t-1}, \mathbf{y}_t)d\mathbf{x}_{t-1}. \tag{7.11}$$

7.4.2 The WCEnKF Algorithm

In this section, the WCEnKF algorithm is described in three stages. The definition starts with the forecast step, in which the previous analysis ensemble state is integrated forward with the model to obtain the forecast ensemble. Two analysis (update) steps are then performed. The first updates, following a Kalman filter-like correction, the forecast ensemble based on the GRACE TWS data; the second update uses information of the water budget closure to perform a second Kalman filter-like correction, leading to the analysis ensemble of interest.

From previous section, it is not possible to analytically compute the integrals in Eqs. (7.5)–(7.11) because of the nonlinearity of the model $\mathcal{M}(.)$. We therefore derive an EnKF solution (Evensen 1994; Hoteit et al. 2015) by applying the standard random sampling properties 1 and 2 listed in Appendix A. Starting at time $t-1$ from an analysis ensemble, $\{\mathbf{x}_{t-1}^{a,(i)}\}_{i=1}^{n}$, the analysis ensemble at next time (t), $\{\mathbf{x}_{t}^{a,(i)}\}_{i=1}^{n}$, can be computed by the following cycles of forecast and update steps.

- *Forecast step.* A forecast ensemble, $\{\mathbf{x}_{t}^{f,(i)}\}_{i=1}^{n}$, is first computed by integrating $\{\mathbf{x}_{t-1}^{a,(i)}\}_{i=1}^{n}$, forward in time with the model:

$$\mathbf{x}_t^{f,(i)} = \mathcal{M}_{t-1}\left(\mathbf{x}_{t-1}^{a,(i)}\right) + \nu^{(i)}, \tag{7.12}$$

where $\nu^{(i)}$ is a random sample from the Gaussian $\mathcal{N}(\mathbf{0}, \mathbf{Q}_t)$.

- *Update with GRACE TWS data (first update).* Once a new observation $\mathbf{y}_t$ is available, new ensemble $\{\tilde{\mathbf{x}}_{t}^{a,(i)}\}_{i=1}^{n}$ and $\{\tilde{\mathbf{x}}_{t-1}^{s,(i)}\}_{i=1}^{n}$ are then computed using an EnKF update of the forecast ensemble and the previous analysis ensemble:

$$\mathbf{y}_t^{f,(i)} = \mathbf{H}\mathbf{x}_t^{f,(i)} + \mathbf{w}^{(i)}; \quad \mathbf{w}^{(i)} \sim \mathcal{N}(\mathbf{0}, \mathbf{R}_t), \tag{7.13}$$

$$\tilde{\mathbf{x}}_t^{a,(i)} = \mathbf{x}_t^{f,(i)} + \mathbf{P}_{\mathbf{x}_t^f}\mathbf{H}^T \underbrace{[\mathbf{H}\mathbf{P}_{\mathbf{x}_t^f}\mathbf{H}^T + \mathbf{R}_t]^{-1}[\mathbf{y}_t - \mathbf{y}_t^{f,(i)}]}_{\boldsymbol{\mu}_t^{(i)}}, \tag{7.14}$$

$$\tilde{\mathbf{x}}_{t-1}^{s,(i)} = \mathbf{x}_{t-1}^{a,(i)} + \mathbf{P}_{\mathbf{x}_{t-1}^a,\mathbf{x}_t^f}\mathbf{H}^T\boldsymbol{\mu}_t^{(i)}. \tag{7.15}$$

The covariance matrices $\mathbf{P}_{\mathbf{x}_t^f}$ and $\mathbf{P}_{\mathbf{x}_{t-1}^a,\mathbf{x}_t^f}$, are evaluated beforehand from the previous analysis and forecast ensembles as,

$$\mathbf{P}_{\mathbf{x}_t^f} = (n-1)^{-1}\mathbf{S}_{\mathbf{x}_t^f}\mathbf{S}_{\mathbf{x}_t^f}^T, \tag{7.16}$$

$$\mathbf{P}_{\mathbf{x}_{t-1}^a,\mathbf{x}_t^f} = (n-1)^{-1}\mathbf{S}_{\mathbf{x}_{t-1}^a}\mathbf{S}_{\mathbf{x}_t^f}^T, \tag{7.17}$$

where $\mathbf{S}_{\mathbf{x}_{t-1}^a}$ and $\mathbf{S}_{\mathbf{x}_t^f}$ are the perturbation matrices (i.e., matrices with n columns formed by the ensemble members minus the ensemble mean). Equations (7.14) and (7.15) are EnKF updates of $\mathbf{x}_t^{f,(i)}$ and $\mathbf{x}_{t-1}^{a,(i)}$, respectively. These updates are achieved based on $\mathbf{y}_t$, with Kalman gains $\mathbf{P}_{\mathbf{x}_t^f}\mathbf{H}^T[\mathbf{H}\mathbf{P}_{\mathbf{x}_t^f}\mathbf{H}^T + \mathbf{R}_t]^{-1}$ (Eq. (7.14)) and $\mathbf{P}_{\mathbf{x}_{t-1}^a,\mathbf{x}_t^f}\mathbf{H}^T[\mathbf{H}\mathbf{P}_{\mathbf{x}_t^f}\mathbf{H}^T + \mathbf{R}_t]^{-1}$ (Eq. (7.15)). The $\tilde{\mathbf{x}}_t^{a,(i)}$ is based on $\mathbf{y}_t$ only, and a second update with $\mathbf{z}_t$ is still required. The index '$\sim$' is used for the first update to distinguish it from the second one.

- *Adjustment with the water budget constraint (second update).* The pseudo-observation, $\mathbf{z}_t$, is then used to update $\{\tilde{\mathbf{x}}_t^{a,(i)}\}_{i=1}^n$, again using an EnKF update, leading to the actual state analysis ensemble of interest:

$$\mathbf{z}_t^{f,(i)} = \mathbf{G}\tilde{\mathbf{x}}_t^{a,(i)} + \mathbf{L}\tilde{\mathbf{x}}_{t-1}^{s,(i)} + \boldsymbol{\xi}_t^{(i)}; \quad \boldsymbol{\xi}_t^{(i)} \sim \mathcal{N}(\mathbf{0}, \boldsymbol{\Sigma}_t), \tag{7.18}$$

$$\mathbf{x}_t^{a,(i)} = \tilde{\mathbf{x}}_t^{a,(i)} + \mathbf{P}_{\tilde{\mathbf{x}}_t^a,\mathbf{z}_t^f}[\mathbf{N}\mathbf{P}_{\eta_t}\mathbf{N}^T + \boldsymbol{\Sigma}_t]^{-1}[\mathbf{z}_t - \mathbf{z}_t^{f,(i)}], \tag{7.19}$$

with $\mathbf{N} = [\mathbf{G}, \mathbf{L}]$, the cross-covariance $\mathbf{P}_{\tilde{\mathbf{x}}_t^a,\mathbf{z}_t^f}$ is evaluated from the ensembles $\{\tilde{\mathbf{x}}_t^{a,(i)}\}_{i=1}^n$ and $\{\mathbf{z}_t^{f,(i)}\}_{i=1}^n$, as in Eq. (7.17), and the covariance $\mathbf{P}_{\eta_t}$ is computed from the augmented state ensemble $\{\eta_t^{(i)}\}_{i=1}^n$, where $\eta_t^{(i)} = [(\tilde{\mathbf{x}}_t^{a,(i)})^T, (\tilde{\mathbf{x}}_{t-1}^{s,(i)})^T]^T$, as in Eq. (7.16). As one can see, Eq. (7.19) translates an EnKF update of $\tilde{\mathbf{x}}_t^{a,(i)}$, based on the pseudo-observation $\mathbf{z}_t$, where gain is $\mathbf{P}_{\tilde{\mathbf{x}}_t^a,\mathbf{z}_t^f}[\mathbf{N}\mathbf{P}_{\eta_t}\mathbf{N}^T + \boldsymbol{\Sigma}_t]^{-1}$, leading to $\mathbf{x}_t^{a,(i)}$, the state analysis ensemble of interest.

The PM Eq. (7.5) estimate is then approximated by the sample mean of the resulting analysis ensemble. As discussed in the introduction, the pseudo-observations are only available at the discharge observations locations, but the Kalman update Eq. (7.18) spreads the information to the whole state vectors. A schematic illustration of the filter algorithm is presented in Fig. 7.3.

Similarly to the standard CEnKF of Pan et al. (2012), the proposed WCEnKF involves one forecast step and two successive update steps. The two filters have the

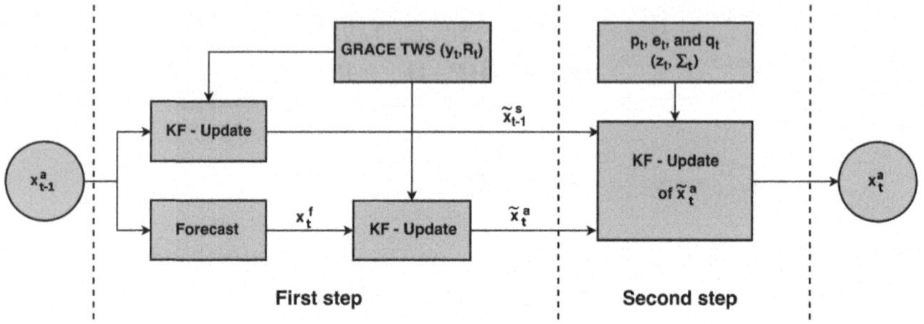

Fig. 7.3 A schematic illustration of the WCEnKF filter's steps applied for data assimilation in this study

same forecast and first update (with observation $\mathbf{y}_t$) steps, and only differ in their second update (adjustment with pseudo-observation $\mathbf{z}_t$)). The state update mechanism Eqs. (7.18) and (7.19) is more general than the one in Pan et al. (2012), as the latter does not involve the OSA smoothing ensemble, $\{\tilde{\mathbf{x}}_{t-1}^{s,(i)}\}_i$ in Eqs. (7.18), (7.19) and assume no noise ($\boldsymbol{\xi}_t^{(i)} = 0$) in Eq. (7.18) and its covariance $\boldsymbol{\Sigma}_t = 0$ in Eq. (7.19). As such, CEnKF can be considered as a direct particular case of WCEnKF. As stated above, accounting for uncertainties in the constraint allows avoiding a perfect pseudo-observation model scenario, which should help mitigating for over-fitting issues. The OSA smoothing terms (e.g., $\tilde{\mathbf{x}}_{t-1}^{s,(i)}$ in Eq. (7.18)) come from the fact that the pseudo-observation, $\mathbf{z}_t$, in the constraint Eq. (7.4) is not only function of $\mathbf{x}_t$ but also of $\mathbf{x}_{t-1}$.

7.4.3 Experimental Setup

All the water fluxes data (including **p**, **e**, and **q**) are accumulated to a monthly scale and used in the monthly assimilation processes. The monthly increment (i.e., the difference between the monthly averaged GRACE TWS and simulated TWS) can be added to each day of the current month, which guarantees that the update of the monthly mean is identical to the monthly mean of the daily updates. In practice, the differences between the predictions and the updated states are added as offsets to the state vectors at the last day of each month to generate the ensembles for the next month assimilation step. Given that not enough information are available to accurately estimate the pseudo-observation error covariance Σ, especially for **q**, to test the sensibility we consider the error values mentioned in Sect. 7.3.3 as *reference errors* and test with three different Σ: (1) the *reference errors* values minus 5% of observation values, (2) *reference errors*, and (3) the *reference errors* plus 5% of observation values. We further assume the observation errors to be spatially uncorrelated. This test allows us to analyze the influence of the pseudo-observations on the final results.

To generate an initial ensemble to start the filtering process, we follow Renzullo et al. (2014) and perturb the meteorological forcing fields. To this end, we assume a Gaussian multiplicative error of 30% for precipitation, an additive Gaussian error of 50 Wm^{-2} for the shortwave radiation, and a Gaussian additive error of 2 °C for temperature (Jones et al. 2007; Renzullo et al. 2014). The initial ensemble is then computed by sampling the above Gaussian distributions (see details in Renzullo et al. 2014). We, then, integrate the resulting ensemble (with 30 members) forward with the model from January 2000 to April 2002 to generate the initial ensembles at the beginning of the study period. An ensemble of 30 members is selected as it was found large enough to obtain sufficient ensemble spread at reasonable computational cost.

We further apply ensemble inflation and localization to enhance the filters performances (e.g., Anderson et al. 2007). These techniques were proven to be useful in dealing with neglected uncertainties in the system and small ensembles (e.g., Hamill and Snyder 2002; Bergemann and Reich 2010). Ensemble inflation with a best case coefficient factor of 1.12 (after testing different values) is applied here to increase the ensemble deviation from the ensemble-mean (Anderson et al. 2007). Local Analysis (LA) (Evensen 2003) is used to restrict the impact of the measurements in the update step to variables located within a certain distance only (5° as suggested by Khaki et al. 2017b). By spatially limiting the influences of observations over large distances in the sample covariance, LA can help mitigating spatial correlation errors and rank deficiency problem during the assimilation (see Khaki et al. 2017b, for more details). This is particularly useful to account for the spatial correlation errors in satellite products, particularly GRACE (Khaki et al. 2017b; Tangdamrongsub et al. 2017).

7.5 Results

We first investigate the effects of different scenarios applied for errors associated with the fluxes in Sect. 7.5.1. In Sect. 7.5.2, we evaluate the performance of WCEnKF against in-situ groundwater measurements over the Mississippi River Basin in the US and the Murray-Darling Basin in Australia. To further assess the behavior of the proposed WCEnKF, we compare its results with the standard EnKF for predicting water storages. Then, in Sect. 7.5.3, we analyze the assimilation results and the performance of the proposed filter in enforcing the balance between water fluxes, e.g., we assess the behaviour of the filters in dealing with water balance problem.

7.5.1 Error Sensitivity Analysis

We first analyze the effects of the different datasets, i.e., both the GRACE TWS and pseudo-observations on the filter estimates. The incorporation of the pseudo-observations in the second update step of the filter modifies the contribution of

Table 7.2 Average correlations and errors between the water storages estimated by WCEnKF and water fluxes observations of **p**, **e** and **q** as well as GRACE TWS data considering three different error values used in the data assimilation process. "Ref" in table refers to the *reference errors* (described in Sect. 7.4.3)

Error level	Correlation				
	p	**e**	**q**	GRACE TWS	Imbalance error (mm)
(1) Ref−5% (observation)	0.78	0.83	0.76	0.77	12.05
(2) Ref+0% (observation)	0.65	0.72	0.69	0.84	18.31
(3) Ref+5% (observation)	0.61	0.63	0.58	0.89	37.24

GRACE TWS data on the state estimations. As such, the three different covariance error matrices (cf. Sect. 7.4.3) of **p**, **e**, and **q** would cause that both the GRACE TWS and pseudo-observations contribute differently. For each grid point, we calculate the correlations between the filter estimations of TWS and the water fluxes **p**, **e**, and **q** as well as the assimilated GRACE TWS data. The results along with the average imbalance errors (from the water balance equation) are presented in Table 7.2. It can be seen that applying the first case with minimum error values, as it is expected, leads to a higher correlation between the filter estimates and other water fluxes. The least imbalance error is also achieved in this case. However, in general, increasing the impact of water fluxes in the second step of the filter decreases the correlation between the estimates and GRACE TWS data. This suggests, as we expected, a trade-off between the effects of observations in the first and second step of the filter according to the values of Σ. In the third scenario, for example, applying pseudo-observations with larger errors leads to smaller correlations with the water flux observations and larger correlation to the GRACE TWS data. Note that we also applied a similar test for **p**, **e**, and **q** with zero error (such that CEnKF), which resulted in the least imbalance error. Nevertheless, this case leads to larger errors compared to groundwater measurements compared with the three scenarios above. Therefore, hereafter, we only present the results associated with the second scenario (with no additional errors on those that are initially assumed). This case is found to lead to better results when groundwater estimates from each scenario are compared to independent groundwater in-situ measurements (details in Sect. 7.5.2).

7.5.2 Assessment Against In-Situ Data

The estimated groundwater storage obtained from each filter is compared to the post-processed in-situ measurements of groundwater changes (cf. Sect. 7.3.4) over the Mississippi Basin and Murray-Darling Basin. To this end, the estimated groundwater storages, as well as model-free run (without data assimilation) are spatially interpolated to the location of the in-situ measurements using the nearest neighbour

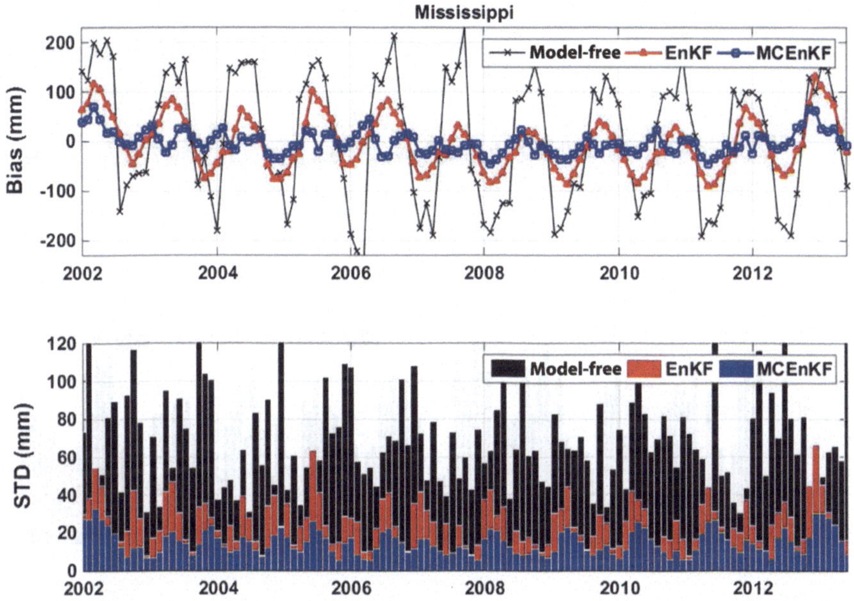

Fig. 7.4 Average bias and STD of the groundwater results from the EnKF and WCEnKF data assimilation filters over the Mississippi Basin with respect to the in-situ groundwater measurements

(the closest four grid values). The groundwater misfits (errors) between the in-situ measurements and those of the EnKF and WCEnKF are then computed. Figures 7.4 and 7.5 plot the resulting bias, namely, differences between groundwater estimated by the filters and in-situ measurements, and STD (of the calculated differences) for the Mississippi Basin and Murray-Darling Basin, respectively.

For both basins, the estimated biases are significantly decreased when the proposed WCEnKF filter is applied. The average estimated bias using WCEnKF is 23.14 mm for the Mississippi Basin and 26.89 mm for the Murray-Darling Basin, indicating an average of 22.10 and 26.38% bias improvements compared to the EnKF. Despite this, we found that the correlation between the filters' estimated groundwater and in-situ groundwater time series are large for both basins. An average of 0.76 (at 95% confidence interval) for both basins is achieved, which means that assimilating only GRACE data (as in the EnKF) is good for estimating annual and inter-annual variations, but not enough to accurately recover their amplitudes. The error reduction using WCEnKF is also noticeable in the STD. WCEnKF decreases the uncertainties in the Mississippi Basin and Murray-Darling Basin by 48.87% and 35.19%, respectively.

For every grid point within each basin, we calculate the Root-Mean-Squared Error (RMSE) and also the correlation between in situ measurements and filters results. Note that cross-correlation is applied to account for lag differences between the time series. We further undertake a significance test for the correlation coefficients using

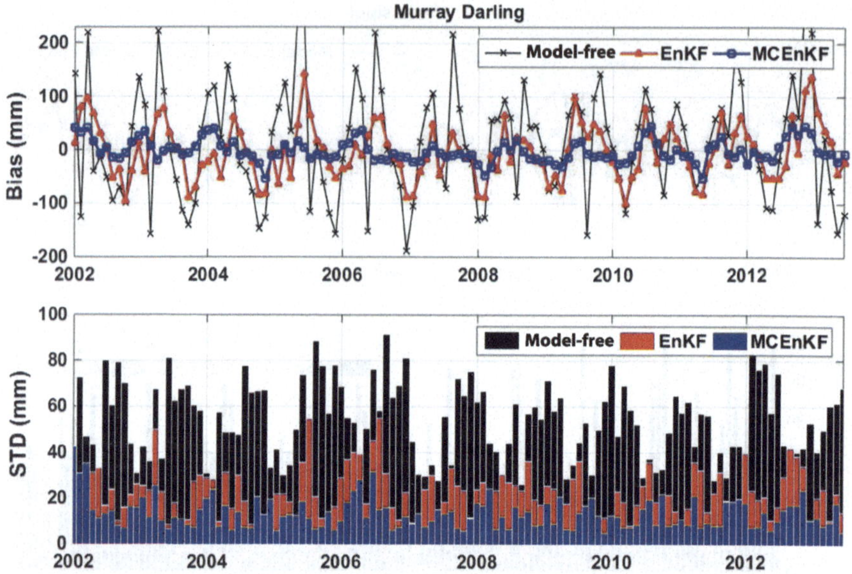

Fig. 7.5 Average bias and STD of the groundwater results from the EnKF and WCEnKF data assimilation filter over the Murray-Darling Basin with respect to the in-situ groundwater measurements

Table 7.3 Summary of the evaluation results from each filter and model-free run against the groundwater in-situ measurements over the Mississippi Basin and Murray-Darling Basin. For each case the RMSE average and its range (±XX) at the 95% confidence interval is presented

	Mississippi Basin		Murray-Darling Basin	
Method	RMSE (mm)	Correlation	RMSE (mm)	Correlation
EnKF	56.74 ± 6.12	0.72	41.58 ± 6.48	0.68
Improvement (%) regarding model-free	38.41	36.11	48.96	47.06
WCEnKF	48.22 ± 5.63	0.84	34.63 ± 5.27	0.79
Improvement (%) regarding model-free	47.66	45.23	57.49	54.43
Improvement (%) regarding EnKF	15.02	14.28	16.71	13.92

t-distribution. The estimated t-value and the distribution at 0.05 significant level are then used to calculate a p-value. The calculated p-values for the correlations in Table 7.3 lie under 5% indicating coefficients are significant. Table 7.3 summarizes these results. The Assimilation of the GRACE data using WCEnKF increases the correlation from 0.72 (EnKF) to 0.84 over the Mississippi Basin and from 0.68 to 0.79 for the Murray-Darling Basin. While both filters significantly improve groundwater estimates with respect to model-free run (48.13 on average), the larger RMSE

Table 7.4 Average RMSE results (with their ranges ±XX at the 95% confidence) by each filter at forecast steps and model-free run compared to the groundwater in-situ measurements over the Mississippi Basin and Murray-Darling Basin. Table also contains correlations between TWS estimated by the methods at forecast steps and water fluxes

Method	RMSE (mm)		Correlation		
	Mississippi Basin	Murray-Darling Basin	**p**	**e**	**q**
Model-free	92.13 ± 12.39	81.46 ± 10.67	0.95	0.86	0.83
EnKF	74.53 ± 8.82	62.71 ± 9.25	0.56	0.53	0.49
WCEnKF	65.48 ± 7.18	47.91 ± 7.95	0.94	0.82	0.85

improvements of 15.02 and 16.71% for the Mississippi Basin and the Murray-Darling Basin, respectively, suggest the enhancement gained from the proposed two-updates filter against the one-update filter.

Furthermore, two analyses are undertaken on the forecast steps to investigate which filter is more efficient in keeping observations effects within the system states. Generally, a filter with better forecasts can perform better during an experiment. We calculate average RMSE of groundwater estimates at forecast steps for the Mississippi and Murray-Darling Basins and compare them with those of model-free run (Table 7.4). It can be seen that both filters reduced RMSE values, while WCEnKF outperforms the EnKF scheme (approximately 12%). We also compute correlations between TWS forecast estimates, both by filters and model-free, and water fluxes (i.e., **p**, **e**, and **q**). A similar analysis as Table 7.3 is done to control the significance of correlation coefficients. Average correlations over the basins of Amazon, Indus, Mississippi, Orange, Danube, St. Lawrence, Murray-Darling, and the Yangtze (cf. Figure 7.1) are listed in Table 7.4. Based on the correlation values, it is evident that WCEnKF achieves larger correlations with respect to the EnKF. The proposed filter obtains improved agreement between the assimilation results and the fluxes.

Furthermore, to statistically investigate the difference between average correlation values, ANOVA (analysis of variance; Nelson 1983; Ullman 1989) method is applied. The method shows how mean values are different. For every flux correlation, the null hypothesis is that the average correlation for the model-free, EnKF, and WCEnKF are equal. ANOVA tests the above hypothesis at 0.05 significance level. Our experiment indicates that the means are not equal, thus, ANOVA in the second step determines which correlations are different (to the level of significance). After implementing the later step, the EnKF result demonstrates a significantly larger difference from the model-free and WCEnKF. In sum, Table 7.4 shows that WCEnKF successfully assimilates data sets into the system, which also leads to a better forecast.

7.5.3 *Water Balance Enforcement*

In the following, we analyze the results of the filter estimates using the second scenario from Sect. 7.5.1 in terms of their relationship to the observations and more importantly water budget closure. Figure 7.6 shows the results for the comparison between the assimilation results and GRACE TWS data. For each grid point, we calculate the average discrepancy and correlation between the two TWS time series. Results indicate that the error between the model and GRACE data is about 26 mm, which is 69% less than those resulting from the free-run (model runs without assimilation) and 13% higher than data assimilation results using the (one-update) EnKF scheme. This means that the application of the second update step, in some cases, decreases the effects of GRACE data by enforcing the balance between water fluxes. Figure 7.6b, in general, suggests a high correlation between the filter estimates and observations. Nevertheless, again, smaller correlations are found in places with a denser discharge stations corresponding to better imbalance control (e.g., central to northern of Asia). Much smaller correlations are observed between GRACE TWS and the model-only results (0.47 on average). Nevertheless, the EnKF provides 11% higher correlation to observations. This is due to the effects of the second update step of the proposed filter.

The above results could be explained by the correlations between the filter estimates and two water fluxes data, i.e., precipitation and evaporation. Indeed, as one can see in Fig. 7.7, the locations where a high correlation is achieved, are places where the second step of the filter affects more due to the availability of discharge data (cf. Fig. 7.1). Approximately 33 and 44% larger correlation coefficients for $\mathbf{p}$ and $\mathbf{e}$, respectively, are achieved in the areas where water balance adjustment is used compared with other areas. This illustrates that forcing water balance condition into the assimilation process increases the agreement between model outputs and other water fluxes on the one hand, and may change the effects of the GRACE data on the model on the other hand.

The average imbalance at each grid point is plotted in Fig. 7.8. The figure clearly demonstrates how the water budget enforcement spatially influences the imbalance between Δs and fluxes. It can be seen that wherever a dense network of water discharge stations exists (cf. Fig. 7.1), e.g., North America, Southeast Asia, and West Australia, a smaller imbalance between all compartments occurs. For other areas, the imbalance is much larger because the second analysis step of WCEnKF cannot be applied due to the lack of discharge data and the method simply performs as the EnKF. Therefore, this highlights the effect of the second step of WCEnKF in dealing with imbalances. This confirms the previous results that the second update step in WCEnKF increases the agreement between the assimilation outputs and the water fluxes, which results in water imbalance decreases.

Table 7.5 summarizes the average correlations between the estimated TWS data and water fluxes, $\mathbf{p}$, $\mathbf{e}$ and $\mathbf{q}$, and the average estimated imbalance errors as suggested by the EnKF and WCEnKF. Note that we only compare the filters' performances over the points in which discharge data is available. WCEnKF successfully increases

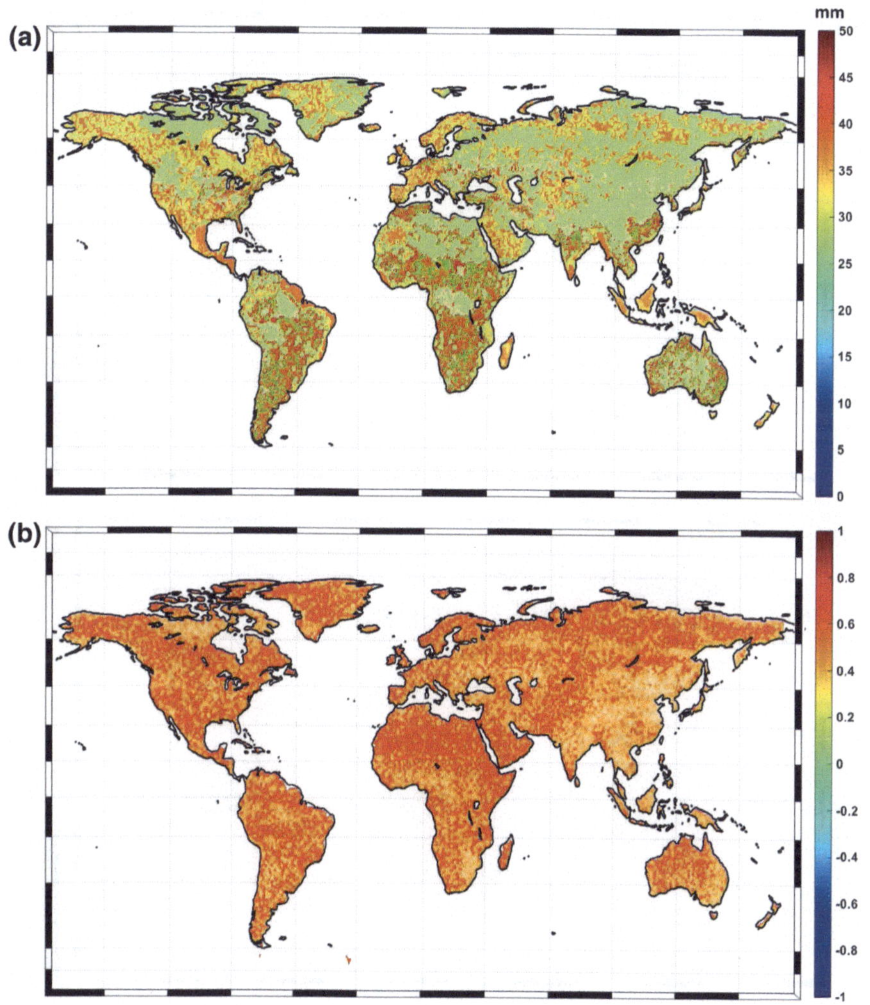

Fig. 7.6 **a**, Temporal average of misfits between the summation of TWS from WCEnKF and the GRACE TWS time series at each grid point, and **b**, The correlation between the two TWS time series

the correlation between the results and water variables of **p**, **e** and **q** with average improvements of 33%, 44%, and 45%, respectively. This leads to a significant imbalance reduction of approximately 82% (suggesting an error of 18.31 mm compared to 62.17 mm for the EnKF).

Next, in order to further investigate the data assimilation results, we focus on the major basins of Amazon, Indus, Mississippi, Orange, Danube, St. Lawrence, Murray-Darling, and the Yangtze (cf. Fig. 7.1). Due to variability of various water fluxes over different areas, these have different characteristics and behaviors with

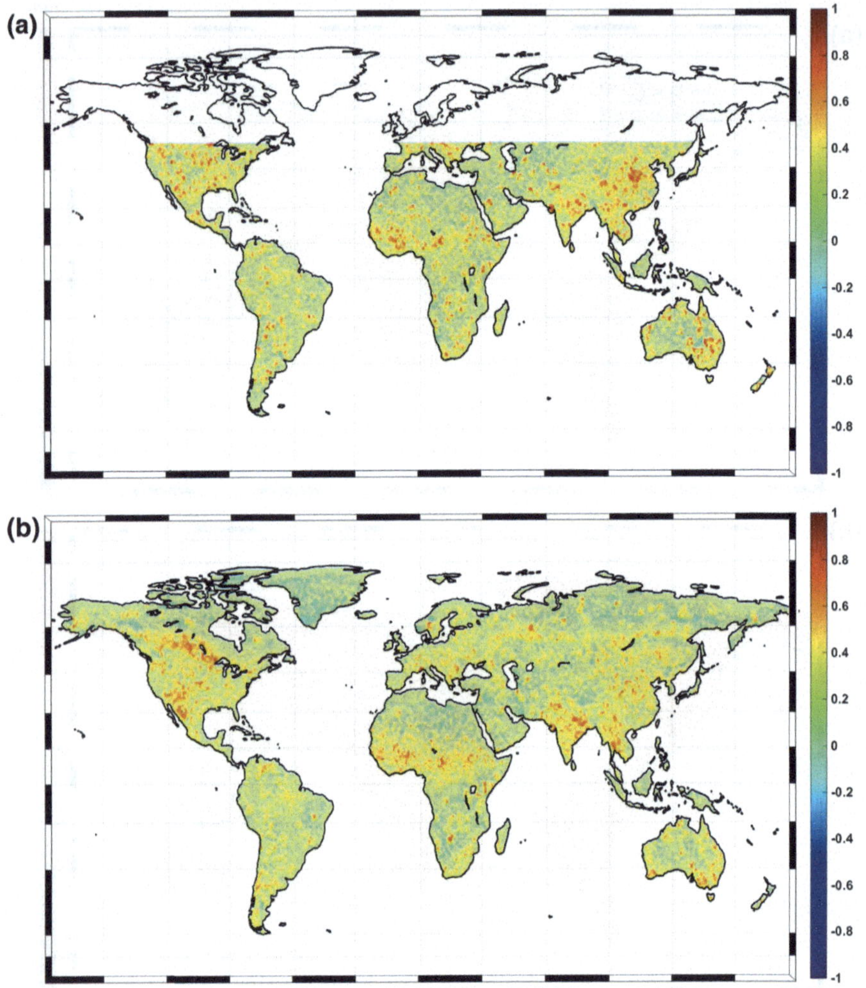

Fig. 7.7 Correlation between the data assimilation results and **p** (**a**) and **e** (**b**) time series at each grid point

various contributions through the second update of the filter (Fig. 7.9). Figure 7.9 illustrates the contribution of each water flux in the water budget closure of the basins. This shows how each variable incorporates in the water balance equation differently over each basin. Generally the larger contributions are found for **p** and **e** for all basins. **q** has a larger contribution over the Amazon Basin and relatively small impacts on the Orange Basin and St. Lawrence Basin. The estimated water storage change (Δ**s**) effects, however, vary significantly between the basins. It is shown in Fig. 7.9 that Δ**s** has larger influences over Mississippi, Danube, and Murray-Darling Basins. The share of Δ**s** in each basin is affected by incorporating **p**, **e** and **q** into the

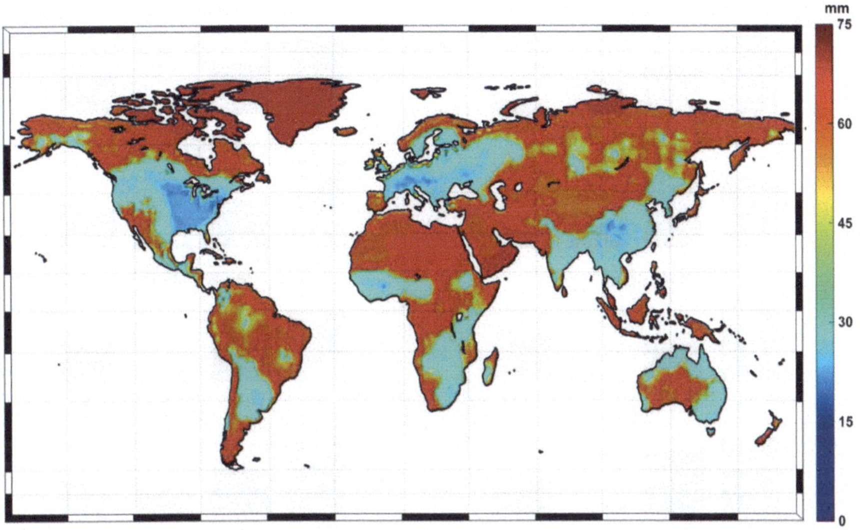

Fig. 7.8 Temporal average of imbalance errors

Table 7.5 Average correlation between the assimilation results (summation of water storages) and the data of **p**, **e** and **q**. The average imbalance errors provided by each filtering method are also indicated

Method	Correlation			Imbalance error (mm)
	p	**e**	**q**	
EnKF	0.32	0.28	0.24	62.17
WCEnKF	0.65	0.72	0.69	18.31
Improvement (%)	50.76	61.11	65.21	70.55

second step of WCEnKF, which is significantly different from the one estimated by the EnKF.

Figure 7.10 presents the average $\Delta\mathbf{s}$ as they result from the EnKF and WCEnKF over each basin. It can be seen that the application of water balance adjustment in the filtering process results in a considerable difference between the estimated TWSs. The larger correlations between the two solutions in the Mississippi Basin (0.50) and St. Lawrence Basin (0.47) indicate less influence of the water budget constraint in these basins. However, the weak agreements between the EnKF and WCEnKF results, with about 0.34 correlation on average, suggest a large impact of water balance enforcement on the process. This remarkable difference is expected to have a large effect on imbalance issue for each basin (Fig. 7.11).

The spatial average time series of imbalance between Δs and fluxes for each basin are shown in Fig. 7.11 for the EnKF and WCEnKF. In all the cases, the new filter successfully decreases the imbalance in comparison to the EnKF. The EnKF

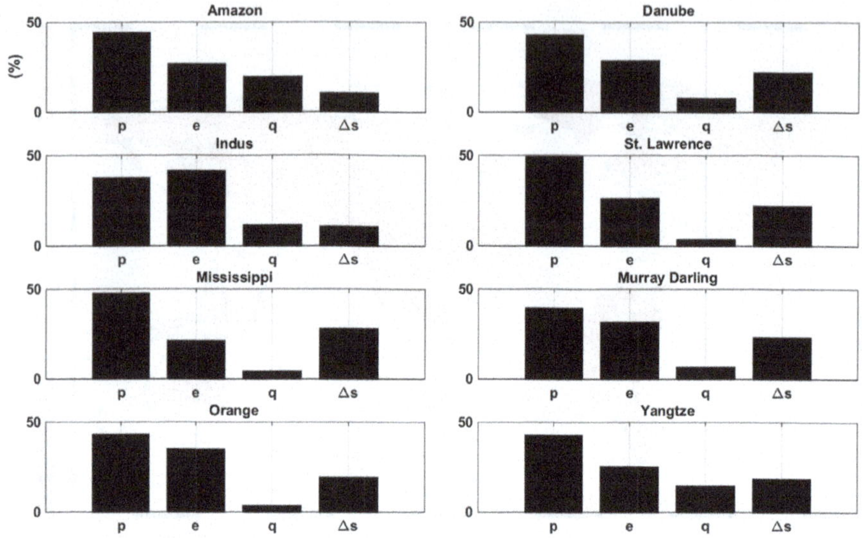

Fig. 7.9 Contributions of each water flux in water budget closure over different basins

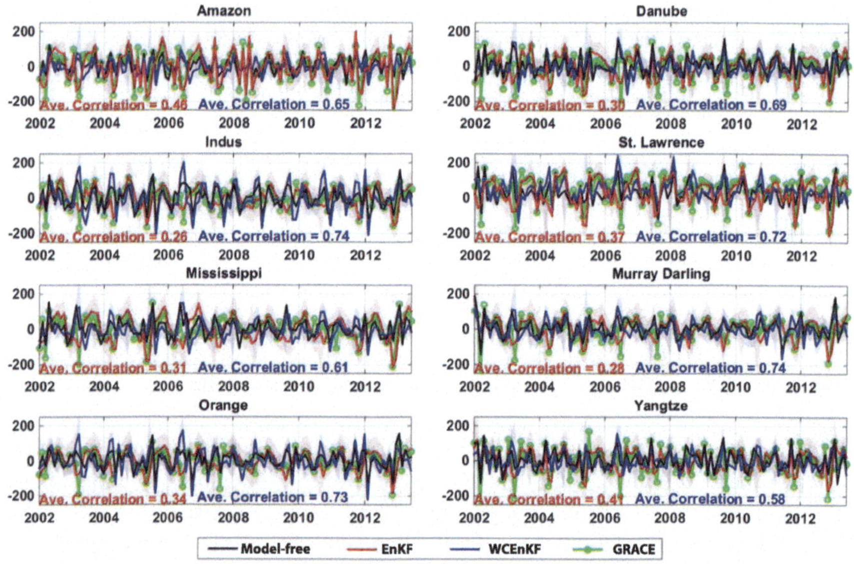

Fig. 7.10 Spatial average time series of **Δs** from each filter over different basins (units are mm). Shaded areas represent ensemble spreads of water storage change time series. Correlation values of WCEnKF and EnKF are depicted on the figure

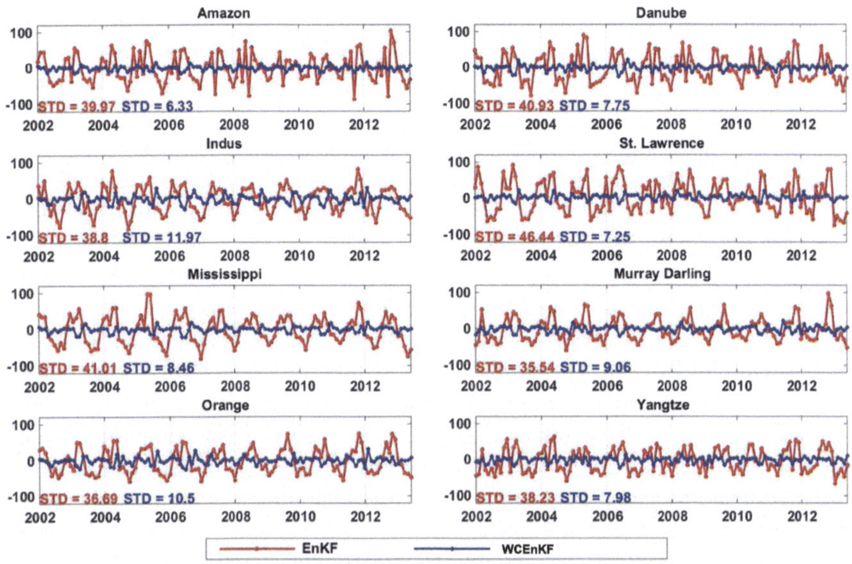

Fig. 7.11 Average imbalance error time series calculated using the EnKF and WCEnKF filters for each basin (units are mm)

results in larger water balance problem in the Mississippi and Danube basins, while the proposed WCEnKF suggests the best performances over these two basins with average imbalance reductions of 87% and 84%, respectively. We also compute the standard deviation (STD) of each time series (cf. Fig. 7.11). The large range of calculated STD in the EnKF (10.9 mm) is reduced to 5.64 mm by applying WCEnKF. Furthermore, the proposed filter appropriately improves disagreement between all compartments, both in terms of magnitudes and STDs. Figure 7.11 further suggests the importance of implementing the water balance adjustment. The absolute (average) imbalance without using this approach is 67.08 mm, and a large part of it is directly connected to the estimated TWS. The WCEnKF data assimilation decreases this value to approximately 14.45 mm, which leads to both better estimation of TWS and higher agreement with the other water fluxes.

7.6 Summary and Conclusions

GRACE TWS data are assimilated into W3RA covering 2002–2012 to improve model outputs and satisfy the terrestrial water budget balance. For that purpose, we propose a two-update weak constrained EnKF (WCEnKF) scheme that enforces water budget closure using the water fluxes. WCEnKF shows a good performance in integrating GRACE TWS data into the system (first update) and constraining the water balance equation (second update). Larger correlations in terms of groundwater estimates are found between assimilation results using the two-update filter (14.10% average) and ground-based observations, compared with those of the model-free. We also achieve 21.12% (on average) groundwater RMSE reductions using WCEnKF compared with EnKF. The application of the proposed filter shows an ability in imposing the water budget closure constraint as demonstrated by higher correlation of the estimated TWS changes to the **p**, **e**, and **q** (0.33, 0.44, and 0.45, respectively), as well as an imbalance reduction, i.e., from 62.17 mm using the traditional EnKF, to 18.31 mm (82.53% improvement).

There are some key factors that affect the performance of WCEnKF. Most importantly errors associated with pseudo-observations can largely alter the results. It is very difficult to achieve spatio-temporal variations of error characteristics of each water budget component. This study assesses three different error scenarios and investigates their impact on the results. However, the assumptions that are made, especially using a fixed uncertainty, might be inappropriate or sometimes strong since various data sets have performed differently within different areas. Therefore, more investigations are still needed to fully assess the filter's capability in terms of data uncertainties, applying multiple data sets for each variable (e.g., **p**, **e**), and using other types of observations such as soil moisture for data assimilation.

Acknowledgements We would like to thank Dr. Natthachet Tangdamrongsub for his useful comments, which contributed to the improvement of this study. M. Khaki is grateful for the research grant of Curtin International Postgraduate Research Scholarships (CIPRS)/ORD Scholarship provided by Curtin University (Australia). This work is a TIGeR publication.

Appendix A. Some Useful Properties of Random Sampling

Property 1 (Hierarchical sampling; Robert et al. 2006). *Assuming that one can sample from* $p(\mathbf{x}_1)$ *and* $p(\mathbf{x}_2|\mathbf{x}_1)$*, then a sample,* $\mathbf{x}_2^*$*, from* $p(\mathbf{x}_2)$ *can be generated by drawing* $\mathbf{x}_1^*$ *from* $p(\mathbf{x}_1)$ *and then* $\mathbf{x}_2^*$ *from* $p(\mathbf{x}_2|\mathbf{x}_1^*)$.

Property 2 (Conditional sampling; Hoffman et al. 1991). *Consider a Gaussian pdf,* $p(\mathbf{x}, \mathbf{y})$*, with* $\mathbf{P}_{xy}$ *and* $\mathbf{P}_y$ *denoting the cross-covariance of* $\mathbf{x}$ *and* $\mathbf{y}$ *and the covariance of* $\mathbf{y}$*, respectively. Then a sample,* $\mathbf{x}^*$*, from* $p(\mathbf{x}|\mathbf{y})$*, can be generated as,* $\mathbf{x}^* = \widetilde{\mathbf{x}} + \mathbf{P}_{xy}\mathbf{P}_y^{-1}[\mathbf{y} - \widetilde{\mathbf{y}}]$*, where* $(\widetilde{\mathbf{x}}, \widetilde{\mathbf{y}}) \sim p(\mathbf{x}, \mathbf{y})$.

Appendix B. Derivation of the WCEnKF Algorithm

Equation (7.12), which computes the forecast ensemble $\{\mathbf{x}_t^{f,(i)}\}_{i=1}^n$ from the previous analysis one, is obtained by applying Property 1 above to the forecast step (7.6). Regarding the first update step (with $\mathbf{y}_t$), one first applies Property 1 on the following formula,

$$p(\mathbf{y}_t|\mathbf{r}_{0:t-1}) = \int \underbrace{p(\mathbf{y}_t|\mathbf{x}_t)}_{\mathcal{N}(\mathbf{H}_t\mathbf{x}_t,\mathbf{R}_t)} p(\mathbf{x}_t|\mathbf{r}_{0:t-1})d\mathbf{x}_t,$$

to sample the observation forecast ensemble, $\{\mathbf{y}_t^{f,(i)}\}_{i=1}^n$, as in Eq. (7.13). Property 2 is then used in Eqs. (7.7) to obtain the ensembles $\{\tilde{\mathbf{x}}_t^{a,(i)}\}_{i=1}^n$ (Eq. (7.14)) and $\{\mathbf{x}_{t-1}^{s,(i)}\}_{i=1}^n$, respectively. For the second update step (with $\mathbf{z}_t$), one first uses Property 1 in Eq. (7.11), with $p(\mathbf{z}_t|\mathbf{x}_t, \mathbf{x}_{t-1}) \overset{(7.4)}{=} \mathcal{N}(\mathbf{G}\mathbf{x}_t + \mathbf{L}\mathbf{x}_{t-1}, \mathbf{\Sigma}_t)$, to obtain the pseudo-observation forecast ensemble $\{\mathbf{z}_t^{f,(i)}\}_{i=1}^n$ (Eq. (7.18)), then Property 2 in Eq. (7.10) to compute the state analysis ensemble $\{\mathbf{x}_t^{a,(i)}\}_{i=1}^n$ (Eq. (7.19)).

Chapter 8
Unsupervised Constraint for Hydrologic Data Assimilation

Addition of water balance uncertainty is necessary to allow for a more realistic water balance control during filtering. This has been framed in a supervised framework, i.e., by assigning approximate error covariance to the water balance observations before filtering, which may not allow for an optimal estimation of corrections (in the second step of the filter) to be applied to results from the first step of the filter.

—M. Khaki (This chapter is presented following Khaki et al. (2018e), "Unsupervised ensemble Kalman filtering with an uncertain constraint for land hydrological data assimilation")

8.1 Summary

The standard ensemble data assimilation schemes often violate the dynamical balances of hydrological models, in particular, the fundamental water balance equation, which relates water storage and water flux changes. The present study aims at extending the recently introduced Weak Constrained Ensemble Kalman Filter (WCEnKF) to a more general framework, namely unsupervised WCEnKF (UWCEnKF), in which the covariance of the water balance model is no longer known, thus requiring its estimation along with the model state variables. This extension is introduced because WCEnKF was found to be strongly sensitive to the (manual) choice of this covariance. The proposed UWCEnKF, on the other hand, provides a more general unsupervised framework that does not impose any (manual, thus heuristic) value of this covariance, but suggests an estimation of it, from the observations, along with the state. The new approach is tested based on numerical experiments of assimilating Terrestrial Water Storage (TWS) from Gravity Recovery and Climate Experiment (GRACE) and remotely sensed soil moisture data into a hydrological model. The experiments are conducted over different river basins, comparing WCEnKF, UWCEnKF, and the standard EnKF. In this setup, the UWCEnKF constrains the system state variables with TWS changes, precipitation, evaporation, and discharge data to balance the summation of water storage simulations. In-situ groundwater and soil moisture

M. Khaki, *Satellite Remote Sensing in Hydrological Data Assimilation*,
https://doi.org/10.1007/978-3-030-37375-7_8

measurements are used to validate the results of the UWCEnKF and to evaluate its performances against the EnKF. Our numerical results clearly suggest that the proposed framework provides more accurate estimates of groundwater storage changes and soil moisture than WCEnKF and EnKF over the different studied basins.

8.2 Introduction

Hydrological models play important roles in environmental studies and are crucial for hydrological applications. Due to a variety of factors, such as model structural errors, data deficiency, and uncertainty in inputs and parameters, the outputs of these models can be far from perfect. Data assimilation techniques offer a framework to improve the models simulations by constraining their outputs to the observations. However, the application of assimilation schemes could introduce an imbalance between water fluxes, namely precipitation $\mathbf{p}$, evaporation $\mathbf{e}$, discharge $\mathbf{q}$, and changes in water storage, $\Delta\mathbf{s}$, through the water balance equation $\Delta\mathbf{s} = \mathbf{p} - \mathbf{e} - \mathbf{q}$. The water balance equation is applied in land hydrological models to describe the relationships between these fluxes (Sokolov and Chapman 1974). The model structure governs variations in the water state changes due to the incoming and outgoing hydrological water fluxes. Data assimilation of any water storages, e.g., soil moisture and/or terrestrial water storage (TWS), breaks the existing balance because the assimilated state does not satisfy the water balance property (Khaki et al. 2017c).

Existing data assimilation methodologies under water budget enforcement rely on a "perfect observations" assumption in the closure constraint (e.g., Pan and Wood 2006; Sahoo et al. 2011; Pan et al. 2012). For example, Pan and Wood (2006) proposed a constrained ensemble Kalman filter (CEnKF) that imposes regional water balance constraint to improve the filtering results. The CEnKF involves two successive EnKF-like updates. The first update uses the observations to update the state forecast, following an EnKF-like step, while the second update imposes the balance constraint via another EnKF-like correction, yet with a different form. Other studies have applied data merging algorithms along with the CEnKF (see, e.g., Sahoo et al. 2011; Pan et al. 2012; Zhang et al. 2016) to provide the flux datasets from various resources for water balance control. Although these improved datasets have resulted in better state estimates over different river basins by incorporating more accurate information about the constraints, the assumption of perfect observations is still problematic. This assumption leads to a strong constraint, which is unrealistic and may cause various issues. Simon and Chia (2002) suggested that even though it does not present any theoretical problems, the assumption can result in a singular covariance matrix, which in practice increases the possibility of numerical issues. Furthermore, by neglecting errors associated with flux observations, one can expect more estimation errors because of the strong water budget enforcement, which could also lead to over-fitting issues (Tangdamrongsub et al. 2017).

In a recent study, Khaki et al. (2017c) proposed a new two-update ensemble Kalman-based scheme, a weak constrained ensemble Kalman filter (WCEnKF), that involves uncertainties in the water budget balance enforcement equation. Unlike previous studies (e.g., Pan and Wood 2006; Sahoo et al. 2011; Pan et al. 2012; Khaki et al. 2017c), water balance uncertainty is added to the equality constraint formulation, which allows for a more realistic water balance control during filtering. This has been framed in a supervised framework, i.e., by assigning approximate error covariance to the water balance observations before filtering, which may not allow for an optimal estimation of corrections (in the second step of the filter) to be applied to results from the first step of the filter. The present study aims to extend the work of Khaki et al. (2017c) to the case where the covariance associated with flux observations is unknown, proposing an unsupervised framework to estimate it along with the hydrology state variable. The proposed Unsupervised WCEnKF (UWCEnKF) introduces an iterative scheme in the second update step of the WCEnKF.

In order to assess the performance of the UWCEnKF, numerical experiments are carried out to assimilate the Gravity Recovery And Climate Experiment (GRACE) derived terrestrial water storage (TWS), as well as soil moisture products from the Advanced Microwave Scanning Radiometer-Earth Observing System (AMSR-E) and Soil Moisture and Ocean Salinity (SMOS) into a hydrological model. Assimilating GRACE TWS data has been performed in a number of previous studies to constrain the mass balance of hydrological models over different river basins (e.g., Zaitchik et al. 2008; van Dijk et al. 2014; Eicker et al. 2014; Reager et al. 2015; Schumacher et al. 2016; Khaki et al. 2018b, c). Several studies already demonstrated a great capability of AMSR-E and SMOS datasets to constrain model estimates through data assimilation (e.g., De Jeu et al. 2008; Renzullo et al. 2014; Leroux et al. 2016; Tian et al. 2017). It has also been shown that simultaneous assimilation of the different datasets generally leads to better results in terms of state estimates (e.g., Zhang et al. 2014; Renzullo et al. 2014; Han et al. 2016; Tian et al. 2017; Lievens et al. 2017) as compared to individual assimilation of the different datasets. This motivates the current study to simultaneously assimilate GRACE TWS and soil moisture observations from AMSR-E and SMOS. We also apply the standard EnKF to compare its results with the proposed UWCEnKF filter. This enables to evaluate the relevance of the proposed approach for enforcing the water budget closure.

We further consider multiple observations of the water components in the water budget equation. This is done to achieve the best estimates of $\mathbf{p}$ and $\mathbf{e}$ over different basins (see Fig. 8.1). Multi-mission products for precipitation and evaporation are used in the data merging approach of Sahoo et al. (2011) to derive a single data set for each observation type (i.e., $\mathbf{p}$ and $\mathbf{e}$). The approach estimates uniform datasets independently for each basin. The merged data, as well as the water discharge measurements from various ground stations, are then applied to constrain the water balance equation in the UWCEnKF's second update. This experiment is undertaken over eight globally distributed basins; Amazon, Indus, Mississippi, Orange, Danube, St. Lawrence, Murray-Darling, and the Yangtze, to better explore the capability of the proposed filter.

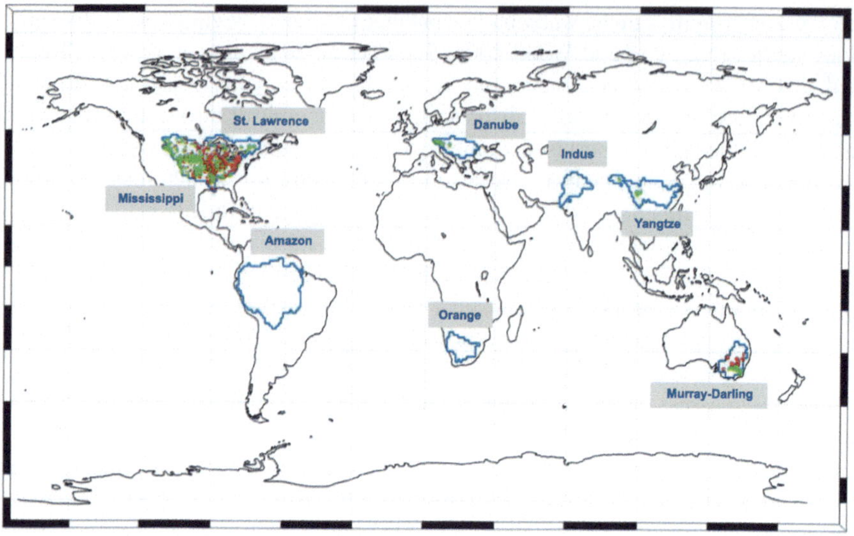

Fig. 8.1 The location of study basins. The figure also contains the distribution of in-situ groundwater (red) and soil moisture (green) gauge stations

8.3 Model and Data

8.3.1 Hydrological Model

Vertical water compartments of the globally distributed World-Wide Water Resources Assessment system (W3RA) model, developed in 2008 by the Commonwealth Scientific and Industrial Research Organisation (CSIRO; Australia), are used to simulate water storages. W3RA is a one-dimensional system that simulates landscape water stored in the vegetation and soil systems (van Dijk 2010). Here, we use the $1^\circ \times 1^\circ$ version of the model to represent the water balance of the soil, groundwater and surface water storage, in which each cell is modeled independently from its neighbors (van Dijk 2010). Groundwater dynamics in the model includes recharge from deep drainage, capillary rise (estimated with a linear diffusion equation), evaporation from groundwater saturated areas, and discharge. The model assumes that redistribution between grid cells can be ignored. Groundwater and river water dynamics are simulated at grid cell level and hence parameters are equal across the grid cell. Meteorological data sets of minimum and maximum temperature, downwelling short-wave radiation, and precipitation products provided by Princeton University (http://hydrology.princeton.edu) are used to force the W3RA model between 2003 and 2013. The model state is composed of the top, shallow and deep root soil water, snow, vegetation, groundwater, and surface water storage.

8.3.2 *Assimilated Observations*

Observations are assimilated in two steps. The first step assimilates GRACE TWS and satellite soil moisture observations, which are used to update the forecast state, while the second step enforces the water balance constraints, based on water flux observations.

8.3.2.1 Data Used in the First Update

GRACE level 2 (L2) gravity field data provided by the ITSG-Grace2016 (Mayer-Gürr et al. 2014) is used to compute monthly TWS after applying a few standard corrections. These include replacing degree 1 (C10, C11, S11) and degree 2 (C20) coefficients by more accurate coefficients from Swenson et al. (2008) and the Satellite Laser Ranging solutions (Cheng and Tapley 2004), respectively. The gravity fields are then converted to $3^\circ \times 3^\circ$ TWS fields (Wahr et al. 1998). Khaki et al. (2017a) showed that implementing GRACE TWS with this spatial resolution exploits better impacts of GRACE TWS mainly because of larger correlation errors in the higher spatial resolution fields, which can be problematic during assimilation (see also Eicker et al. 2014; Schumacher et al. 2016). Colored/correlated noise and leakage errors are reduced using the Kernel Fourier Integration (KeFIn) filter, as proposed by Khaki et al. (2018a). The KeFIn filter works through a two-step post-processing algorithm: in the first step it mitigates the measurement noise and the aliasing of unmodelled high-frequency mass variations, and in the second step it decreases the leakage errors. Note that, here, rather using model outputs, fixed signal to noise ratio is applied during the KeFIn filtering (see Khaki et al. 2018a, for details). The application of the KeFIn filter was shown in Khaki et al. (2018a) to outperform a number of existing GRACE filtering techniques, e.g., land-grid-scaling method applied in Mass Concentration blocks (Mascons) products justifying its use in the current study.

Furthermore, soil moisture products from the Advanced Microwave Scanning Radiometer for EOS (AMSR-E) and ESA's Soil Moisture Ocean Salinity (SMOS) Earth Explorer mission are used to update soil storage variations. AMSR-E measures surface brightness temperature that corresponds to surface soil moisture content of 2 cm depth (Njoku et al. 2003). SMOS, on the other hand, measures microwave emissions from Earth's surface at about 5 cm depth. Here we use descending passes (see, e.g., De Jeu and Owe 2003) of gridded Level-3 land surface product AMSR-E (Njoku 2004) between 2003 and 2011, and Level 3 CATDS (Centre Aval de Traitement des Donnees SMOS) on ascending passes (see, e.g., Draper et al. 2009) for the period of 2011–2013. These passes are selected due to their higher agreement with in-situ measurements (see also Jackson and Bindlish 2012; Su et al. 2013). Both data products are rescaled to a monthly $1^\circ \times 1^\circ$ scale for the present study. Cumulative distribution function (CDF) matching (Reichle and Koster 2004; Drusch et al. 2005) is applied to rescale the observations and remove the bias between the model simulations and observations. These measurements are mainly used to constrain the model variability, and not its absolute values. CDF matching relies on the assumption that the difference between observed soil moisture and that of the model is stationary and guarantees that the statistical distribution of both time series is the same (Draper et al. 2009; Renzullo et al. 2014).

8.3.2.2 Data Used in the Second Update

Multiple data sets are used for flux net observations. Details of these products are outlined in Table 8.1. For precipitation, we use the Tropical Rainfall Measuring Mission (TRMM-3B43; Huffman et al. 2007), NOAA CPC Morphing Technique (CMORPH; Joyce et al. 2004), the Global Precipitation Climatology Project (GPCP) Version 2.3 (Adler et al. 2003), Global Precipitation Climatology Centre (GPCC; Schneider et al. 2008), and CPC unified gauge dataset (Chen et al. 2002). TRMM-3B43, CMORPH, and GPCP are used to generate the merged precipitation for data assimilation, while GPCC and CPC are applied for uncertainty analysis (cf. Sect. 8.5.1). Evaporation data are collected from MODIS Global Evapotranspiration Project (MOD16; Mu et al. 2007), Global Land Evaporation Amsterdam Model (GLEAM; Miralles et al. 2011), ERA-interim (Simmons et al. 2007), and Variable Infiltration Capacity (VIC) land surface model (Liang et al. 1994). Similar to precipitation, an uncertainty analysis is undertaken for evaporation with respect to ERA-interim and VIC products. All of these products are rescaled into a monthly $1^\circ \times 1^\circ$ spatial resolution. Various data sources are considered for discharge (see Table 8.1) to achieve the maximum amount of coverage within the basins of Amazon, Indus, Mississippi, Orange, Danube, St. Lawrence, Murray-Darling, and Yangtze (Fig. 8.1).

8.3.3 In-Situ Measurements

Monthly in-situ groundwater and soil moisture measurements are used to validate the results. The groundwater stations are located in the Mississippi, St. Lawrence, and Murray-Darling basins. Specific yield values provided by the literature (e.g., Gutentag et al. 1984; Strassberg et al. 2007; Seoane et al. 2013; Khaki et al. 2017c) are used to convert well measurements into groundwater storage anomalies. We further use in-situ soil moisture measurements over the Mississippi, St. Lawrence, Danube, Yangtze, and Murray-Darling basins to assess the estimated soil moisture. These data are collected from the International Soil Moisture Network (ISMN) and the moisture-monitoring network. It is worth mentioning that the temporal averages from the in-situ time series are removed before using them to validate the assimilation results. The distribution of both groundwater and soil moisture in-situ products are displayed in Fig. 8.1. Details of the datasets are outlined in Table 8.1.

Table 8.1 A summary of the datasets used in this study

Product	Platform	Reference
Terrestrial water storage (TWS)	GRACE	Mayer-Gürr et al. (2014)
Soil moisture	AMSR-E	Njoku (2004)
Soil moisture	SMOS	Draper et al. (2009)
Precipitation (**p**)	TRMM-3B42	Huffman et al. (2007)
Precipitation (**p**)	CMORPH	Joyce et al. (2004)
Precipitation (**p**)	GPCP	Adler et al. (2003)
Precipitation (**p**)	GPCC	Schneider et al. (2008)
Precipitation (**p**)	CPC	Chen et al. (2002)
Evapotranspiration (**e**)	MOD16	Mu et al. (2007)
Evapotranspiration (**e**)	GLEAM	Miralles et al. (2011)
Evapotranspiration (**e**)	ERA-interim	Simmons et al. (2007)
Evapotranspiration (**e**)	VIC	Liang et al. (1994)
Water discharge (**q**)	GRDC	http://www.bafg.de/GRDC/EN/Home/homepage_node.html
Water discharge (**q**)		http://www.hydrosciences.fr/sierem/consultation/choixaccess.asp?lang=en
Water discharge (**q**)	USGS	https://waterdata.usgs.gov/nwis/sw
Water discharge (**q**)		http://www.bom.gov.au/waterdata/
Water discharge (**q**)	NRFA	http://nrfa.ceh.ac.uk/data/
Water discharge (**q**)		http://www.ore-hybam.org/
Water discharge (**q**)		http://www.hydrology.gov.np/new/bull3/index.php/hydrology/home/main
Hydrological model	W3RA	http://www.wenfo.org/wald/data-software/
Groundwater in-situ measurements	NSW	http://waterinfo.nsw.gov.au/pinneena/gw.shtml
Groundwater in-situ measurements	USGS	https://water.usgs.gov/ogw/data.html
Soil moisture in-situ measurements	OzNet	Smith et al. (2012)
Soil moisture in-situ measurements	ISMN	https://ismn.geo.tuwien.ac.at/

8.4 Methodology

8.4.1 Problem Formulation

Our discrete-time state-space system is represented as,

$$\begin{cases} \mathbf{x}_t = \mathcal{M}_{t-1}(\mathbf{x}_{t-1}) + \nu_t, \\ \mathbf{y}_t = \quad \mathbf{H}_t \mathbf{x}_t + \mathbf{w}_t, \end{cases} \tag{8.1}$$

where $\mathbf{x}_t \in \mathbb{R}^{n_x}$ and $\mathbf{y}_t \in \mathbb{R}^{n_y}$ stand for the system state and the observation at time t and of sizes n_x and n_y, respectively. In system (8.1), $\mathcal{M}_{t-1}(.)$ is a nonlinear operator integrating the system state from time $t-1$ to t, and $\mathbf{H}_t$ is the observational (design) operator at time t, which is linear in our application. Note, however, that the proposed scheme can be easily extended to the nonlinear case (Liu and Xue 2002). The model process noise, $\nu = \{\nu_t\}_{t=0}^T$, and the observation process noise, $\mathbf{w} = \{\mathbf{w}_t\}_{t=0}^T$, are assumed to be independent in time, jointly independent, and independent of the initial state, shown by $\mathbf{x}_0$. Furthermore, ν_t and $\mathbf{w}_t$ are assumed to be Gaussian with zero means and covariances $\mathbf{Q}_t$ and $\mathbf{R}_t$, respectively. The model time step, t, is considered to be equal to the assimilation time step. More details about the state-space formulation (i.e., about the structures of $\mathbf{x}_t$, $\mathbf{y}_t$, $\mathcal{M}_t$ and $\mathbf{H}_t$) of our application can be found in Khaki et al. (2017c).

The ensemble Kalman filter update step does not constrain the water fluxes and this likely distorts their balance ($\Delta\mathbf{s} = \mathbf{p} - \mathbf{e} - \mathbf{q}$). This was enforced by Khaki et al. (2017c), up to a weak constraint:

$$\mathbf{d}_t = -\mathbf{x}_t + \mathbf{x}_{t-1} + \mathbf{p}_t - \mathbf{e}_t - \mathbf{q}_t + \boldsymbol{\xi}_t, \tag{8.2}$$

accounting for the uncertainty in the different water fluxes data through a noise term $\boldsymbol{\xi}_t$, which we assume here to be Gaussian with zero mean and covariance, $\boldsymbol{\Sigma}$, and independent of $\boldsymbol{\xi}_{t'\neq t}$, $\{\nu_t\}_{t=0}^T$, $\{\mathbf{w}_t\}_{t=0}^T$ and $\mathbf{x}_0$. Considering Eq. (8.2), one can see that changes in the water storage at two successive time steps is equal to the difference between precipitation and summation of evaporation and discharge up to uncertainties in the involved data. The constraint in Eq. (8.2) can be rewritten as another observation equation in the state-space formulation, Eq. (8.3), which also involves the state at the previous time,

$$\mathbf{z}_t = \mathbf{G}\mathbf{x}_t + \mathbf{L}\mathbf{x}_{t-1} + \boldsymbol{\xi}_t, \tag{8.3}$$

where $\mathbf{z}_t \overset{\text{def}}{=} \mathbf{d}_t - \mathbf{p}_t + \mathbf{e}_t + \mathbf{q}_t$ plays the role of a "pseudo-observation", $\mathbf{L}$ is an $n_z \times n_x$ identity matrix, and $\mathbf{G} = -\mathbf{L}$ (here, $n_z = n_x$). Define $\mathbf{r}_t = [\mathbf{y}_t^T, \mathbf{z}_t^T]^T$ and $\mathbf{r}_{0:t} = \{\mathbf{r}_0, \mathbf{r}_1, \cdots, \mathbf{r}_t\}$. In the state-space system (8.1)–(8.3), a generic filtering algorithm has been recently introduced by Khaki et al. (2017c), recursively computing the analysis pdf of the state $\mathbf{x}_t$ from the history of the augmented observations, $\mathbf{r}_{0:t}$, $p(\mathbf{x}_t|\mathbf{r}_{0:t})$. The computation of $p(\mathbf{x}_t|\mathbf{r}_{0:t})$ from $p(\mathbf{x}_{t-1}|\mathbf{r}_{0:t-1})$ proceeds in a succession of a forecast step and two Bayesian update steps. The forecast step consists of moving from $p(\mathbf{x}_{t-1}|\mathbf{r}_{0:t-1})$ to the forecast pdf, $p(\mathbf{x}_t|\mathbf{r}_{0:t-1})$, based on the state transition pdf $p(\mathbf{x}_t|\mathbf{x}_{t-1})$ (which is described by the state model). The resulting forecast pdf is then updated, based on the likelihood of the observations, $p(\mathbf{y}_t|\mathbf{x}_t)$ (which is represented by the observation model), resulting in an unconstrained analysis pdf[1], $p(\mathbf{x}_t|\mathbf{r}_{0:t-1}, \mathbf{y}_t)$. The latter is, in turn, updated in the second Bayesian step, based on the likelihood of the pseudo-observation, $p(\mathbf{z}_t|\mathbf{x}_{t-1,t})$ (which is represented by

[1] The term *unconstrained* comes from the fact that these pdfs are not based on the pseudo-observation, $\mathbf{z}_t$, that "represents" the equality constraint.

the constraint Eq. (8.3)), leading to the desirable analysis pdf at the current time t, $p(\mathbf{x}_t|\mathbf{r}_{0:t})$. Details about these steps can be found in (Khaki et al. 2017c).

In a supervised framework, where the parameters of the constrained state-space system (including $\boldsymbol{\Sigma}$) are known, the above generic algorithm was implemented by Khaki et al. (2017c) through Monte Carlo approximation of the posterior mean (PM) estimate of the state and its covariance, which led to the ensemble Kalman-type WCEnKF. Khaki et al. (2017c) noticed that the WCEnKF is sensitive to the choice of $\boldsymbol{\Sigma}$, which can strongly affect the filter behaviors. Here, we design a more general unsupervised framework in which $\boldsymbol{\Sigma}$ is an unknown diagonal covariance matrix, which thereby needs to be estimated concurrently with the state.

8.4.2 The Unsupervised Weak Constrained Ensemble Kalman Filter (UWCEnKF)

8.4.2.1 The Generic Algorithm

The UWCEnKF shares the same forecast and first update steps as the WCEnKF, but computes the posterior distribution of both state and pseudo-observation noise covariance in the second update step, instead of only that of the state. In a Bayesian framework, this consists in viewing the covariance, $\boldsymbol{\Sigma}$, as another random variable with a given prior pdf; the goal is then to compute its posterior pdf jointly with the state[2], $p(\mathbf{x}_{t-1}, \mathbf{x}_t, \boldsymbol{\Sigma}|\mathbf{r}_{0:t})$. However, the statistical dependencies between the states, $\mathbf{x}_{t-1:t}$, and the covariance, $\boldsymbol{\Sigma}$, makes its computation quite tricky. One way to overcome this difficulty is to resort to the variational Bayesian (VB) approach and approximate $p(\mathbf{x}_{t-1}, \mathbf{x}_t, \boldsymbol{\Sigma}|\mathbf{r}_{0:t})$ with a separable pdf $q(\mathbf{x}_{t-1}, \mathbf{x}_t, \boldsymbol{\Sigma}|\mathbf{r}_{0:t}) = q(\mathbf{x}_{t-1}, \mathbf{x}_t|\mathbf{r}_{0:t})q(\boldsymbol{\Sigma}|\mathbf{r}_{0:t})$, under the Kullback-Leibler divergence (KLD) minimization criteria (Jaakkola and Jordan 2000; Smidl and Quinn 2008; Ait-El-Fquih and Hoteit 2015; Ait-El-Fquih et al. 2016). This reads,

$$\begin{aligned} q(\mathbf{x}_{t-1}, \mathbf{x}_t, \boldsymbol{\Sigma}|\mathbf{r}_{0:t}) &= \underset{\phi(\mathbf{x}_{t-1}, \mathbf{x}_t, \boldsymbol{\Sigma}|\mathbf{r}_{0:t})}{\operatorname{argmin}} \ \mathrm{KLD}\left(\phi(\mathbf{x}_{t-1}, \mathbf{x}_t, \boldsymbol{\Sigma}|\mathbf{r}_{0:t})||p(\mathbf{x}_{t-1}, \mathbf{x}_t, \boldsymbol{\Sigma}|\mathbf{r}_{0:t})\right), \\ &= \underset{\phi(\mathbf{x}_{t-1}, \mathbf{x}_t, \boldsymbol{\Sigma}|\mathbf{r}_{0:t})}{\operatorname{argmin}} \ \mathbb{E}_{\phi(\mathbf{x}_{t-1}, \mathbf{x}_t, \boldsymbol{\Sigma}|\mathbf{r}_{0:t})}\left[\ln\left(\frac{\phi(\mathbf{x}_{t-1}, \mathbf{x}_t, \boldsymbol{\Sigma}|\mathbf{r}_{0:t})}{p(\mathbf{x}_{t-1}, \mathbf{x}_t, \boldsymbol{\Sigma}|\mathbf{r}_{0:t})}\right)\right], \end{aligned} \tag{8.4}$$

where $\mathbb{E}_{\phi(u)}[f(u)]$ denotes the expected value of $f(u)$ with respect to (w.r.t.) the pdf $\phi(u)$. The solution of Eq. (8.4) can be obtained from (the proof can be found for instance in Smidl and Quinn 2006, pp. 28–31):

[2] For the sake of clarity, the inclusion of both $\mathbf{x}_t$ and $\mathbf{x}_{t-1}$ in the joint posterior pdf of interest is due to the fact that both these states appear in the pseudo-observation model Eq. (8.3), which necessitates estimating both of them.

$$q(\mathbf{x}_{t-1}, \mathbf{x}_t|\mathbf{r}_{0:t}) \propto \exp\left(\mathbb{E}_{q(\boldsymbol{\Sigma}|\mathbf{r}_{0:t})}\left[\ln\left(p(\mathbf{x}_{t-1}, \mathbf{x}_t, \boldsymbol{\Sigma}, \mathbf{r}_{0:t})\right)\right]\right), \tag{8.5}$$

$$q(\boldsymbol{\Sigma}|\mathbf{r}_{0:t}) \propto \exp\left(\mathbb{E}_{q(\mathbf{x}_{t-1},\mathbf{x}_t|\mathbf{r}_{0:t})}\left[\ln\left(p(\mathbf{x}_{t-1}, \mathbf{x}_t, \boldsymbol{\Sigma}, \mathbf{r}_{0:t})\right)\right]\right). \tag{8.6}$$

According to Eqs. (8.5) and (8.6), the independence that is inserted between the marginal posteriors, $q(\mathbf{x}_{t-1}, \mathbf{x}_t|\mathbf{r}_{0:t})$ and $q(\boldsymbol{\Sigma}|\mathbf{r}_{0:t})$, is partially compensated by the fact that each of these pdfs remains dependent on the expected value of $\ln(p(\mathbf{x}_{t-1}, \mathbf{x}_t, \boldsymbol{\Sigma}, \mathbf{r}_{0:t}))$ w.r.t. the other. However, this property of "cyclic" dependence between $q(\mathbf{x}_{t-1}, \mathbf{x}_t|\mathbf{r}_{0:t})$ and $q(\boldsymbol{\Sigma}|\mathbf{r}_{0:t})$ makes it impossible to exactly evaluate these pdfs, or any of their statistics, such as for instance their means, which are taken as the PM estimates of the states and the covariance, $\boldsymbol{\Sigma}$, respectively. A standard approximation is to proceed with cyclic iterations between (8.5) and (8.6), evaluating one pdf after the other, until convergence is reached (Smidl and Quinn 2008; Sato 2001; Massoud et al. 2018). Based on the factorization,

$$p(\mathbf{x}_{t-1}, \mathbf{x}_t, \boldsymbol{\Sigma}, \mathbf{r}_{0:t}) \propto p(\mathbf{z}_t|\mathbf{x}_{t-1}, \mathbf{x}_t, \boldsymbol{\Sigma})p(\mathbf{x}_{t-1}, \mathbf{x}_t|\mathbf{r}_{0:t-1}, \mathbf{y}_t)q(\boldsymbol{\Sigma}|\mathbf{r}_{0:t-1}), \tag{8.7}$$

which stems from the conditional independence properties of the state-space system (8.1)–(8.3), the iterative form of Eqs. (8.5) and (8.6) becomes,

$$q^{(\ell)}(\mathbf{x}_{t-1}, \mathbf{x}_t|\mathbf{r}_{0:t}) \propto \exp\left(\mathbb{E}_{q^{(\ell-1)}(\boldsymbol{\Sigma}|\mathbf{r}_{0:t})}\left[\ln\left(p^{(\ell-1)}(\mathbf{z}_t|\mathbf{x}_{t-1}, \mathbf{x}_t, \boldsymbol{\Sigma})\right)\right]\right) p(\mathbf{x}_{t-1}, \mathbf{x}_t|\mathbf{r}_{0:t-1}, \mathbf{y}_t), \tag{8.8}$$

$$q^{(\ell)}(\boldsymbol{\Sigma}|\mathbf{r}_{0:t}) \propto \exp\left(\mathbb{E}_{q^{(\ell)}(\mathbf{x}_{t-1},\mathbf{x}_t|\mathbf{r}_{0:t})}\left[\ln\left(p^{(\ell-1)}(\mathbf{z}_t|\mathbf{x}_{t-1}, \mathbf{x}_t, \boldsymbol{\Sigma})\right)\right]\right) q(\boldsymbol{\Sigma}|\mathbf{r}_{0:t-1}), \tag{8.9}$$

where $p^{(\ell)}(.)$ and $q^{(\ell)}(.)$ respectively denote the pdfs $p(.)$ and $q(.)$ at iteration ℓ. As can be seen below (cf. Sect. 8.4.2.2), iterating over the pdfs Eqs. (8.8) and (8.9) amounts in practice to iterate over their (approximate) parameters, thereby leading to an unsupervised ensemble-based filtering scheme, which iterates in its second step over the PM estimates of the states and the pseudo-observation noise covariance.

8.4.2.2 Practical Implementation

For the sake of simplicity, we first focus on the case of a homogeneous noise with a covariance matrix,

$$\boldsymbol{\Sigma} = \lambda \times \mathbb{I}_{n_z}, \tag{8.10}$$

where λ is the variance value and $\mathbb{I}_{n_z}$ denotes the $n_z \times n_z$ identity matrix. The more general inhomogeneous case will be discussed later. The prior probability distribution $p(\lambda)$ is chosen as an inverse-Gamma distribution (as a natural choice for variances), with shape and scale parameters $\hat{\alpha}_0$ and $\hat{\beta}_0$, respectively (Smidl and Quinn 2006). In the case of non-informative priors, one could take $\hat{\alpha}_0 = \hat{\beta}_0$ relatively small. At each iteration $(\ell - 1) \rightarrow (\ell)$, inserting in Eqs. (8.8) and (8.9) the Gaussian pdf,

$$p^{(\ell-1)}(\mathbf{z}_t|\mathbf{x}_{t-1},\mathbf{x}_t,\boldsymbol{\Sigma}) = \mathcal{N}_{\mathbf{z}_t}(\mathbf{G}\mathbf{x}_t + \mathbf{L}\mathbf{x}_{t-1}, \boldsymbol{\Sigma}^{(\ell-1)}),$$

one obtains a posterior $q^{(\ell)}(\lambda|\mathbf{r}_{0:t})$ that is also an inverse-Gamma distribution with parameters, $\hat{\alpha}_t$ and $\hat{\beta}_t^{(\ell)}$, given in Eqs. (8.17) and (8.18) below. Likewise, $q^{(\ell)}(\mathbf{x}_{t-1},\mathbf{x}_t|\mathbf{r}_{0:r})$ is Gaussian with an ensemble representation given in Eqs. (8.14)-(8.16).

The UWCEnKF Starting at time $t-1$ from an analysis ensemble, $\{\mathbf{x}_{t-1}^{a,(i)}\}_{i-1}^{m}$, and shape and scale parameters $(\hat{\alpha}_{t-1}, \hat{\beta}_{t-1})$ of the inverse-Gamma posterior pdf $p(\lambda|\mathbf{r}_{0:t-1})$, these at the next time t can be computed following a succession of a forecast and two update steps. The forecast step, which computes the forecast ensemble, $\{\mathbf{x}_t^{f,(i)}\}_{i-1}^{m}$, and the first update step (with $\mathbf{y}_t$), which computes the unconstrained analysis and smoothing ensembles, $\{\tilde{\mathbf{x}}_t^{a,(i)}\}_{i-1}^{m}$ and $\{\tilde{\mathbf{x}}_{t-1}^{s,(i)}\}_{i-1}^{m}$, are identical to those in Khaki et al. (2017c), namely,

$$\mathbf{x}_t^{f,(i)} = \mathcal{M}_{t-1}(\mathbf{x}_{t-1}^{a,(i)}) + \nu^{(i)}, \tag{8.11}$$

$$\tilde{\mathbf{x}}_t^{a,(i)} = \mathbf{x}_t^{f,(i)} + \mathbf{P}_{\mathbf{x}_t^f}\mathbf{H}^T \underbrace{[\mathbf{H}\mathbf{P}_{\mathbf{x}_t^f}\mathbf{H}^T + \mathbf{R}_t]^{-1}[\mathbf{y}_t + \epsilon^{(i)} - \mathbf{H}\mathbf{x}_t^{f,(i)}]}_{\mu_t^{(i)}}, \tag{8.12}$$

$$\tilde{\mathbf{x}}_{t-1}^{s,(i)} = \mathbf{x}_{t-1}^{a,(i)} + \mathbf{P}_{\mathbf{x}_{t-1}^a,\mathbf{x}_t^f}\mathbf{H}^T \times \mu_t^{(i)}, \tag{8.13}$$

where $\mathbf{P}_{\mathbf{x}_t^f}$ is the sample forecast error covariance and $\mathbf{P}_{\mathbf{x}_{t-1}^a,\mathbf{x}_t^f}$ represents the sample cross-covariance between the previous analysis and current forecast errors, $\nu^{(i)} \sim \mathcal{N}(\mathbf{0},\mathbf{Q}_t)$, and $\epsilon^{(i)} \sim \mathcal{N}(\mathbf{0},\mathbf{R}_t)$.

As for the second update step (with $\mathbf{z}_t$), which applies the adjustment to enforce the water budget balance constraint, it involves iterations to compute Eqs. (8.8) and (8.9). Let $\hat{\alpha}_t = \hat{\alpha}_{t-1} + \frac{n_z}{2}$, the iteration begins with the initialization $\hat{\lambda}_t^{(0)} = \frac{\hat{\beta}_{t-1}}{\hat{\alpha}_t}$ and correspondingly $\hat{\boldsymbol{\Sigma}}_t^{(0)} = \hat{\lambda}_t^{(0)} \times \mathbb{I}_{n_z}$. For $\ell = 0 \cdots L$, the state members are first updated as,

$$\mathbf{z}_t^{f,(i,\ell)} = \mathbf{G}\tilde{\mathbf{x}}_t^{a,(i)} + \mathbf{L}\tilde{\mathbf{x}}_{t-1}^{s,(i)} + \boldsymbol{\xi}_t^{(i,\ell)}; \quad \boldsymbol{\xi}_t^{(i,\ell)} \sim \mathcal{N}(\mathbf{0},\hat{\boldsymbol{\Sigma}}_t^{(\ell)}), \quad i = 1,\cdots,m, \tag{8.14}$$

$$\mathbf{x}_t^{a,(i,\ell)} = \tilde{\mathbf{x}}_t^{a,(i)} + \mathbf{P}_{\tilde{\mathbf{x}}_t^a,\mathbf{z}_t^{f,\ell}} \underbrace{[\mathbf{M}\mathbf{P}_{\eta_t}\mathbf{M}^T + \hat{\boldsymbol{\Sigma}}_t^{(\ell)}]^{-1}[\mathbf{z}_t - \mathbf{z}_t^{f,(i,\ell)}]}_{\nu_t^{(i,\ell)}}, \quad i = 1,\cdots,m, \tag{8.15}$$

$$\mathbf{x}_{t-1}^{s,(i,\ell)} = \tilde{\mathbf{x}}_{t-1}^{s,(i)} + \mathbf{P}_{\tilde{\mathbf{x}}_{t-1}^s,\mathbf{z}_t^{f,\ell}} \times \nu_t^{(i,\ell)}, \quad i = 1,\cdots,m, \tag{8.16}$$

where $\mathbf{M} \stackrel{\text{def}}{=} [\mathbf{G},\mathbf{L}]$; $\mathbf{P}_{\tilde{\mathbf{x}}_t^a,\mathbf{z}_t^{f,\ell}}$ and $\mathbf{P}_{\tilde{\mathbf{x}}_{t-1}^s,\mathbf{z}_t^{f,\ell}}$ are the sample cross-covariances computed using the ensembles $\{\tilde{\mathbf{x}}_t^{a,(i)}\}_{i=1}^{m}$, $\{\tilde{\mathbf{x}}_{t-1}^{s,(i)}\}_{i=1}^{m}$ and $\{\mathbf{z}_t^{f,(i,\ell)}\}_{i=1}^{m}$; and $\mathbf{P}_{\eta_t}$ is the sample covariance of the ensemble $\{\boldsymbol{\eta}_t^{(i)}\}_{i=1}^{m}$ with $\boldsymbol{\eta}_t^{(i)} \stackrel{\text{def}}{=} [(\tilde{\mathbf{x}}_t^{a,(i)})^T, (\tilde{\mathbf{x}}_{t-1}^{s,(i)})^T]^T$. Based on the resulting ensembles, the observation noise variance is then updated as,

$$\hat{\beta}_t^{(\ell+1)} = \hat{\beta}_{t-1} + \frac{1}{2}[||\mathbf{z}_t - \mathbf{G}\hat{\mathbf{x}}_t^{a,(\ell)} - \mathbf{L}\hat{\mathbf{x}}_{t-1}^{s,(\ell)}||^2 + \text{Trace}\,(\mathbf{M}\mathbf{P}_{\gamma_t^\ell}\mathbf{M}^T)], \quad (8.17)$$

$$\hat{\lambda}_t^{(\ell+1)} = \hat{\beta}_t^{(\ell+1)}/\hat{\alpha}_t, \quad (8.18)$$

$$\hat{\boldsymbol{\Sigma}}_t^{(\ell+1)} = \hat{\lambda}_t^{(\ell+1)} \times \mathbb{I}_{n_z}, \quad (8.19)$$

where $\hat{\mathbf{x}}_t^{a,(\ell)}$ and $\hat{\mathbf{x}}_{t-1}^{s,(\ell)}$ are the (empirical) means of the ensembles $\{\mathbf{x}_t^{a,(i,\ell)}\}_{i=1}^m$ and $\{\mathbf{x}_{t-1}^{s,(i,\ell)}\}_{i=1}^m$, respectively; and $\mathbf{P}_{\gamma_t^\ell}$ is the sample covariance of the ensemble $\{\boldsymbol{\gamma}_t^{(i,\ell)}\}_{i=1}^m$ with $\boldsymbol{\gamma}_t^{(i,\ell)} \overset{\text{def}}{=} [(\mathbf{x}_t^{a,(i,\ell)})^T, (\mathbf{x}_{t-1}^{s,(i,\ell)})^T]^T$. The $\hat{\boldsymbol{\Sigma}}_t^{(L)}$ and $\{\mathbf{x}_t^{a,(i,L)}\}_{i=1}^m$ are then considered as the analysis covariance and state estimates, respectively, that will be used in the next assimilation cycle. In our numerical experiments, only few iterations (less than 10) were needed to reach convergence based on the variance estimate. Note that instead of pre-setting the number of iterations, L, on may use an alternative stopping criteria based, for instance, on the relative squared error norm (RSEN) of the estimated state and/or variance(s), or the evidence lower bound (ELB), defined as Blei et al. (2017),

$$\mathcal{E}_1 = \mathbb{E}_{q(\xi_t,\boldsymbol{\Sigma}|\mathbf{r}_{0:t})}\big[\ln\,(p(\mathbf{v}_t, \boldsymbol{\Sigma}, \mathbf{r}_t|\mathbf{r}_{0:t-1}))\big] - \mathbb{E}_{q(\xi_t,\boldsymbol{\Sigma}|\mathbf{r}_{0:t})}[\ln\,(q(\mathbf{v}_t, \boldsymbol{\Sigma}|\mathbf{r}_{0:t}))], \quad (8.20)$$

with $\mathbf{v}_t = [\mathbf{x}_t^T, \mathbf{x}_{t-1}]^T$. Note that it is not possible to use the KLD as this requires the knowledge of the target pdf, $p(\mathbf{v}_t, \boldsymbol{\Sigma}|\mathbf{r}_{0:t})$, which, indeed, is not known. Furthermore, minimizing the KLD amounts to maximizing the ELB (Blei et al. 2017). However, a problem occurs in practice with ELB (8.20) in case of large dimensional systems (i.e., when $n_x > m$). In such a case, the covariance $\mathbf{P}_{\gamma_t}$, whose inverse is involved in the expression of the (assumed Gaussian) pdf, $q(\mathbf{v}_t|\mathbf{r}_{0:t})$, is a low-rank matrix, and thus not invertible. To overcome this limitation, we propose to remove the variable, $\mathbf{v}_t$, from the ELB, by rather using pdfs that are conditional on this variable (i.e., for which $\mathbf{v}_t$ is a fixed known value). Since the iterations' process occurs in the second update step (i.e., which uses $\mathbf{z}_t$), we assign to $\mathbf{v}_t$ the mean $\hat{\boldsymbol{\eta}}_t$ of $\{\boldsymbol{\eta}_t^{(i)}\}_{i=1}^m$, which, indeed, is an approximation of $\mathbb{E}_{q(\xi_t|\mathbf{r}_{0:t-1},\mathbf{y}_t)}[\mathbf{v}_t]$ (i.e., the unconstrained analysis mean of $\mathbf{v}_t$). The resulting ELB reads,

$$\begin{aligned}\mathcal{E}_2 &= \mathbb{E}_{q(\boldsymbol{\Sigma}|\mathbf{r}_{0:t})}\big[\ln\big(p(\boldsymbol{\Sigma}, \mathbf{r}_t|\mathbf{r}_{0:t-1}, \hat{\boldsymbol{\eta}}_t)\big)\big] - \mathbb{E}_{q(\boldsymbol{\Sigma}|\mathbf{r}_{0:t})}[\ln\,(q(\boldsymbol{\Sigma}|\mathbf{r}_{0:t}))],\\ &\approx \text{cte} + \mathbb{E}_{q(\boldsymbol{\Sigma}|\mathbf{r}_{0:t})}\big[\ln\big(p(\mathbf{z}_t|\boldsymbol{\Sigma}, \hat{\boldsymbol{\eta}}_t)\big)\big] + \mathbb{E}_{q(\boldsymbol{\Sigma}|\mathbf{r}_{0:t})}\big[\ln\,(q(\boldsymbol{\Sigma}|\mathbf{r}_{0:t-1}))\big] - \mathbb{E}_{q(\boldsymbol{\Sigma}|\mathbf{r}_{0:t})}[\ln\,(q(\boldsymbol{\Sigma}|\mathbf{r}_{0:t}))],\end{aligned} \quad (8.21)$$

where the term "cte" encompasses all the terms that do not depend on Σ. This suggests that the convergence of the proposed scheme can be monitored based either on the change in $\mathcal{E}_2$ only, the change in RSEN of the state only, the change in $\mathcal{E}_2$ and RSEN of the state, or, as stated above, the change in RSEN of both state and $\boldsymbol{\Sigma}$. Finally, based on the Gaussian expression of $p(\mathbf{z}_t|\boldsymbol{\Sigma}, \hat{\boldsymbol{\eta}}_t)$ and the inverse-Gamma expression of $q(\boldsymbol{\Sigma}|\mathbf{r}_{0:t-1})$ and $q(\boldsymbol{\Sigma}|\mathbf{r}_{0:t})$, one readily shows that Eq. (8.21) at iteration $(\ell) \to (\ell+1)$ is given as,

$$\mathcal{E}_2^{(\ell)} \approx \text{cte} + \frac{\hat{\alpha}_t}{\hat{\beta}_t^{(\ell+1)}} \left[\hat{\beta}_t^{(\ell+1)} - \hat{\beta}_{t-1} - \|\mathbf{z}_t - \mathbf{M}\hat{\boldsymbol{\eta}}_t\|^2/2 \right] - \ln(\hat{\beta}_t^{(\ell+1)}), \quad (8.22)$$

where cte gathers the terms that do not vary with iterations (i.e., independent of (ℓ)).

The adaptation of the algorithm above to the case of an inhomogeneous noise with a covariance is straightforward,

$$\boldsymbol{\Sigma} = \text{diag}\left(\lambda^1, \ldots, \lambda^{n_z}\right), \quad (8.23)$$

where diag($\mathbf{v}$) denotes a diagonal matrix with diagonal $\mathbf{v}$. More specifically, Eqs. (8.11)–(8.16) that compute the state ensembles are kept unchanged, and only those related to the noise variance will be updated (i.e., Eqs. (8.17)–(8.19) for each λ^j). Each variance λ^j, $j = 1, \ldots, n_z$, is estimated separately from the others, λ^k, $k \neq j$, by a direct application of Eqs. (8.17)–(8.19) and (8.22), which, correspond to the $n_z \times 1$ vectorial model (8.3), on the scalar (marginal) model,

$$\mathbf{z}_{t,j} = \mathbf{G}(j,:)\mathbf{x}_t + \mathbf{L}(j,:)\mathbf{x}_{t-1} + \boldsymbol{\xi}_{t,j}, \quad (8.24)$$

where $\mathbf{z}_{t,j}$ and $\boldsymbol{\xi}_{t,j}$ respectively denote the jth component of $\mathbf{z}_t$ and $\boldsymbol{\xi}_t$ (i.e., $\boldsymbol{\xi}_{t,j} \sim \mathcal{N}(\mathbf{0}, \lambda^j)$), and $\mathbf{G}(j,:)$ and $\mathbf{L}(j,:)$ are the jth rows of $\mathbf{G}$ and $\mathbf{L}$, respectively. A schematic illustration of this algorithm is presented in Fig. 8.2.

8.5 Experimental Setup

8.5.1 Data Merging

A single product for each water flux term of precipitation (**p**) and evaporation (**e**) is required to close the water balance in the second update step of UWCEnKF. One can use only one data product for each flux components, e.g., only TRMM-3B43 for **p** for the filtering process. However, this may introduce errors because various products are subject to a different rate of uncertainty over different areas. Alternatively, the different data products for each component can be merged into a unique **p** and **e** to better represent the water balance over the globally distributed basins (Sahoo et al. 2011). Here, we merge various datasets of precipitation and evaporation prior to data assimilation. To this end, we follow Sahoo et al. (2011) and merge the data considering their relative error levels w.r.t. non-satellite products. This combination is done in a way that satellite-based products are merged to be used in data assimilation while other products are only applied for the merging objective. For **p**, the average of GPCC and CPC unified gauge over each basin is assumed as the truth and is used to estimate the error level of each satellite-based product, i.c, TRMM-3B43, CMORPH, and GPCP. A similar strategy is applied for evaporation, where ERA-interim and VIC products are used to quantify the error level associated with the data

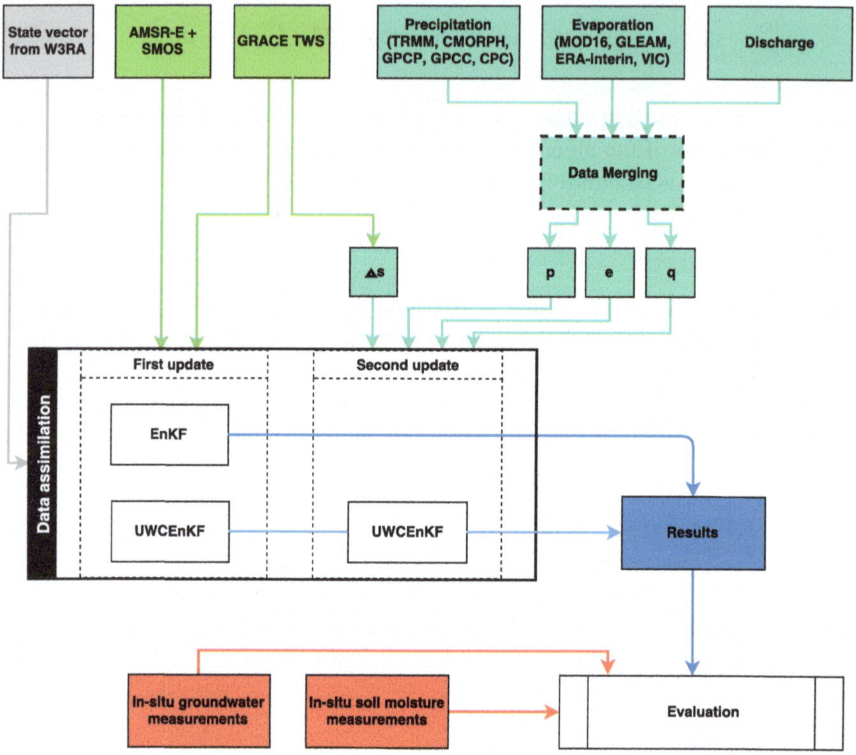

Fig. 8.2 A schematic illustration of the UWCEnKF steps applied for data assimilation, as well as data merging process

of MOD16 and GLEAM outputs that are based on satellite products (Miralles et al. 2011). It is worth mentioning that a more robust merging process can be achieved by involving ground-based measurements as a reference rather than ERA-interim and VIC. Obtaining and analyzing such an enhanced evaporation dataset from in-situ stations over all tested basins is however very difficult and is out of the scope of this study. Therefore, we use these model outputs to merge satellite-based datasets into a single **e**. Once the references are calculated, we use a multiplicative error model to estimate the offset, scale parameter, and error variance for each data product. These variances are then used to compute the observations weights as,

$$w_i = \frac{1}{\sigma_i^2} / \sum_{k=1}^{n_p} \frac{1}{\sigma_k^2}. \tag{8.25}$$

For each data product (i), using the error variances of that specific product σ_i^2 and all products (σ_k^2) in the same data type (with the total number of n_p), weight w_i can be calculated. Equation (8.25) is applied for both precipitation and evaporation to provide merged data with reduced error (Luo et al. 2007; Sahoo et al. 2011). Note that the above approach is applied only to merge the various data products and to obtain uniform precipitation and evaporation datasets prior to assimilation. The estimated errors (e.g., σ_i^2 in Eq. (8.25)) are used only for this objective and are not related to the water flux error covariance calculation in the filtering procedure (cf. Sect. 8.4.2).

8.5.2 Data Assimilation

To start the assimilation process, the initial ensemble is generated by perturbing the forcing fields. To this end, we use Monte Carlo sampling to perturb the precipitation, shortwave radiation, and temperature field considering a Gaussian multiplicative error of 30% for precipitation, an additive Gaussian error of 50 Wm^{-2} for the shortwave radiation, and a Gaussian additive error of 2 °C for temperature (Jones et al. 2007). The system state includes top soil, shallow soil, deep soil water, snow, vegetation, surface, and groundwater storages. Except for groundwater and surface storage, all the other components are simulated with two hydrological response units (HRU) of tall, e.g., deep-rooted vegetation and short, e.g., shallow-rooted vegetation. This leads to a state vector of dimension $(2 \times 5 + 1 + 1) \times 1695$ (corresponding to 1695 grid points over all basins).

All observations, including GRACE TWS, satellite soil moisture data, and water fluxes are assimilated monthly. The monthly increment is then be added to each day of the current month, which guarantees that the update of the monthly mean is identical to the monthly mean of the daily updates. Here, the differences between the predictions and the updated state variables are added as offsets to the state variables at the last day of each month to generate the ensembles for the next month assimilation step (see Eicker et al. 2014, for more details). The observation operator aggregates different water storages at each grid point to update with GRACE TWS and scales the top-layer soil storage by the field capacity value to provide a relative wetness for updating with the soil moisture products of AMSR-E and SMOS (Renzullo et al. 2014).

In addition, observation error covariances for the first update step are required. Full error information about the Stokes' coefficients are used to construct the TWS error covariance matrix. This is done by converting GRACE spherical harmonic error coefficients to TWS error covariances following Khaki et al. (2017b). Since such an information is not available for soil moisture products, we assume their error covariances to be uncorrelated with standard deviations of 0.04 $m^3\ m^{-3}$ for SMOS (as suggested by Leroux et al. 2016) and 0.05 $m^3\ m^{-3}$ for AMSR-E (as suggested by De Jeu et al. 2008). We further apply two common auxiliary techniques of ensemble variance inflation and covariance localization to mitigate for the ensemble spread collapse and rank deficiency (Anderson 2001; Houtekamer and Mitchell 2001). These include an ensemble inflation with a coefficient factor of 1.12 and Local Analysis (LA) with a localization length scale of 5° (see Khaki et al. 2017a, for more details).

8.6 Results

The results are discussed in three parts. UWCEnKF implementation is first presented and discussed in Sect. 8.6.1.1. The validation of the proposed approach against in-situ groundwater and soil moisture measurements is then presented in Sect. 8.6.2. The relevance of the second update step in UWCEnKF and its overall effects on the assimilation system performance is finally analyzed in Sect. 8.6.3. UWCEnKF estimates are also compared with the results of WCEnKF and EnKF. UWCEnKF is tested with both constant (*Structure in* Eq. (8.10), indicated by UWCEnKF-1) and spatially varying (*Structure in* Eq. (8.23), indicated by UWCEnKF-2) error variances for the water balance equation. While UWCEnKF-1 assigns a fixed error variance to water fluxes at all points, different values for individual points are calculated by UWCEnKF-2.

8.6.1 Implementation Results

8.6.1.1 Iteration Impacts

We first study the sensitivity of UWCEnKF-1, and UWCEnKF-2 to the iteration procedure. As mentioned, in contrast with WCEnKF, which assumes that these uncertainties are known, UWCEnKF estimates the error covariance through an iteration process. To show how this iteration works, we compare the convergence of UWCEnKF-1 and UWCEnKF-2, based on Eq. (8.22), in Fig. 8.3. The average evolutions of $\mathcal{E}_2^{(\ell+1)} - \mathcal{E}_2^{(\ell)}$ (the difference between Eq. (8.22) in each two successive iterations) from both filters for $\ell = 0 \ldots 10$ are shown in this figure. After few iterations, generally less than 8, both UWCEnKF-1 and UWCEnKF-2 converge. Faster convergence and lower differences $\mathcal{E}_2^{(\ell+1)} - \mathcal{E}_2^{(\ell)}$ are also generally achieved by UWCEnKF-2 compared to UWCEnKF-1. It can be seen that after 5 iterations, UWCEnKF-2 decreases to a value below the selected arbitrary threshold of $\mathcal{E}_2^{(\ell+1)} - \mathcal{E}_2^{(\ell)} = 10\,\text{mm}$. This is due to the fact that UWCEnKF-2 enables more degree of freedom in the optimization process by using different error variance for each grid point as compared to UWCEnKF-1, which tries to fit a single value for the entire domain.

In order to demonstrate the relevance of the UWCEnKF, we compare its results against those of the WCEnKF wit hvarious preselected values of error variances. The sensitivity of the WCEnKF to the choice of $\mathbf{\Sigma}$ can be seen in Fig. 8.4. The various implementations of the WCEnKF result in different performances in terms of imbalance and the Root-Mean-Squared Error (RMSE), which is calculated based on the assimilation results and groundwater in-situ measurements over the Murray-Darling Basin. The estimated groundwater time series from the WCEnKF and UWCEnKF are spatially interpolated to the nearest gauge stations. The difference between in-situ and filtered time series are then used to calculate the RMSE.

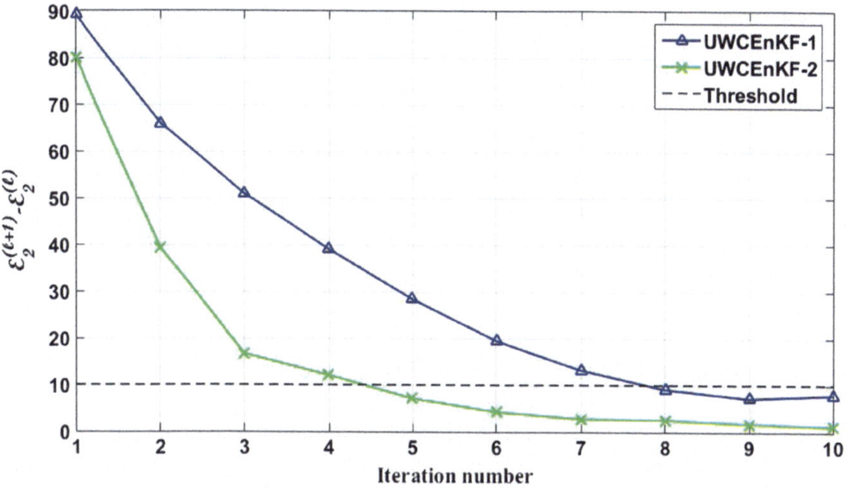

Fig. 8.3 Average $\mathcal{E}_2^{(\ell+1)} - \mathcal{E}_2^{(\ell)}$ estimates (unit is mm) from UWCEnKF-1 and UWCEnKF-2 filters during assimilation in each iteration (for $\ell = 0 \ldots 10$). The threshold value (10 mm) is chosen arbitrary based on a trial and error procedure

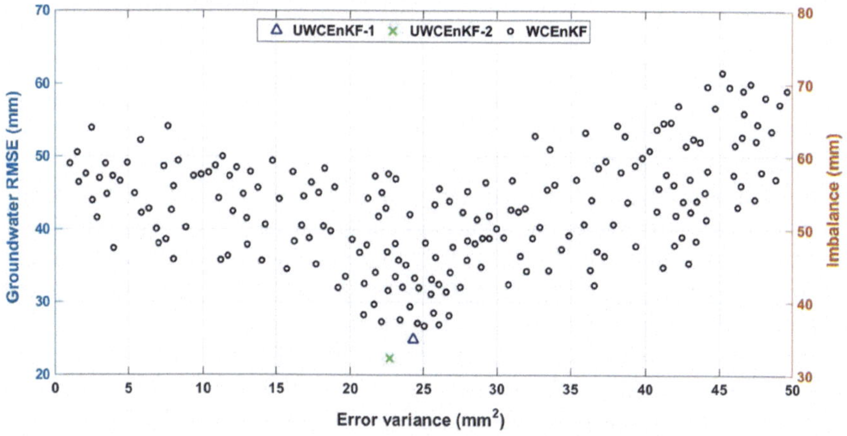

Fig. 8.4 Average groundwater RMSE and imbalance for various implementations of the WCEnKF filter using different error variance assumed (circles) considering different error variance. UWCEnKF-1 and UWCEnKF-2 results are indicated by triangle and cross, respectively

Each circle in Fig. 8.4 refers to the average results of an independent implementation of WCEnKF. It can be seen that the results of this filter largely vary depending on the selection of the error variance. Overall, lower imbalance and RMSE are obtained by assuming 20 to 30 mm^2. UWCEnKF-1 and UWCEnKF-2, on the other hand, achieve better results, shown by the triangle and cross, respectively, in a single implementation. The optimization algorithms used in UWCEnKF cause this independence of the error variance choice. It can also be seen that WCEnKF can achieve comparable results to that of UWCEnKF-1 in few cases. UWCEnKF-2, however, generally leads to the minimum RMSE and imbalance.

8.6.1.2 Spatial and Temporal Balance Error Variance

The performance of the proposed UWCEnKF in estimating water balance error variance and their effects on the imbalance between water fluxes are discussed in this section and is further compared with WCEnKF results. Both spatial and temporal variabilities are examined. Figure 8.5 shows the temporally averaged error variances assigned to the observations for WCEnKF, as well as those estimated by UWCEnKF-1 and UWCEnKF-2 over the Amazon Basin. It can be seen that UWCEnKF-1 and UWCEnKF-2 estimate different errors at each iteration. The error variance maps in WCEnKF, on the other hand, is fixed to what has been assigned prior to data assimilation. After eight iterations, it is observed that the error estimated by UWCEnKF-1 is closer to the average of UWCEnKF-2 results (34.70 mm^2), i.e., 41.19 mm^2 for UWCEnKF-1 and, in comparison to 68.74 mm^2 for WCEnKF. This indicates that both UWCEnKF-1 and UWCEnKF-2 result in uncertainties with close magnitude for water balances and the implemented algorithms allow for such an adjustment during iteration steps. Furthermore, Fig. 8.5 depicts the spatial variability characteristics of error variances estimated by UWCEnKF-2. This property allows for more flexibility for error adjustment in UWCEnKF-2. These flexibilities in the UWCEnKF filtering method, as illustrated in Fig. 8.6, result in a smaller imbalance.

The better performances of UWCEnKF-1 and UWCEnKF-2 compared to WCEnKF in minimizing imbalance errors are clear in Fig. 8.6, where each map shows the estimated imbalance corresponding to Fig. 8.5 setups. Figure 8.6 shows that the iteration algorithm effectively reduces imbalance errors, even after only few iterations (e.g., four). In addition, it can be seen that the applied algorithm in UWCEnKF provides the opportunity for error variances to be adjusted with no supervision as in WCEnKF. UWCEnKF-2, with more flexibility for such adjustment than UWCEnKF-1 (cf. Figure 8.5), leads to the smaller imbalance, that is ∼6 mm (absolute average of all values) against ∼13 mm (on average) for UWCEnKF-1. This larger improvement for UWCEnKF-2 results is achieved by estimating different error variance values over each grid point, and correspondingly applying different rate of adjustments (based on the estimated water balance uncertainty) from the equality constraint to the points.

An example of the abovementioned spatially varying error variance in UWCEnKF-2 can be seen in Fig. 8.7. Figure 8.7a depicts the average imbalance

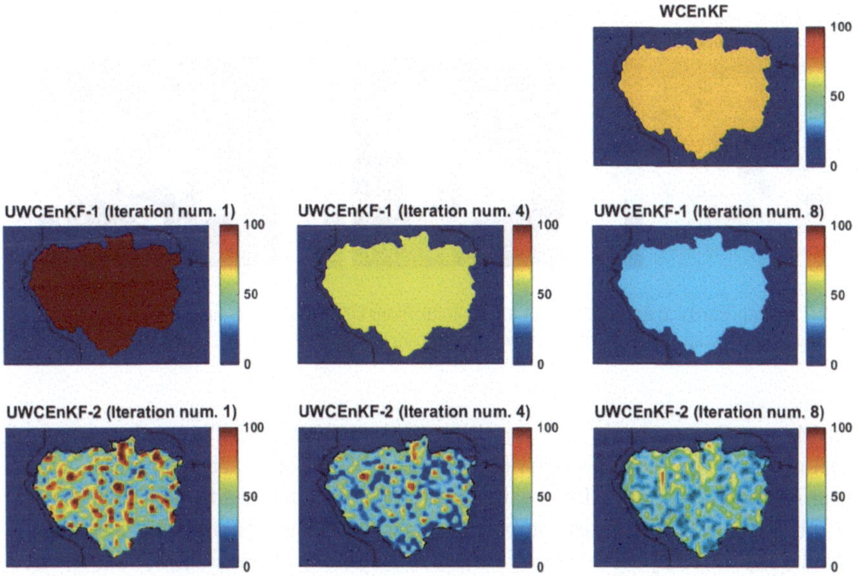

Fig. 8.5 Spatial variability of error variances estimated by WCEnKF, UWCEnKF-1, and UWCEnKF-2. The corresponding results for different iterations are also demonstrated for WCEnKF-1 and UWCEnKF-2

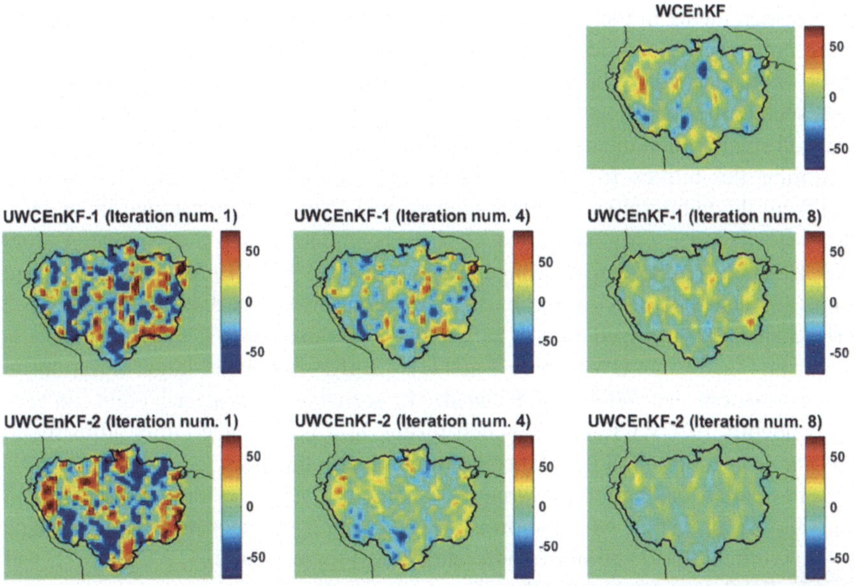

Fig. 8.6 Spatial variability of imbalances from WCEnKF, UWCEnKF-1, and UWCEnKF-2 corresponding to the errors presented in Fig. 8.5

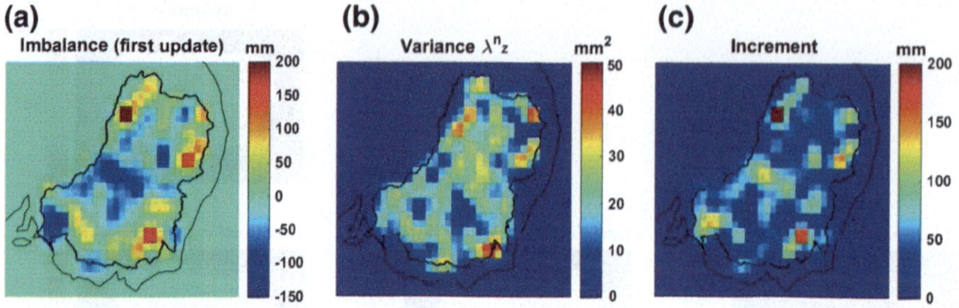

Fig. 8.7 Temporarily averaged maps of imbalances from UWCEnKF-2's first update (**a**), estimated error variance (**b**), and increments applied in the second analysis step of UWCEnKF-2 (**c**)

over Murray-Darling basin after jointly assimilating GRACE TWS and satellite soil moisture in the first analysis step of UWCEnKF. It is worth mentioning that we find larger impacts of GRACE TWS data (approximately 7.5 times for all the basins) on the imbalance between fluxes compared to the satellite soil moisture products, which could be explained by the fact that contrary to the soil moisture assimilation, GRACE data influences all compartments. The temporally averaged estimated variances are displayed in Fig. 8.7b. It can be seen that both estimated maps exhibit similar spatial patterns in some areas. One can also see in Fig. 8.7b that, in general, a larger variance is estimated over the areas with larger imbalance. Figure 8.7c shows the average applied increments in the second analysis step of UWCEnKF-2 to account for the above imbalances. It is clear that larger increments are applied over the areas with larger imbalances, e.g., the north, southeast, and southwest parts of the basin. The areas such as the central parts, which display smaller imbalance in Fig. 8.7a, are also assigned smaller increments as shown in Fig. 8.7c.

Similar flexibilities for error variance estimation in UWCEnKF can also be seen from the temporal variabilities of error variances as demonstrated in Fig. 8.8. The water balance error variances at each assimilation step are estimated from UWCEnKF-1 for the entire Orange Basin and from UWCEnKF-2 for each grid point (green shaded area) of the basin. The figure also plots that of UWCEnKF-2 derived spatially averaged values, as well as errors used in WCEnKF. Again, it is clear from Fig. 8.8 that UWCEnKF-1 and UWCEnKF-2 allow for larger variations in error estimations than WCEnKF. It can also be seen that errors at each point can vary independently in UWCEnKF-2, which results in a better uncertainty adjustment. This can help for optimal imbalance minimization in the filter.

Both spatial and temporal variabilities of error variances are summarized in Fig. 8.9 over all basins, which shows variation ranges of water balance covariance in time (vertical lines) and space (horizontal lines) for WCEnKF, UWCEnKF-1, and UWCEnKF-2. In contrast to WCEnKF and UWCEnKF-1, spatial variabilities can be observed in UWCEnKF-2 results. As discussed, this helps for a better error adjustment during the filtering process. In terms of temporal variations, both UWCEnKF-1 and UWCEnKF-2 perform comparably well representing a larger range of changes than WCEnKF over all basins. The unsupervised error estimation algorithm in

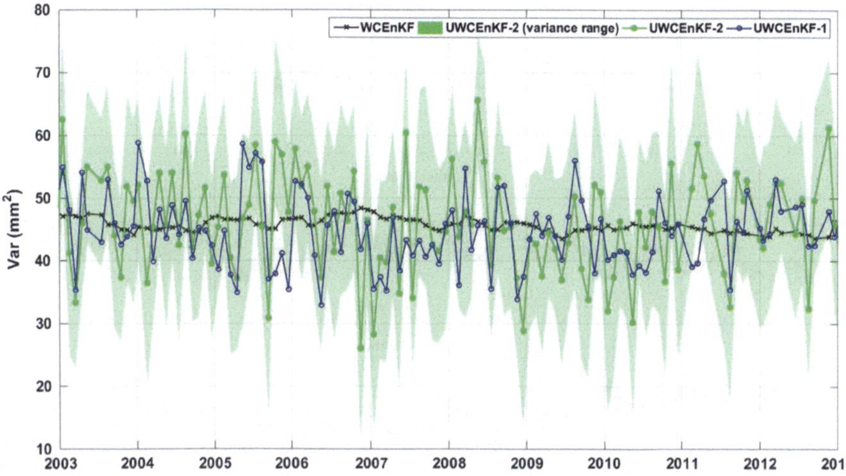

Fig. 8.8 Average water balance variances estimated by UWCEnKF-1 and UWCEnKF-2. The plots also contains the assigned variance values for WCEnKF implementation

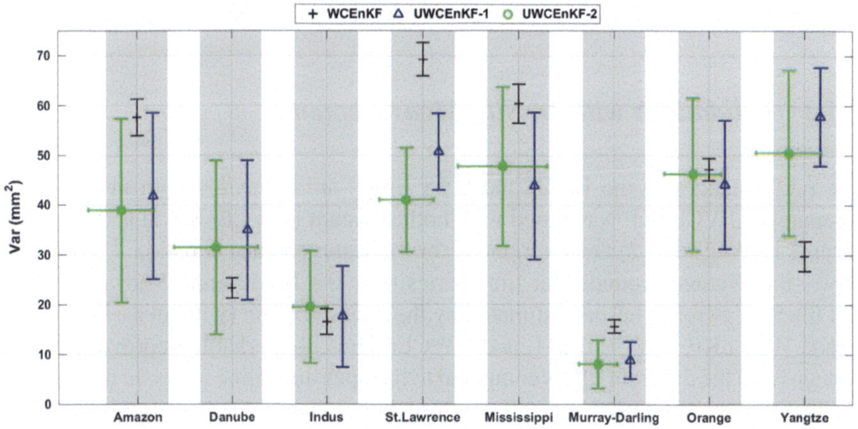

Fig. 8.9 Variation ranges of water balance covariance in time (vertical lines) and space (horizontal lines) for WCEnKF, UWCEnKF-1, and UWCEnKF-2

UWCEnKF enables to estimate an "optimal" water balance error calculation, which as it will be shown in Sect. 8.6.3 (cf. Fig. 8.15) leads to smaller imbalance errors. In cases where assigned error to WCEnKF is close to what is calculated by UWCEnKF, e.g., Indus Basin, the final achieved imbalance from the filters are also close. In other cases with larger differences between assigned and estimated errors, there are larger discrepancies in imbalances.

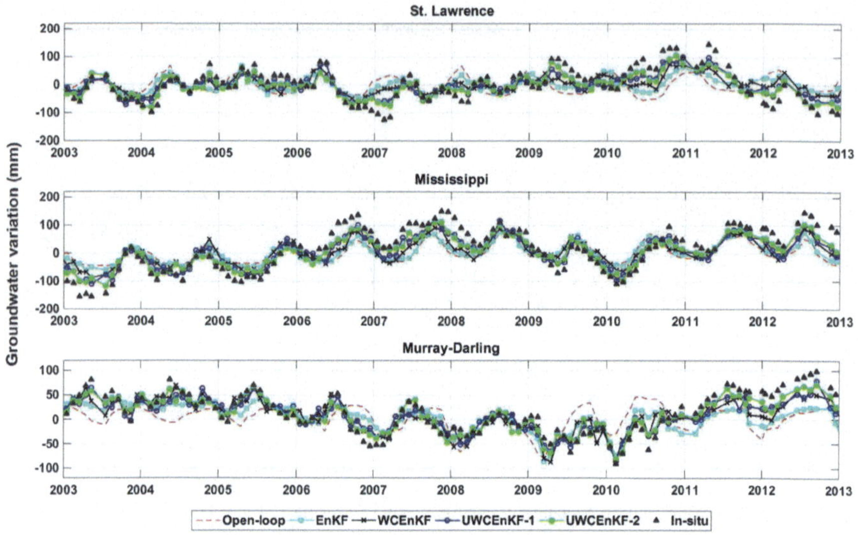

Fig. 8.10 Average groundwater variation time series by the open-loop run, EnKF, WCEnKF, UWCEnKF-1, and UWCEnKF-2 over St. Lawrence, Mississippi, and Murray-Darling basins

8.6.2 Validations with In-Situ Measurements

The performances of the EnKF and UWCEnKF are compared with in-situ measurements. UWCEnKF was tested with both constant (UWCEnKF-1) and spatially varying (UWCEnKF-2) error variances for the water balance equation. Figure 8.10 shows the average groundwater time series over the Mississippi, Murray-Darling and the St. Lawrence basins, estimated by the open-loop run (without assimilation), EnKF, WCEnKF, UWCEnKF-1, and UWCEnKF-2. Remarkable improvement can be seen from the different filters compared to the open-loop time series. In this regard, WCEnKF and UWCEnKF generally perform better than EnKF. This is more evident when a considerable trend exists in the time series, e.g., within the Murray-Darling basin after 2009 and St. Lawrence between 2010 and 2012. It can also be seen that UWCEnKF groundwater time series in most of the times better match to those of in-situs. A clear example of this can be found in Murray-Darling basin 2011–2013. Furthermore, comparing UWCEnKF-1 and UWCEnKF-2, better agreements between in-situ and estimated groundwater changes are achieved for UWCEnKF-2 over all three basins, particularly in the Mississippi basin.

To better monitor how UWCEnKF improves the groundwater estimates, their results are compared with in-situ measurements and against those of EnKF. RMSE and standard deviation (STD) are calculated for groundwater error time series, i.e., the difference between in-situ and filtered groundwater time series, at the location of each in-situ station. Figures 8.11 and 8.12 display the results over the Murray-Darling and Mississippi basins, respectively.

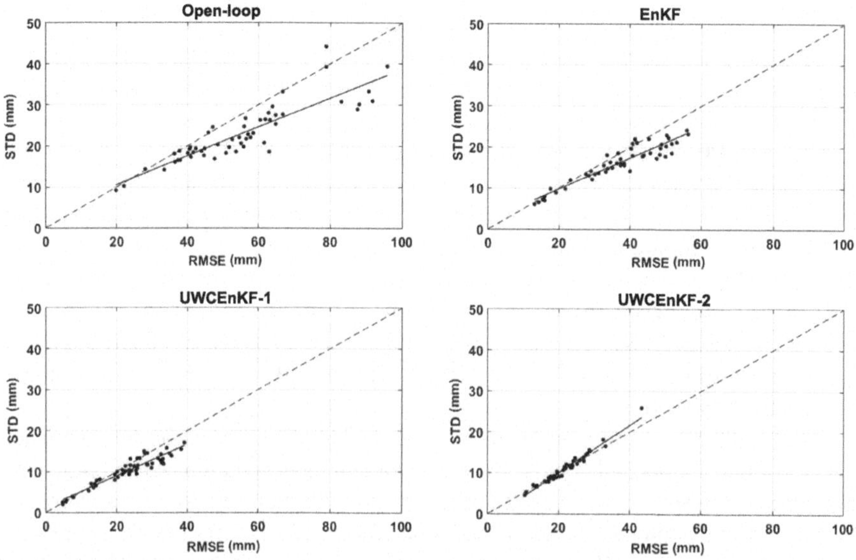

Fig. 8.11 Average RMSE and STD of the groundwater results from the EnKF, UWCEnKF-1, and UWCEnKF-2 filters over the Murray-Darling basin regarding the in-situ groundwater measurements

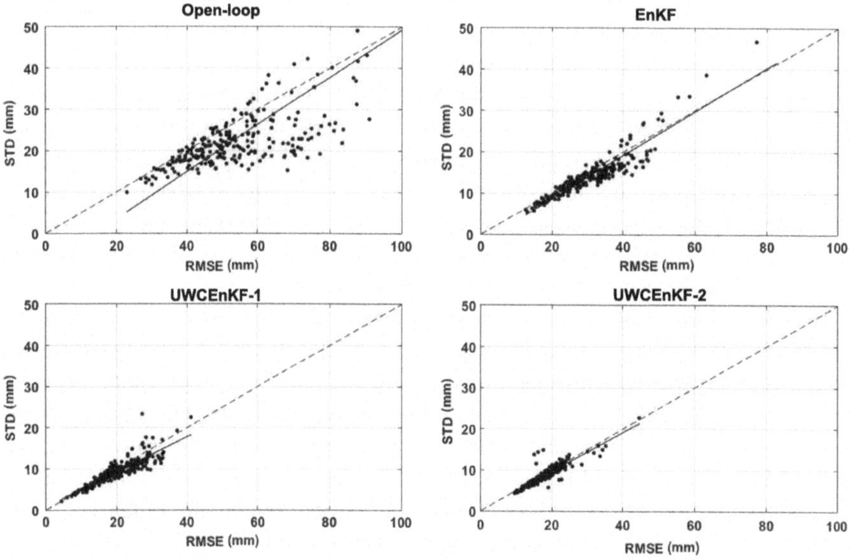

Fig. 8.12 Average RMSE and STD of the groundwater results from the EnKF, UWCEnKF-1, and UWCEnKF-2 filters over the Mississippi basin regarding the in-situ groundwater measurements

One can see that the filters successfully reduce RMSE and STD w.r.t. the open-loop run. This indicates the relevance of assimilation for decreasing state estimate errors. The groundwater estimate improvements are different for each filter. UWCEnKF-1 and UWCEnKF-2 suggest more (18% on average) error reduction than EnKF. Overall, more pronounced error reductions are achieved over the Mississippi basins, which could be attributed to larger model errors within the basin. Slightly better performances (~4%) in terms of groundwater error reduction are obtained with UWCEnKF-2 compared to UWCEnKF-1. We also compute the correlations (at 0.05 significance level) between the filtered and in-situ groundwater time series. Similarly, larger correlations result from the filter estimates compared to the open-loop run, namely, 14% from EnKF, 26% for UWCEnKF-1, and 29% for UWCEnKF-2. The correlation results also confirm that UWCEnKF provides better estimates of the groundwater time series.

In-situ soil moisture measurements are also used to assess the assimilation impact on soil storage. To this end, similar to groundwater assessment, filtered soil moisture time series at the stations' locations are compared with their in-situ counterpoints at different layers. Figure 8.13 shows root-zone soil moisture variation time series as estimated by the various filters, as well as in-situ measurements over the Mississippi, Murray-Darling, St. Lawrence, Danube, and the Yangtze basins. It can be seen that all filters decrease the misfits between estimated and measured soil moisture variations. In some cases, however, UWCEnKF, and to a lesser degree WCEnKF, performs better, e.g., Mississippi (2009), Murray-Darling (2004 and 2008), and Danube (2006). There are also various occasions during which the WCEnKF and UWCEnKF-1 results are very close, such as St. Lawrence 2010–2012 and Yangtze 2005–2006. This can be explained by the fact that both methods use a single error variance value for water balance uncertainties, so whenever a good approximation is used to assign this value prior to data assimilation in WCEnKF, close to what is estimated in UWCEnKF-1, the corresponding state estimates seen to be also close. UWCEnKF-2, on the other hand, performs relatively better, being more successful in matching soil moisture estimates to the in-situ soil moisture variations.

The correlation results between the monthly soil moisture estimates for all filters w.r.t. the monthly in-situ measurements are presented in Table 8.2. Note that different soil moisture estimates of various soil layers are compared to soil moisture measurements at corresponding layers and their average are reported in the table. For instance, the model top layer is compared with 0–8 cm measurements over the Murray-Darling basin and 0–10 cm over Mississippi basin, summations of the model top, shallow, and a small portion of deep-root soil layers are tested against 0–30 cm and 0–50 cm measurements over the Murray-Darling and Mississippi basins, respectively, and summations of the model's soil layers are compared to 0–90 cm (for Murray-Darling) and 0–100 cm (for Mississippi) soil measurements. Due to a difference between the soil moisture estimates (i.e., column water storage measured in mm) and the in-situ measurements (i.e., volumetric soil moisture), only a correlation analysis is conducted. Additionally, in order to statistically assess the results, a significance test for the correlation coefficients is applied based on the t-distribution. The estimated

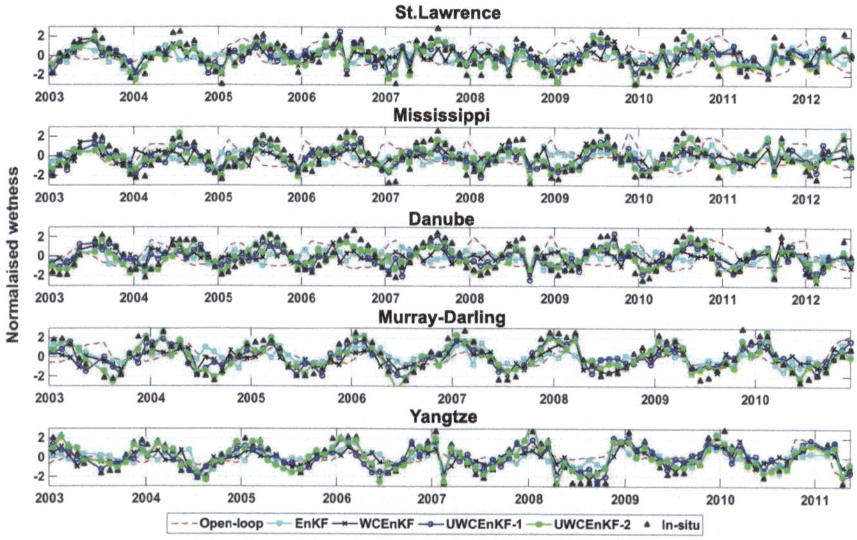

Fig. 8.13 Average soil moisture variation time series by the open-loop run, EnKF, WCEnKF, UWCEnKF-1, and UWCEnKF-2 over St. Lawrence, Mississippi, Danube, Yangtze, and Murray-Darling basins

Table 8.2 Average correlations between in-situ and soil moisture estimates from various methods. Improvements in the assimilation results are calculated as [(*assimilation - open-loop run*)/*open-loop run*] × 100 (%)

Basin	Open-loop	EnKF	WCEnKF	UWCEnKF-1	UWCEnKF-2
Danube	0.67	0.74	0.79	0.81	0.82
St. Lawrence	0.69	0.72	0.76	0.84	0.87
Mississippi	0.72	0.81	0.85	0.86	0.88
Murray-Darling	0.76	0.83	0.86	0.89	0.91
Yangtze	0.73	0.75	0.78	0.80	0.81
Improvemts (%)	–	7.85	13.22	17.75	20.28

t-value and the distribution at 0.05 significant level are used to calculate the p-value, which is assumed to be significant if it lies under 5%.

The results indicate that assimilation significantly improves soil storages regardless of the applied filter. All the filters have positive effects on soil moisture estimates. UWCEnKF performs better than both WCEnKF and EnKF with respectively 6 and 11% higher correlations with the in-situ measurements. It can also be seen that in some cases, e.g., Mississippi basin, the filters generally perform comparably, especially WCEnKF and UWCEnKF-1. This indicates that WCEnKF is capable of improving soil moisture estimates as UWCEnKF subject to using an accurate water

Table 8.3 Average correlations between the filtered water discharge and independent observations over different basins

Basin	Open-loop	EnKF	UWCEnKF-1	UWCEnKF-2
Amazon	73.62	78.04	95.26	96.58
Danube	76.13	76.28	90.77	90.60
Indus	77.08	74.71	84.48	85.37
St. Lawrence	68.55	80.65	87.41	89.17
Mississippi	71.91	73.78	94.29	93.32
Murray-Darling	79.36	83.12	96.31	96.89
Orange	69.47	71.82	93.42	94.05
Yangtze	71.15	75.49	92.69	93.91

balance uncertainty because this is the only difference between the two approaches. The largest improvement with an average 20.28% for all basins is achieved by UWCEnKF-2, better than UWCEnKF-1 (17.75% on average) and noticeably larger than EnKF (7.85%).

We further examine the assimilation results against independent discharge data over different basins. It is worth mentioning that these discharge datasets are not assimilated. The average correlations between the estimated water discharge time series and those from the in-situ data over each basin are presented in Table 8.3. Improvements are achieved for all assimilation experiments w.r.t. the open-loop run. The EnKF increases the correlation by 4% (on average), while UWCEnKF-1 and UWCEnKF-2 increase the correlation by approximately 23respectively. Again, UWCEnKF provides better results than EnKF over all basins. The largest correlation values are obtained for the Murray-Darling and Amazon basins, while the largest correlation improvements are achieved over the Orange, Amazon, and the Yangtze basins.

8.6.3 Impact of the Equality Constraint

To further investigate the relevance of the second analysis step of UWCEnKF, we calculate correlations between the filters estimates and assimilated observations at the forecast and analysis steps for all basins. The average correlations improvements w.r.t. the open-loop run are plotted in Fig. 8.14. As expected, larger correlations are obtained in the analysis step. In general, applying EnKF results in larger correlations between the estimates and assimilated observations (e.g., GRACE TWS and AMSR-E+SMOS) because during the EnKF assimilation the full magnitude of the update is applied to the variables regardless of the water balance. However, the WCEnKF and UWCEnKF take into account the water balance in a second update, which leads to the most improvements regarding $\mathbf{p}$, $\mathbf{e}$, and $\mathbf{q}$. This is due to the fact that the first update

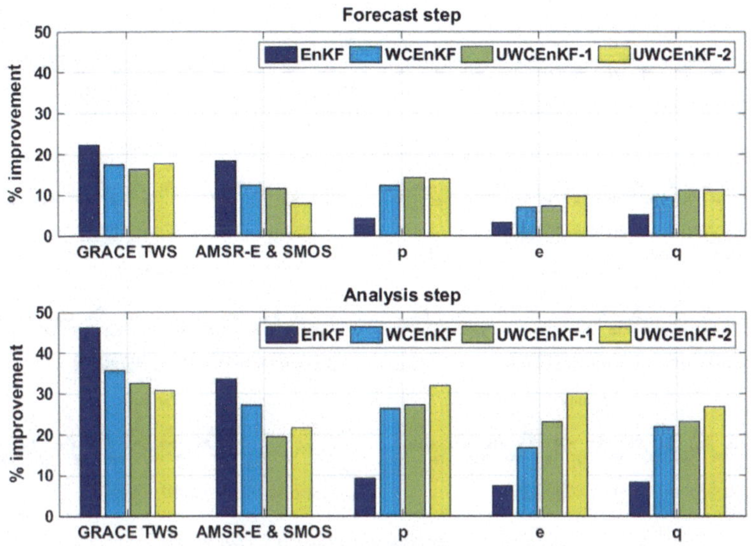

Fig. 8.14 Average correlation improvements of filtered TWS time series to GRACE TWS, **p**, **e**, and discharge **q** with respect to open-loop run in forecast and analysis steps. For AMSR-E+SMOS correlation, filtered top soil storage estimates are used

in the WCEnKF and UWCEnKF corrects the state variables with the observations, and the second update corrects the water balance. This suggests that water budget constraint slightly degrades the effects of observations in the (second) update step in both WCEnKF and UWCEnKF filters, which is generally due to the observation overfitting problem, when no constraint is applied (e.g., standard EnKF) in data assimilation (see also Tangdamrongsub et al. 2017; Khaki et al. 2017c). Furthermore, there is a degree of disagreement between TWS changes and other flux observations (e.g., precipitation, evaporation, and discharge), which could be attributed to different sources of uncertainties in the observations (see, e.g., Aires 2014; Munier et al. 2014). The water budget constraint applied to data assimilation (i.e., the second update of UWCEnKF) accounts for this effect by further correcting the estimated states from the first update step based on GRACE TWS. The second step partly removes the artifacts from data assimilation of GRACE in the first step. It can clearly be seen that UWCEnKF provides higher correlations to the flux observations than WCEnKF. This improvement is more pronounced by using UWCEnKF-2. UWCEnKF's both variants remarkably increase the correlations between TWS estimates and water fluxes compared to EnKF. Overall, a better performance is observed for UWCEnKF-2 in comparison to UWCEnKF-1.

The results of water budget closure resulting from each filter for every basin are shown in Fig. 8.15. UWCEnKF-1 and UWCEnKF-2 clearly reduce water budget imbalances for all basins compared to WCEnKF and especially EnKF. It can also be seen that UWCEnKF-2 better enforces the balance between water components

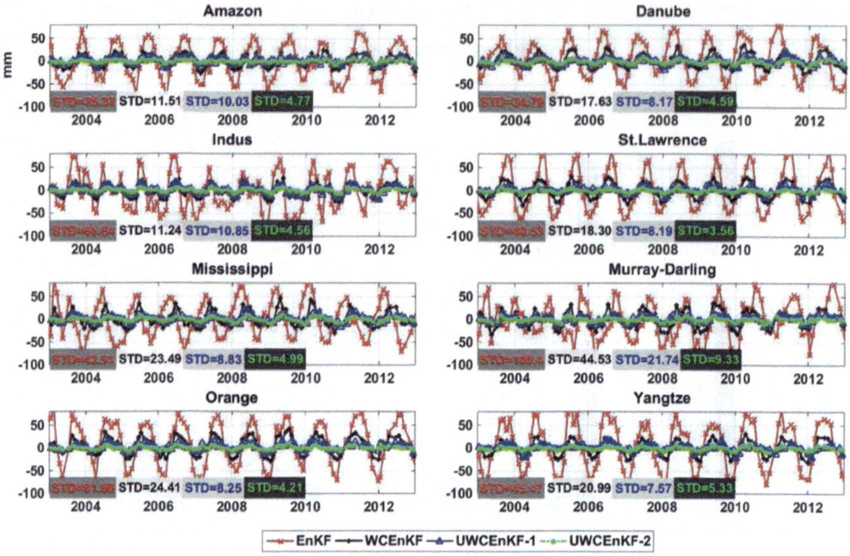

Fig. 8.15 Average water budget imbalance time series calculated using EnKF, WCEnKF, and UWCEnKF variants for each basin (units are mm)

after assimilation. The absolute imbalance from UWCEnKF-2 is 15.28 mm, 8.26% smaller than UWCEnKF-1, 17.84% smaller than WCEnKF, and 36.47% smaller than EnKF. Note that these average values are computed for all basins. The imbalance reductions can also be seen from the reported STD values for each time series in Fig. 8.15. In all basins, the largest STD results from the EnKF and the least from the UWCEnKF-2. In some cases such as Indus, and to a lesser degree Amazon, WCEnKF performs comparably to UWCEnKF-1. UWCEnKF-2, on the other hand, achieves the largest water budget imbalance reduction, in terms of amplitude and STD, which confirms the results of Fig. 8.14, as well as the validation results against in-situ measurements.

8.7 Conclusions

This study introduced an Unsupervised Weak Constrained Ensemble Kalman Filter (UWCEnKF) to mitigate for water budget imbalance while accounting for uncertainties in the inputs of the water balance components. UWCEnKF is an extension of the previously proposed Weak Constrained Ensemble Kalman Filter (WCEnKF) to a more general (unsupervised) framework, in which the covariance associated with the water balance model is estimated along with the system state. Numerical experiments were carried out to assess the performance of the UWCEnKF against WCEnKF, as well as the standard Ensemble Kalman Filter (EnKF). The filters' results examinations against available in-situ measurements indicated that UWCEnKF performs best in terms of groundwater error reduction and soil moisture estimate improvements. In general, UWCEnKF reduced groundwater errors (w.r.t. groundwater in-situ measurements) by 18% (on average), and 11% (on average) more than EnKF and WCEnKF, respectively. UWCEnKF-2 also achieved 4% (on average) smaller groundwater RMSE than UWCEnKF-1. Furthermore, UWCEnKF increased the correlation values between soil moisture estimates and those of the in-situ measurements by 6% more than WCEnKF and 12% more than EnKF. Again, UWCEnKF-2 performed better than UWCEnKF-1 with larger soil moisture correlations w.r.t. the in-situ soil moisture measurements, i.e., 20.28% against 17.75%. UWCEnKF also achieved larger correlations to independent discharge datasets, e.g., respectively 6 and 11% larger correlations with the in-situ measurements than WCEnKF and EnKF. The experiments results also suggested that the UWCEnKF using spatially varying error variances for the water balance equation provides better groundwater and soil moisture estimates than applying a constant error variance. A similar performance was also obtained for the water budget imbalance reduction, where the prior variant better mitigated the imbalance problem than the latter case.

Overall, UWCEnKF achieved maximum correlations with the flux observations, both during the forecast and analysis steps. The largest imbalance reduction was also obtained using UWCEnKF. More specifically, the absolute imbalance for UWCEnKF-2 is 15.28 mm, 8.26% smaller than UWCEnKF-1, 17.84% smaller than WCEnKF, and 36.47% smaller than EnKF. These results demonstrate the relevance of the new proposed unsupervised scheme, which is straightforward to implement and computationally not intensive. Future work will consider extending the proposed framework to jointly estimate the model biases with the state and the observation error variance.

Acknowledgements M. Khaki is grateful for the research grant of Curtin International Postgraduate Research Scholarships (CIPRS)/ORD Scholarship provided by Curtin University (Australia). This work is a TIGeR publication.

Part VI
Data-Driven Approach

Classical data assimilation methods, despite their success in hydrologic modelling, have their limitations. These include, but not limited to, their requirement of intensive computations for high-dimensional systems (Tandeo et al. 2015), as well as the availability of a physical model (see, e.g. Palmer 2001; Reichle and Koster 2005). The later is essential for data assimilation to run the system advance in time, i.e. in the forecast step. Nevertheless, these methods cannot be applied when the model is either unavailable or partially available (e.g. Sauer 2004; Lguensat et al. 2017; Berry and Harlim 2016). In such cases, data-driven (non-parametric) approaches have been used (e.g. Tandeo et al. 2015; Dreano et al. 2015; Hamilton et al. 2016). The idea behind these alternatives is replacing the actual model with a data-driven non-parametric reconstruction of the system's dynamics, which can sometimes reduce process complexity leading to more efficient filtering. This is then allowing us to update the system, usually built from the system trajectories, based on additional observations.

The recently proposed data-driven approach for hydrologic application is proposed in the chapter. The approach is based on the Kalman-Takens method to reconstruct the model for more complicated state observation transition systems in hydrology. Various aspects of the method are assessed such as its computational efficiency, its ability to effective sampling and stability during the process. All these factors are crucial to determine a good assimilation method. Furthermore, it is important to compare the data-driven approach with standard data assimilation techniques, particularly in terms of the computation burden. Moreover, it is important to explore the methods' performance in dealing with multiple assimilation observation such as GRACE TWS and satellite soil moisture simultaneously. This is particularly of interest knowing that joint data assimilation can indeed lead to a more model improvement compared to single observation data assimilation (e.g. Tian et al. 2017; Khaki et al. 2018c, d; Girotto et al. 2019). Soil moisture data for this objective can be derived from various sources such as AMSR-E (e.g. Draper et al. 2009; Mladenova et al. 2011) and SMOS (e.g. Peischl et al. 2012; Jing et al. 2018) soil moisture products with global coverage and high temporal resolution.

Chapter 9
Non-parametric Hydrologic Data Assimilation

"Classically, data assimilation can be used to improve imperfect models by integrating available observations with the underlying physical model". On the other hand, "data-driven methods offer comparable solutions when the physics underlying the models are unknown".

—M. Khaki (This chapter is presented following Khaki et al. (2018f), "Non-parametric data assimilation scheme for land hydrological applications")

9.1 Summary

Data assimilation, which relies on explicit knowledge of dynamical models, is a well-known approach that addresses models' limitations due to various reasons, such as errors in input and forcing datasets. This approach, however, requires intensive computational efforts, especially for high dimensional systems such as distributed hydrological models. Alternatively, data-driven methods offer comparable solutions when the physics underlying the models are unknown. For the first time in a hydrological context, a non-parametric framework is implemented here to improve model estimates using available observations. This method uses Takens delay-coordinate method to reconstruct the dynamics of the system within a Kalman filtering framework, called the Kalman-Takens filter. A synthetic experiment is undertaken to fully investigate the capability of the proposed method by comparing its performance with that of a standard assimilation framework based on an adaptive unscented Kalman filter (AUKF). Furthermore, using terrestrial water storage (TWS) estimates obtained from the Gravity Recovery And Climate Experiment (GRACE) mission, both filters are applied to a real case scenario to update different water storages over Australia. In-situ groundwater and soil moisture measurements within Australia are used to further evaluate the results. The Kalman-Takens filter successfully improves the estimated water storages at levels comparable to the AUKF results, with an average RMSE reduction of 37.30% for groundwater and 12.11% for soil moisture estimates.

M. Khaki, *Satellite Remote Sensing in Hydrological Data Assimilation*,
https://doi.org/10.1007/978-3-030-37375-7_9

Additionally, the Kalman-Takens filter, while reducing estimation complexities, requires a fraction of the computational time, i.e., ∼8 times faster compared to the AUKF approach.

9.2 Introduction

A precise study of terrestrial water storage (TWS) changes is essential to better understand the spatio-temporal variations of water resources and their effects on the hydrological cycles. In this regard, hydrological models become valuable tools for simulating hydrological processes at global (e.g., Döll et al. 2003; Huntington 2006; Coumou and Rahmstorf 2012; van Dijk et al. 2013) and regional (e.g., Chiew et al. 1993; Wooldridge and Kalma 2001; Christiansen et al. 2007; Huang et al. 2016) scales. These models are formulated based on physical/conceptual principles to represent 'reality' and are still being developed to accurately simulate all complex hydrological processes, including interactions between water cycle components (e.g., surface and sub-surface water exchange). These models, however, can be subject to various sources of uncertainties, e.g., errors in input and forcing data, and imperfect accounting for the physical underlying dynamics, such as those used to simulate evapotranspiration (van Dijk et al. 2011; Vrugt et al. 2013).

Classically, data assimilation can be used to improve imperfect models by integrating available observations with the underlying physical model. Many studies have implemented data assimilation techniques in the fields of ocean and atmospheric sciences (e.g., Bennett 2002; Hoteit et al. 2002; Kalnay 2003; Schunk et al. 2004; Lahoz 2007; Zhang et al. 2012; Hoteit et al. 2012; Tardif et al. 2015; Zhao et al. 2017) and hydrology (e.g., Seo et al. 2003; Vrugt et al. 2005; Weerts et al. 2006; Rasmussen et al. 2015; Kumar et al. 2016; Girotto et al. 2016, 2017; Schumacher et al. 2018). Data assimilation is often used to improve model simulations of soil moisture (e.g., Entekhabi et al. 1994; Calvet et al. 1998; Montaldo et al. 2001; Reichle et al. 2002; De Lannoy et al. 2007, 2009; Kumar et al. 2009, 2015; Brocca et al. 2010; Renzullo et al. 2014; Lievens et al. 2015; De Lannoy et al. 2015), TWS (e.g., Zaitchik et al. 2008; van Dijk et al. 2014; Tangdamrongsub et al. 2015; Schumacher et al. 2016, 2018; Khaki et al. 2017b, 2018b, c), evapotranspiration and sensible heat fluxes (e.g., Schuurmans et al. 2003; Pipunic et al. 2008; Irmak and Kamble 2009; Yin et al. 2014), surface water and river discharge (e.g., Bras and Restrepo-Posada 1980; Awwad et al. 1994; Young 2002; Madsen and Skotner 2005; Vrugt et al. 2006; Andreadis et al. 2007; Neal et al. 2009; Giustarini et al. 2011; Lee et al. 2011; McMillan et al. 2013; Li et al. 2015). Standard data assimilation techniques have their limitations though, e.g., the general requirement of intensive computations for high dimensional systems in realistic applications (Tandeo et al. 2015). Furthermore, when a physical model (i.e., model's underlying equations) is not available, the application of a traditional data assimilation framework that relies on these equations for forecasting can be limited (see, e.g., Palmer 2001; Reichle and Koster 2005; Hersbach et al. 2007; Arnold et al. 2013).

A number of studies employ data-driven (non-parametric) approaches to produce accurate statistical simulations (e.g., Sauer 2004; Tandeo et al. 2015; Dreano et al. 2015; Hamilton et al. 2016; Lguensat et al. 2017). Hamilton et al. (2015) and Berry and Harlim (2016) considered the case when models are partially known. In other cases with completely unknown systems, e.g., no available information about the physics of the underlying models and correspondingly their equations, the application of data assimilation becomes rather complicated. Hamilton et al. (2016) developed a new model-free filter based on the non-parametric Takens approach and Kalman filtering when the physical model is not available. The main idea of Takens' theorem is that the model equations can be replaced by a data-driven non-parametric reconstruction of the system's dynamics. The filter implements Takens' method for attractor reconstruction within the Kalman filtering framework, allowing for a model-free approach to filter noisy data (Hamilton et al. 2016). Takens method has been used in various studies for non-parametric time series predictions (see, e.g., Packard et al. 1980; Takens 1981; Sauer et al. 1991; Sauer 2004). This technique replaced the model with a delay coordinate embedding scheme and has been shown by Hamilton et al. (2016) to not only obtain comparable results to a standard Kalman filter-based framework, but also may perform better when model errors are significant. A similar idea has been used by Tandeo et al. (2015) and Lguensat et al. (2017) to simulate the dynamics of complex systems using a non-parametric sampler. They applied an Analog Data Assimilation (AnDA) scheme that reconstructs the system's dynamics in a fully data-driven manner. While AnDA does not require knowledge of the dynamical model, it assumes that a representative catalog of trajectories of the system is available. They show that the data-driven method performs well without using the physical model.

The main motivation of this study, therefore, is to apply for the first time the Kalman-Takens method in a hydrological context and investigate its capability to enhance a hydrological model's estimates. Its performance is then compared with that of a traditional data assimilation system. The motivation behind selecting the Kalman-Takens method is that it does not use the model's equations, and requires less computational burden to predict high-dimensional systems compared to other existing methods (e.g., Hamilton et al. 2015; Tandeo et al. 2015; Berry and Harlim 2016). This study extends the Kalman-Takens approach to enable its application to a more complicated state observation transition systems, e.g., for a case of updating various variables (e.g., soil moisture and groundwater) using only TWS observations. The proposed scheme exploits model trajectories for these variables as the training data and is then applied to assimilate TWS data derived from the Gravity Recovery And Climate Experiment (GRACE) satellite mission into the hydrological system states over Australia for the period 2003–2013. It should be pointed out here that the use of model trajectory in this method, and the reliance of data-driven on data in general, results in updating observable state variables only. This, however, is different in a standard data assimilation, which can further update other variables subject to availability of the physical model.

GRACE TWS data have been assimilated in many studies, where they have proved to be highly capable of improving the performance of hydrological models (e.g.,

Zaitchik et al. 2008; van Dijk et al. 2014; Eicker et al. 2014; Reager et al. 2015; Schumacher et al. 2018). Nevertheless, GRACE data assimilation has always been challenging due to the unique characteristics of its measurements, such as the coarser spatio-temporal resolution compared to most of the existing hydrological models (Khaki et al. 2017a). A successful data assimilation method should be able to account for these limitations in GRACE products while vertically spreading their information into various water compartments (see, e.g., Schumacher et al. 2016; Khaki et al. 2017a). Khaki et al. (2017b) showed that assimilating GRACE data can significantly improve the hydrological model performance over Australia (see also Khaki et al. 2017c; Tian et al. 2017). In order to benchmark the performance of the proposed data-driven technique, its outputs are compared to those of a standard data assimilation framework based on an adaptive unscented Kalman filter (AUKF, Berry et al. 2013). The results of both methods are evaluated against in-situ measurements, as well as through a synthetic experiment to fully investigate their efficiency in assimilating GRACE TWS data.

9.3 Model and Data

9.3.1 W3RA

The $1^\circ \times 1^\circ$ version of the World-Wide Water Resources Assessment (W3RA; http://www.wenfo.org/wald/data-software/) model from the Commonwealth Scientific and Industrial Research Organisation (CSIRO) is chosen for the study. The model is designed to simulate landscape water stores and describe the water balance of the soil, groundwater and surface water stores in which each cell is modeled independently of its neighbors (van Dijk 2010; Renzullo et al. 2014). The model's forcing includes daily meteorological fields of minimum and maximum temperature, short-wave radiation, and precipitation from Princeton University (Sheffield et al. 2006). The model state vector in our experiment is composed of storages of the top, shallow root and deep root soil layers, groundwater, and surface water for the period of January 2003 to December 2012.

9.3.2 GRACE TWS

For the same period, GRACE level 2 (L2) Stokes' coefficients (up to degree and order 90) and associated full error information are obtained from the ITSG-Grace2014 gravity field model (Mayer-Gürr et al. 2014). Three degree 1 coefficients (C10, C11, and S11) and degree 2 and order 0 (C20) coefficient are replaced by those of Swenson et al. (2008) and that of Cheng and Tapley (2004), respectively. Further, we apply the DDK2 smoothing filter (Kusche et al. 2009) to mitigate a colored/correlated noise in the coefficients (see also Khaki et al. 2018a), and thereafter convert them into $1^\circ \times 1^\circ$ TWS fields following Wahr et al. (1998). The mean TWS for the study period is taken from the W3RA model and is aggregated to the GRACE TWS change time series to reach absolute values related to W3RA (Zaitchik et al. 2008). Error information of ITSG-Grace2014 is used to construct an observation error covariance matrix (Eicker et al. 2014; Schumacher et al. 2016).

9.3.3 *In-Situ Measurements*

In-situ groundwater and soil moisture measurements are used to evaluate the performance of the proposed data assimilation framework. Groundwater data is provided from the New South Wales Government (NSW) within the Murray-Darling Basin, which includes 70% of Australia's irrigated area, covers an area of over one million square kilometers, and extends over much of the central and south-eastern parts of Australia (Mercer et al. 2007). The data is rescaled to a monthly temporal scale to be consistent with GRACE and time series of groundwater storage anomalies. Considering that a specific yield for converting well-water levels to variations in groundwater storage (Rodell et al. 2007; Zaitchik et al. 2008) is not available, we use the value of 0.13 specific yield obtained from the range between 0.115 and 0.2 as suggested by the Australian Bureau of Meteorology (BOM) and Seoane et al. (2013).

Furthermore, in-situ soil moisture products are acquired from the moisture-monitoring network, known as the OzNet network (http://www.oznet.org.au/), over the Murrumbidgee catchment (Smith et al. 2011) and rescaled to the same temporal scale as above. The data contains long-term records of measured volumetric soil moisture at various soil depths at 57 locations across the Murrumbidgee catchment area. Soil measurements at 0–8 cm, the 0–30 cm, and 0–90 cm layers are used to assess the estimated soil moisture results of the proposed assimilation framework. The results can be evaluated using representative soil moisture sites within the basin. Here, we use an analysis suggested by De Lannoy et al. (2007) to acquire the representative soil moisture in-situ measurements (see other methods, in e.g., Famiglietti et al. 2008; Orlowsky and Seneviratne 2014; Nicolai-Shaw et al. 2015). The method is based on relative differences $d_{m,n}$ for site m and time step n, which can be calculated as De Lannoy et al. (2007),

$$d_{m,n} = \frac{SM_{m,n} - \overline{SM}_n}{\overline{SM}_n}, \tag{9.1}$$

where $SM_{m,n}$ is the soil moisture measurement at m and n, and $\overline{SM}_n$ represents the spatially averaged soil moisture. Once $d_{m,n}$ is calculated for each site, the temporally average difference ($\bar{d}_m$) and its standard deviation ($STD(d_m)$) are computed. The most representative site is then the one with $\bar{d}_m$ and $STD(d_m)$ closer to 0.

9.4 Methodology

9.4.1 *Adaptive Unscented Kalman Filter (AUKF)*

Consider the following nonlinear system,

$$\mathbf{x}_t = \mathbf{f}(\mathbf{x_{t-1}}) + \mathbf{v_{t-1}}, \tag{9.2}$$

$$\mathbf{y}_t = \mathbf{h}(\mathbf{x_t}) + \mathbf{u_t}, \tag{9.3}$$

where $\mathbf{f}$, the system dynamics, describes the evolution of state vector, $\mathbf{x}$, over time (t) and $\mathbf{h}$, the observation function, maps $\mathbf{x_t}$ to the observations, $\mathbf{y_t}$. $\mathbf{v}_{t-1}$ represent the process noise, which is assumed to be Gaussian with mean 0 and covariance $\mathbf{Q}$. $\mathbf{u}_t$ indicates observation noise with covariance $\mathbf{R}$, which is assumed to be known (see Sect. 9.3). In the present study, $\mathbf{x}$ consists of different water storages including top, shallow and deep soil water, vegetation, snow, surface, and groundwater storages while $\mathbf{y}$ represents the GRACE TWS data.

For nonlinear systems, the unscented Kalman filter (UKF) (Julier and Uhlmann 1997, 2004; Julier et al. 2000; Simon 2006; Wan and van der Merwe 2001; Terejanu 2009) can be used for state estimation. The UKF approximates the propagation of the mean and covariance of a random variable through a nonlinear function using a deterministic sampling approach that generates an ensemble of state values known as sigma points. Given the current state and covariance estimates $\mathbf{x}^a_{t-1}$ and $\mathbf{P}^a_{t-1}$ at step t of the filter, $2L+1$ sigma points (where L is the dimension of the state vector) are generated by,

$$\mathbf{x}^0_{t-1} = \mathbf{x}^a_{t-1}, \tag{9.4}$$

$$\mathbf{x}^i_{t-1} = \mathbf{x}^a_{t-1} + \left(\sqrt{(L+\lambda)\mathbf{P}^a_{t-1}}\right)_i \quad i = 1, \ldots, L, \tag{9.5}$$

$$\mathbf{x}^{i+L}_{t-1} = \mathbf{x}^a_{t-1} - \left(\sqrt{(L+\lambda)\mathbf{P}^a_{t-1}}\right)_i \quad i = 1, \ldots, L, \tag{9.6}$$

with $\left(\sqrt{(L+\lambda)\mathbf{P}^a_{t-1}}\right)_i$ being the ith column of the matrix square root (e.g., lower triangular Cholesky factorization, Wan and van der Merwe 2000) of $(L+\lambda)\mathbf{P}^a_{t-1}$. The corresponding weights to the above sigma points defined as,

$$w^0_s = \frac{\lambda}{(L+\lambda)}, \tag{9.7}$$

$$w^0_c = \frac{\lambda}{(L+\lambda)} + (1-\alpha^2+\beta), \tag{9.8}$$

$$w^i_s = w^i_c = \frac{1}{2(L+\lambda)} \quad i = 1, \ldots, 2L, \tag{9.9}$$

where $\sum_{i=0}^{2L} w_s^i = \sum_{i=0}^{2L} w_c^i = 1$. In Eqs. 9.5–9.9, λ is the scaling parameter, which can be calculated as $\lambda = \alpha^2(L+\kappa) - L$. The scaling factor α determines the spread of the sigma points around $\mathbf{x}_{t-1}^a$, and κ is a secondary scaling parameter usually set to 0 (the specific value of kappa is not critical, see e.g., Julier and Uhlmann 1997; Van der Merwe 2004). β is employed to incorporate a prior knowledge about the noise distribution (e.g., the optimal choice for Gaussian distribution is $\beta = 2$, e.g., Wan and van der Merwe 2001).

Between these factors, the selection of α has larger impacts on the ensemble spreads and controls the "size" of the sigma-point distribution. α determines how the sigma points can be scaled towards or away from the mean of the prior distribution. For example, $\alpha = 1$ and correspondingly $\lambda = 0$ leads the distance between $\mathbf{x}_{t-1}^a$ and the sigma points to be proportional to $\sqrt{L}$. Positive values of λ (for $\alpha > 1$) scales the sigma points further from $\mathbf{x}_{t-1}^a$ while negative values of λ (for $\alpha < 1$) scales the sigma points towards $\mathbf{x}_{t-1}^a$. In other words, the larger values for this scaling factor causes a larger spread in the sigma points while smaller values result in more concentration around prior distribution (Van der Merwe 2004). Ideally α should be a small number, e.g., $1e-4 \leq \alpha \leq 1$ (Song and He 2005) to avoid sampling non-local effects when the nonlinearities are strong. However, optimal sets of this factor along with κ and β are generally problem specific and can be optimized arbitrary. For the current study, the values of parameters are assumed as $\alpha = 0.5$, $\kappa = 0$, and $\beta = 2$. Nevertheless, it is found that the implemented AUKF is not very sensitive to the parameter selection as long as they result in a numerically well-behaved set of sigma-points and weights (see also Van der Merwe 2004).

The sigma points are advanced forward one time step using model $\mathbf{f}$ and observed using the function $\mathbf{h}$,

$$\mathbf{x}_t^{f,j} = \mathbf{f}(\mathbf{x}_{t-1}^{\mathbf{j}}), \quad \mathbf{j} = \mathbf{0}, \ldots, \mathbf{2L}, \tag{9.10}$$

$$\mathbf{y}_t^{f,j} = \mathbf{h}(\mathbf{x}_t^{\mathbf{f},\mathbf{j}}), \quad \mathbf{j} = \mathbf{0}, \ldots, \mathbf{2L}. \tag{9.11}$$

The transformed points ($\mathbf{x}_t^{f,j}$ and $\mathbf{y}_t^{f,j}$) are then used to calculate their respective forecast means and covariance matrices,

$$\mathbf{x}_t^f = \sum_{j=0}^{2L} w_s^j \mathbf{x}_t^{f,j}, \tag{9.12}$$

$$\mathbf{y}_t^f = \sum_{j=0}^{2L} w_s^j \mathbf{y}_t^{f,j}, \tag{9.13}$$

$$\mathbf{P}_t^f = \sum_{j=0}^{2L} w_c^j \left(\mathbf{x}_t^{f,j} - \mathbf{x}_t^f\right)\left(\mathbf{x}_t^{f,j} - \mathbf{x}_t^f\right)^T + \mathbf{Q}_{t-1}, \tag{9.14}$$

$$\mathbf{P}_{\mathbf{y}_t^f} = \sum_{j=0}^{2L} w_c^j \left(\mathbf{y}_t^{f,j} - \mathbf{y}_t^f\right)\left(\mathbf{y}_t^{f,j} - \mathbf{y}_t^f\right)^T + \mathbf{R}_t, \tag{9.15}$$

as well as the cross covariance between $\mathbf{x}_t^f$ and $\mathbf{y}_t^f$,

$$\mathbf{P}_{\mathbf{x}_t^f,\mathbf{y}_t^f} = \sum_{j=0}^{2L} w_c^j \left(\mathbf{x}_t^{f,j} - \mathbf{x}_t^f\right)\left(\mathbf{y}_t^{f,j} - \mathbf{y}_t^f\right)^T. \tag{9.16}$$

In the analysis step of the filter, the measurements (e.g., GRACE-derived TWS) are used to correct the forecasted state and respective covariance matrix using the Kalman update equations,

$$\mathbf{x}_t^a = \mathbf{x}_t^f + \mathbf{K}(\mathbf{y}_t - \mathbf{y}_t^f), \tag{9.17}$$

$$\mathbf{K} = \mathbf{P}_{\mathbf{x}_t^f,\mathbf{y}_t^f}\mathbf{P}_{\mathbf{y}_t^f}^{-1}, \tag{9.18}$$

$$\mathbf{P}_t^a = \mathbf{P}_{\mathbf{x}_t^f} - \mathbf{K}\mathbf{P}_{\mathbf{y}_t^f}\mathbf{K}^T. \tag{9.19}$$

where $\mathbf{K}$ is the Kalman gain.

Critical to the success of the UKF is the selection of the filter noise covariances, and in particular the process noise covariance matrix $\mathbf{Q}$. Here, we use the method of Berry et al. (2013) to adaptively estimate this covariance matrix. We refer to this as the *adaptive unscented Kalman filter* (AUKF). Building on the method of Mehra (1990, 1992), the general idea of Berry et al. (2013) is to use the increment, $\varepsilon_t = \mathbf{y}_t - \mathbf{y}_t^f$, to estimate the noise covariance at each time step. The method begins by forming an empirical estimate $\mathbf{Q}_{t-1}^e$ for $\mathbf{Q}$,

$$\mathbf{P}_{t-1}^e = \mathbf{F}_{t-1}^{-1}\mathbf{H}_{t-1}^{-1}\varepsilon_{t-1}\varepsilon_{t-1}^T\mathbf{H}_{t-1}^{-T} + \mathbf{K}_{t-1}\varepsilon_{t-1}\varepsilon_{t-1}^T\mathbf{H}_{t-1}^{-T}, \tag{9.20}$$

$$\mathbf{Q}_{t-1}^e = \mathbf{P}_{t-1}^e - \mathbf{F}_{t-2}\mathbf{P}_{t-2}^a\mathbf{F}_{t-2}^T, \tag{9.21}$$

where $\mathbf{P}_{t-1}^e$ is an empirical estimate of the background covariance. In Eqs. 9.20 and 9.21, $\mathbf{F}$ and $\mathbf{H}$ are local linearizations of the nonlinear dynamic models $\mathbf{f}$ and $\mathbf{h}$, respectively, and are estimated using a linear regression on the ensembles (see Eq. 7 in Berry et al. (2013), for details regarding this linearization). It is worth mentioning that we must store linearizations $\mathbf{F}_{t-2}$, $\mathbf{F}_{t-1}$, $\mathbf{H}_{t-1}$, $\mathbf{H}_t$, increments ε_{t-1}, ε_t, analysis covariance $\mathbf{P}_{t-2}^a$, and Kalman gain $\mathbf{K}_{t-1}$ from the $t-1$ and $t-2$ steps of the filter. To form a stable estimate of $\mathbf{Q}$, the noisy estimate $\mathbf{Q}_{t-1}^e$ is combined using an exponentially weighted moving average,

$$\mathbf{Q}_t = \mathbf{Q}_{t-1} + (\mathbf{Q}_{t-1}^e - \mathbf{Q}_{t-1})/\tau, \tag{9.22}$$

where τ is the window of the moving average. Berry et al. (2013); Hamilton et al. (2016) provide additional details on the estimation of noise covariance.

9.4.2 Kalman-Takens Method

The main idea of the Kalman-Takens method is to replace the model-based forecast in the AUKF with the advancement of the dynamics non-parametrically, thus requiring no knowledge of $\mathbf{f}$ (in Eq. 9.2). We provide a brief description of the method below, specifically highlighting modifications in adopting the algorithm to our problem. Full details of the methodology can be found in Hamilton et al. (2016, 2017).

In the present study, we consider a different setup to implement the Kalman-Takens filter for a more complicated state observation transition systems. While the data available are gridded GRACE TWS, our interest is in estimating the different water variables (i.e., top, shallow and deep soil water, vegetation, snow, surface, and groundwater). These variables with no independent observation available, are provided by the W3RA model and are used to produce delay-coordinate vectors. We generate a synthetic set of model trajectories (open-loop run) for these variables to serve as the training data for the Kalman-Takens filter. The training data represents the state of the system. It is also used to generate a local proxy $\tilde{\mathbf{f}}$ for the unknown model $\mathbf{f}$ (cf. Eq. 9.2), which is not available in the non-parametric framework, so Eq. 9.10 for advancing the ensemble forward in time in AUKF is not implementable. This brings us to Eq. 9.23, which defines the delay-coordinate vector $\mathbf{z}$ at each step of the filter using the historical state variables from the open-loop run by,

$$\mathbf{z}_t = [\mathbf{x^o}_t, \mathbf{x^o}_{t-1}, \ldots, \mathbf{x^o}_{t-d}], \tag{9.23}$$

where d is the number of temporal delays. $\mathbf{x^o}$ contains the open-loop top, shallow and deep soil moisture, vegetation, snow, surface, and groundwater. Once the delay coordinate is created, the assimilation procedure can be applied. At each AUKF step, an ensemble of delay vectors is formed and advanced non-parametrically using a local approximation $\tilde{\mathbf{f}}$. This nonparametric prediction helps to build local models for predicting the dynamics at the forecast step (Hamilton et al. 2017). Given the above current delay-coordinate, the non-parametric advancement starts by locating the N nearest neighbors (i.e., points located within a given Euclidean distance; not only adjacent points), within a set of training data,

$$\begin{aligned}
\mathbf{z}_t^1 &= [\mathbf{x^o}^1_t, \mathbf{x^o}^1_{t-1}, \ldots, \mathbf{x^o}^1_{t-d}], \\
\mathbf{z}_t^2 &= [\mathbf{x^o}^2_t, \mathbf{x^o}^2_{t-1}, \ldots, \mathbf{x^o}^2_{t-d}], \\
&\vdots \\
\mathbf{z}_t^N &= [\mathbf{x^o}^N_t, \mathbf{x^o}^N_{t-1}, \ldots, \mathbf{x^o}^N_{t-d}].
\end{aligned} \tag{9.24}$$

The known $\mathbf{z}^1_{t+1}, \mathbf{z}^2_{t+1}, \ldots, \mathbf{z}^N_{t+1}$ (based on $\mathbf{x^o}^1_{t+1}, \mathbf{x^o}^2_{t+1}, \ldots, \mathbf{x^o}^N_{t+1}$), are used in a local model to predict $\mathbf{z}_{t+1}$. The local model $\tilde{\mathbf{f}}$, which can be generated using a weighted average of the nearest neighbors (Hamilton et al. 2016; Lagergren et al. 2018) can be written as,

$$\mathbf{z}_{t+1} = \omega_1 \mathbf{z}_{t+1}^1 + \omega_2 \mathbf{z}_{t+1}^2 + \cdots + \omega_N \mathbf{z}_{t+1}^N, \tag{9.25}$$

$$\omega_i = \frac{\mathrm{e}^{-(d_i/\sigma)^2}}{\sum_{j=1}^N \mathrm{e}^{-(d_j/\sigma)^2}}, \tag{9.26}$$

where d_i is the distance of the jth neighbour to $\mathbf{z}_t$ and σ is a bandwidth parameter, which controls the contribution of each neighbor in the local model (here $\sigma = 2$). The above prediction is applied to estimate the delay coordinate vector at $t + 1$.

The process of building a local model for forecasting the delay-coordinate vector is repeated for each sigma point in the ensemble. After $\tilde{\mathbf{f}}$ has been defined, the remainder of the AUKF update scheme is implemented. Important to the Kalman-Takens method is the selection of d (the number of delays) and neighbors N. Here, we consider different values of N and d and set them based on the filter performance, which is described in Sect. 9.5. The assumption of using the model trajectory rather than observations for generating delay vectors allows us to reconstruct the system representing various water storage compartments. The same assumption is made by Lguensat et al. (2017), where trajectories of the system and not the physical model is available. In fact, we hypothesize that the available model outcomes can be used for the non-parametric sampling of the dynamics and updated by the GRACE TWS (as a summation of all the water variables at each grid point). This means that one can essentially correct state variables of the system, without having data for each individually, using the data-driven framework. The application of this method can address some severe limitations in traditional data assimilation such as large computational cost. A summary of the proposed approach is illustrated in Fig. 9.1.

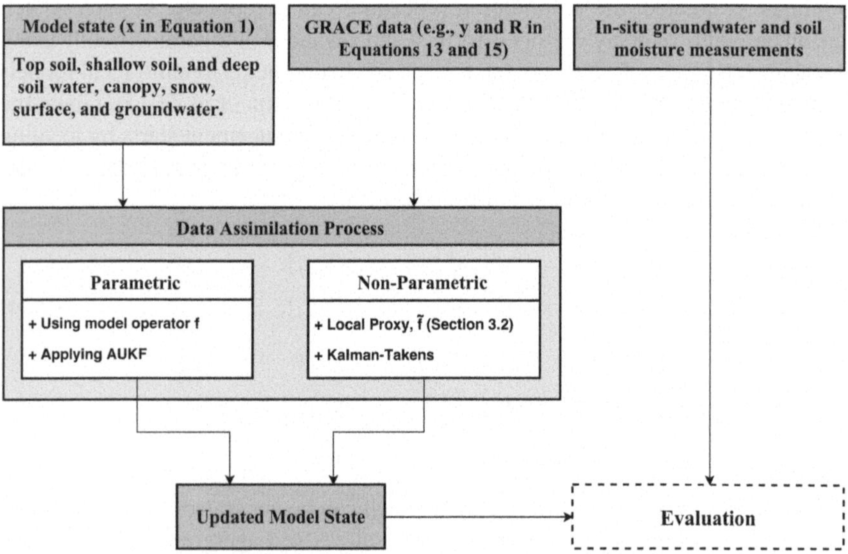

Fig. 9.1 A schematic illustration of the data integration process implemented for this study

9.4.3 Synthetic Experiment

A synthetic experiment is undertaken to assess the efficiency of the proposed data assimilation schemes in simulating physical processes. One important problem with hydrological models, and specifically W3RA, is their limitations in simulating anthropogenic impacts on the water cycle. For example, excessive groundwater extractions, which can largely affect sub-surface water storages, are not modeled in W3RA and a successful data assimilation process should be able to correct for this drawback by taking the advantage of additional observations. Here, we choose to test both AUKF and Kalman-Takens filters to improve upon model simulations between 2003 and 2013 over Iran (32.4279 °N, 53.6880 °E). The rationale behind choosing Iran for this synthetic analysis, and not Australia, is that a remarkable water storage decline is reported over this region, mainly due to anthropogenic impacts, which cannot be detected by W3RA (see Khaki et al. 2018c). A major part of the negative water storage trend is due to human impacts (see details in Forootan et al. 2017; Khaki et al. 2018c). Synthetic observations are produced using the WaterGAP Global Hydrology Model (WGHM; Döll et al. 2003; Müller et al. 2014) monthly TWS outputs, which contain the anthropogenic impacts (Khaki et al. 2018c), at two different spatial resolution of $1^\circ \times 1^\circ$ and $3^\circ \times 3^\circ$. This can help to test whether data assimilation can account for human impacts on water storage and also to investigate the effect of spatial resolution on the final results. WGHM TWS estimates are assumed as our observations after rescaling into $1^\circ \times 1^\circ$ and $3^\circ \times 3^\circ$ and perturbing using Monte Carlo sampling of multivariate normal distributions with the errors representing the GRACE level 2's standard errors. The data assimilation is implemented using both filtering methods at the aforementioned spatial scales.

9.4.4 Evaluation Metrics

To evaluate the assimilation results against in-situ groundwater and soil moisture measurements, three metrics, (i) the Root-Mean-Squared Errors (RMSE), (ii) standard deviation (STD), and (iii) Nash-Sutcliffe coefficient (NSE) are used. Groundwater and soil moisture in-situ measurements from various stations are spatially averaged to the location of the nearest model grid points and are compared with their respective estimates. To this end, using the variation time series of in-situ data and the results of assimilation techniques, RMSE, STD, and NSE are calculated by,

$$RMSE = \sqrt{\frac{1}{n}\sum_{i=1}^{n}(x_i - z_i)^2}, \tag{9.27}$$

$$STD = \sqrt{\frac{1}{n}\sum_{i=1}^{n}(x_i - \bar{x})^2}, \tag{9.28}$$

$$NSE = 1 - \left[\frac{\sum_{i=1}^{n} (x_i - z_i)^2}{\sum_{i=1}^{n} (z_i - \bar{z})^2} \right], \tag{9.29}$$

where x_i is the predicted value (for n samples) and z_i represents the measured in-situ value. In Eqs. 9.27–9.29, $\bar{x}$ and $\bar{z}$ are the average of the predicted and measured values, respectively. Furthermore, to statistically assess the significance of the results, the student t-test is applied. The estimated t-value and the distribution at 0.05 significant level are used to calculate p-values.

9.5 Results

9.5.1 Synthetic Experiment

The results of synthetic experiment, which is chosen to assess the capability of the two data assimilation schemes in improving model's simulation of physical processes are presented in this section. TWS variations from W3RA (open-loop; model integration without assimilation), AUKF and Kalman-Takens filters (with $N = 14$ and $d = 11$, see Sect. 9.5.2 for details), as well as synthetic observations, are displayed in Fig. 9.2, where the time series represent spatially averaged TWS variations over the entire Iran. The trend lines corresponding to each time series are also depicted in the figure. As can be clearly seen, W3RA's open-loop run does not correctly capture the negative trend in the TWS time series as visible in the observations. Assimilation results, on the other hand, successfully reproduce the negative trend. Except for few cases, e.g., 2009 and 2011, Kalman-Takens performs closely to AUKF. The assimilation trend lines also show that the filtered results capture the existing trend of the observations. In addition to the trends, there are larger correlations between AUKF (14% on average)

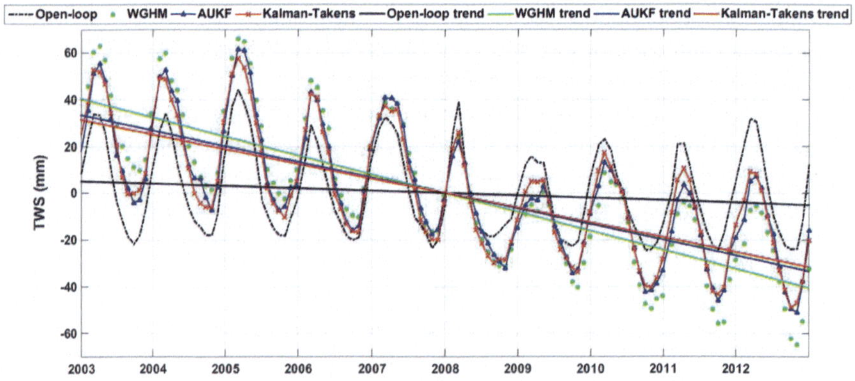

Fig. 9.2 Average TWS variation time series over Iran from AUKF, Kalman-Takens, open-loop run, and WGHM with corresponding trend lines

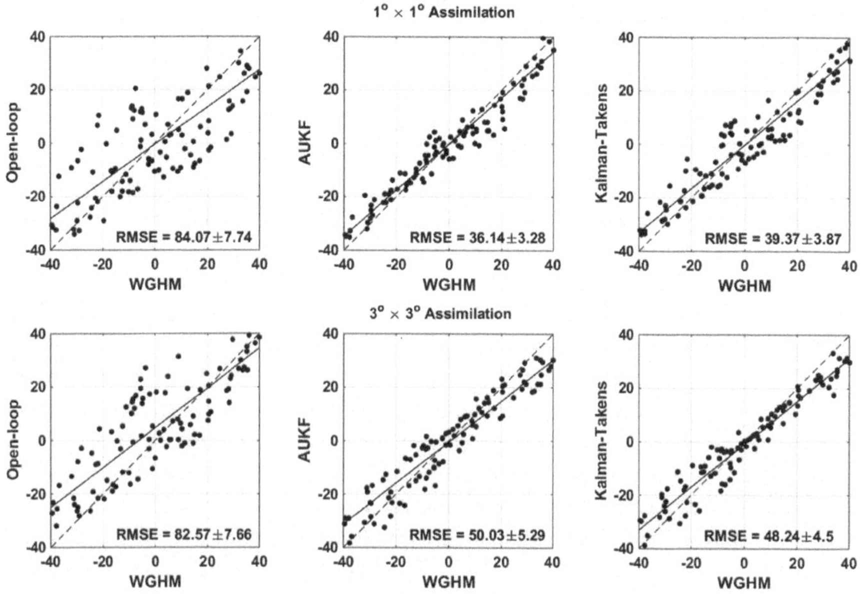

Fig. 9.3 Scatter plots of open-loop, AUKF, and Kalman-Takens TWS estimates with respect to WGHM TWS at the two spatial resolution of 1° × 1° and 3° × 3°. The presented average RMSE values for each method is calculated based on the original WGHM TWS (before perturbation using GRACE errors). In each sub-figure reference (dashed) and fitted (solid) lines are illustrated

and Kalman-Takens (12% on average) with the observations compared to the open-loop results. An evaluation of the assimilation results against the original WGHM TWS, i.e., before perturbation using GRACE noises, are shown in Fig. 9.3.

Figure 9.3 shows the scatter plot of the open-loop, AUKF, and Kalman-Takens TWS estimates against WGHM at the two spatial resolutions of 1° × 1° and 3° × 3° to assess the filters' performances at various spatial scales. Note that temporal assessment is also investigated in Sect. 9.5.2. It can be seen that at both spatial scales, there are larger agreements between the filtered results and WGHM. There are also smaller RMSEs after filtering, which suggests the capability of both methods to improve model simulations even in case of remarkable human impact. While every assimilation scenario leads to smaller RMSE than the open-loop run, the least RMSEs are achieved at 1° × 1° resolution. This shows that assimilating TWS observations at a finer resolution can provide better estimates regardless of the filtering method. It can also be seen that both AUKF and Kalman-Takens provide comparable results at both spatial scales, leading to approximately 48% RMSE reduction. The filters' comparable results at 3° × 3° spatial resolution suggests their similar performance for downscaling TWS observations into the 1° × 1° W3RA resolution.

9.5.2 Assessment with In-Situ Data

Independent groundwater and soil moisture in-situ measurements within the Murray-Darling Basin in Australia are used to evaluate the results. This is done by comparing the AUKF and Kalman-Takens estimates of groundwater and soil moisture with those of the in-situ measurements. Note that further analysis is undertaken to assess the impacts of the filters on non-assimilated variables and the results are provided in the supplementary material. Before comparing AUKF and Kalman-Takens results against in-situ measurements, we investigate the effect of various setups in the Kalman-Takens performance. Different scenarios are considered regarding the number of neighbors N (i.e., 2–40) and also the number of delays d (i.e., 1–25). To reach the best setup amongst these values, we compare the results of each scenario to the in-situ groundwater measurements. Figure 9.4 shows the average absolute groundwater errors resulting from each case. Increasing the number of neighbors can improve the approximation of training data for a particular point to a certain extent (due to the existing spatial correlations). However, selecting N too large can cause a rapid growth of errors, which is related to the effect of over-smoothing the training step. This is different for delays d, where much larger errors are present for smaller values that underestimate temporal variabilities in the data. From our numerical investigations, it can be seen that applying the Kalman-Takens filter with $N = 14$ and $d = 11$ provides the best result. It is worth mentioning that we use these setups of the Kalman-Takens filter throughout this study.

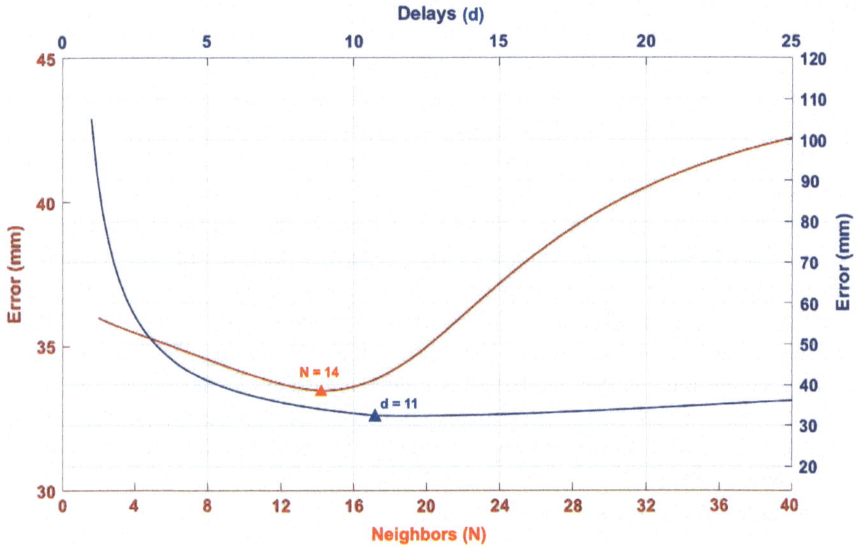

Fig. 9.4 Estimated average errors from different scenarios considered based on the number of neighbors N and delays d. The best estimates are achieved by applying the Kalman-Takens method using $N = 14$ and $d = 11$

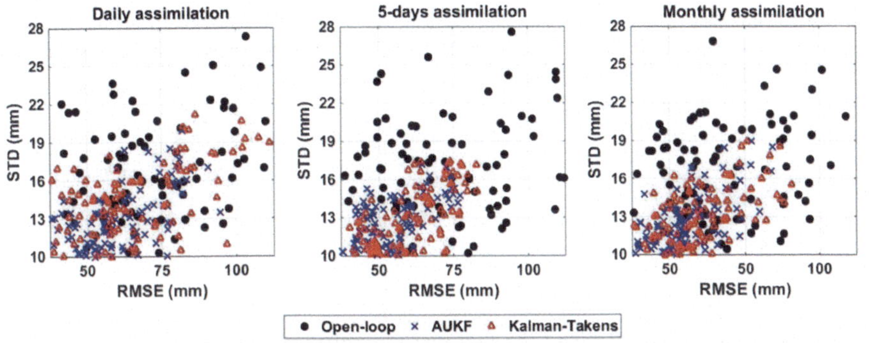

Fig. 9.5 Average groundwater RMSE and STD of from the Kalman-Takens filter, AUKF, and open-loop run computed using groundwater in-situ measurement. The results are presented for assimilation with three different temporal scales (i.e., daily, 5-day, and monthly)

The comparison between the open-loop run, AUKF, and Kalman-Takens results are depicted in Fig. 9.5, which displays scatter plot of each filter's RMSE and STD calculated using in-situ groundwater measurements. Three different temporal evaluations are considered to further investigate the effect of temporal downscaling on the results. The GRACE TWS data (with approximately 30 days temporal scale) and associated errors are interpolated into a daily and 5-daily samples (see also Tangdamrongsub et al. 2015; Khaki et al. 2017a) using the spline interpolation between consecutive months. The assimilation is then undertaken on a daily, 5-day, and monthly basis. Figure 9.5 indicates that both AUKF and Kalman-Takens filters result in smaller RMSE and STD compared to the open-loop run for all the three temporal scales. In the daily and to a lesser degree 5-day assimilation cases, AUKF performs slightly better than the Kalman-Takens, with smaller RMSE and STD, which could be attributed to the contribution of the model equations for spreading TWS information between different variables after assimilation. Nevertheless, the performance of the non-parametric filter is satisfactory for both cases and comparable to that of AUKF. Interestingly, the performances of the two filters are even closer when assimilating monthly data. As a general result, this demonstrates that temporal downscaling of GRACE TWS data is recommended for data assimilation purpose regardless of the filtering method used. The average RMSE values for the 5-day assimilation using AUKF and Kalman-Takens filters are 51.28 ($\sim$13% smaller than daily and $\sim$7% smaller than monthly) and 53.61 ($\sim$16% smaller than daily and $\sim$5% smaller than monthly), respectively. Based on the above evaluation, it can be concluded that different temporal scales have similar effects on both filters, where the AUKF and Kalman-Takens filters perform better for the 5-day assimilation case.

More detailed statistics are provided in Table 9.1 to better compare the performances of the implemented filters against in-situ groundwater measurements. The evaluation is undertaken using RMSE and NSE metrics (see Sect. 9.4.4) based on the 5-day assimilation case. Note that in this table, basin-scale results are provided in addition to the results of the grid-based evaluation. Considering the coarse

Table 9.1 Summary of statistical values derived from the implemented methods using the groundwater in-situ measurements. The reduction of the RMSE value of the AUKF and Kalman-Takens filters are calculated in relation to the RMSE of the open-loop run

	Grid-based evaluation			Basin scale evaluation		
Metric	Open-loop	AUKF	Kalman-Takens	Open-loop	AUKF	Kalman-Takens
RMSE (mm)	74.57	51.28	53.61	69.40	45.16	46.96
NSE	0.51	0.77	0.75	0.58	0.82	0.81
RMSE reduction (%)	–	31.23	28.11	–	34.93	32.33

spatial resolution of W3RA and the fact that a number of groundwater stations can be found in each grid cell, basin-averaged assessment is performed as an alternative examination. The spatially averaged open-loop results and those from filters over the Murray-Darling Basin are tested against basin-average groundwater time series. Results of Table 9.1 confirm the behavior seen in Fig. 9.5. While smaller RMSEs are obtained from AUKF for both grid- and basin-based tests, the application of the Kalman-Takens method significantly decreases groundwater RMSE values (30.22% on average). Also larger NSE values are obtained by both filters compared to the open-loop run. These results prove a high capability of the Kalman-Takens for improving state estimates, very close to the AUKF performance. Table 9.1 also indicates that the Kalman-Takens approach can be used as traditional data assimilation to reduce noise in the final state variables, which are the results of solving complex inverse problems, e.g., groundwater estimates are improved from GRACE-derived TWS. This is also true for soil moisture estimates (cf. Table 9.2).

We use different soil moisture layers from in-situ measurements including 0–8 cm (compared to the model top soil moisture layer), 0–30 cm (compared to the summation of the model top and shallow soil moisture layers), and 0–90 cm (compared to the summation of the model top, shallow, and deep soil moisture layers) for evaluating the results. Note that considering the difference between W3RA states (i.e., column water storage measured in mm) and the OzNet measurements (i.e., volumetric soil moisture) and the fact that converting the model outputs into volumetric units may introduce a bias (Renzullo et al. 2014), only NSE analysis is carried out and the results are provided in Table 9.2. Similar improvements as for groundwater evaluation are also found by comparing the filters estimates against OzNet soil moisture measurements (Table 9.2). Larger NSE values are found from data assimilation filters for all three soil layers. Average NSE from the Kalman-Takens method is 0.73, ~12.3% larger than the open-loop run, and slightly smaller than AUKF results (0.74). Table 9.2 confirms that the capability of the Kalman-Takens method for improving the soil moisture estimates similar to AUKF (13.7% on average). The largest improvements for both filters are achieved in the root zone (0–90 cm) moisture layer. Table 9.2 suggests that AUKF better reflects the GRACE observations, especially at this layer. This, however, does not necessarily lead to better approximations in the shallow soil moisture layer, where the non-parametric approach shows higher improvements compared to AUKF.

Table 9.2 Summary of NSE values estimated using state estimates derived from implemented methods and the soil moisture in-situ measurements at different layers. The improvements (in %) are calculated based on the increased correlation by applying the methods with respect to the open-loop run

	Method	0–8 cm	0–30 cm	0–90 cm
	Open-loop	0.59	0.64	0.72
	AUKF	0.63	0.71	0.89
	Kalman-Takens	0.61	0.73	0.85
Improvements (%)	AUKF	6.77	10.94	23.61
	Kalman-Takens	3.39	14.06	18.05

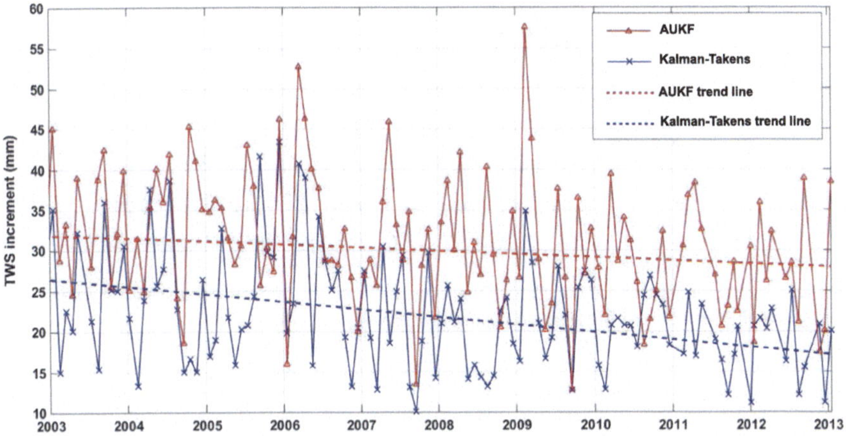

Fig. 9.6 An average TWS increment time series of AUKF and the Kalman-Takens filter on state vectors during the process. Both methods decrease the increment as assimilation proceeds forward in time

9.5.3 Assessing the Performance of AUKF and Kalman-Taken Filters

Here, we compare the performances of the AUKF and Kalman-Takens filters from various perspectives including increment applied, state covariance, computational efficiency, and water storage forecasting. Fig. 9.6 shows the increments implemented by each filtering technique during the study period. We estimate average increment (i.e., ε discussed in Sect. 9.4.1) at all grid points for AUKF and the Kalman-Takens approach. One can see how the filters deal with the GRACE TWS observations in the update steps. Both methods decrease the increment as assimilation proceeds forward in time. This is found to be smaller for the Kalman-Takens method (see the trend lines in Fig. 9.6) compared to AUKF. In fact, AUKF integrates the ensemble members through the model $\mathbf{f}$, while the non-parametric approach uses the local proxy $\tilde{\mathbf{f}}$. Consequently, larger misfits between the Kalman-Takens method forecast estimates

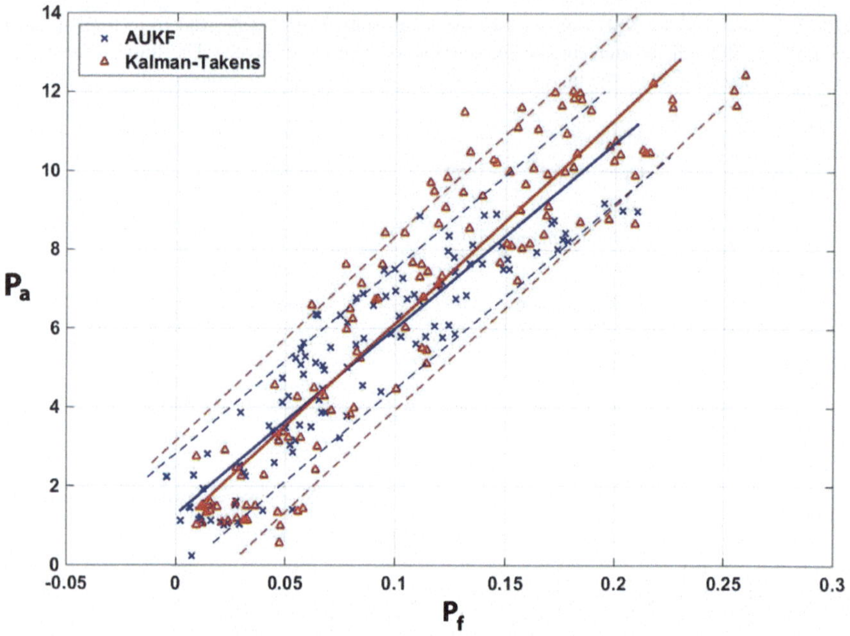

Fig. 9.7 An average estimated covariance matrices of P_f and P_a corresponding to 95% confidence level (dashed lines) at each filtering step using the implemented filters

and observations can be expected. Nevertheless, Fig. 9.6 shows that the local proxy performs comparably to **f** in most of the time. In addition to increments, the difference between the filter's forecasting also affects the estimated error covariances, especially forecast covariance matrix (cf. Fig. 9.7).

P_f and P_a are calculated at assimilation steps for both filters. The average of the matrices' diagonal elements are displayed in Fig. 9.7. Despite the filters different performances in Fig. 9.6, both methods perform very similar in dealing with the error covariances. The distribution of scattered error points from the Kalman-Takens filter and the corresponding trend line largely matches that of AUKF, which demonstrate that the filters have comparable uncertainty estimates. This indicates the ability of the Kalman-Takens method, which not only improves the model states but is also competitive with the traditional data assimilation system.

9.5.3.1 Filters Efficiency

Computational complexity is important for data assimilation methods, especially when dealing with a high dimensional system, such as in hydrological studies. Therefore, a good data assimilation filter requires balancing between processes undertaken to achieve accurate estimates and computational efficiency. While the

Kalman-Takens filter's capability for improving state estimates have already been demonstrated (cf. Sects. 9.5.1 and 9.5.2), its potential for decreasing the computational cost is examined here. This is done by comparing the computation time of the AUKF and Kalman-Takens filtering methods from various perspectives including forecasting, analysis steps, and filtering over the entire study period. Importantly, the following computation time estimates have been obtained using identical hardware. In the forecast step, the average computation time (for 794 grid points within Australia) is considerably lower for the Kalman-Takens filter, e.g., 6.12 s against 8.57 s for AUKF. This is due to the fact that the Kalman-Takens filter exploits the proxy model ($\tilde{f}$), which is based on a local approximation and requires much less computation than a physics-based model. The average computational time at the analysis steps is 5.74 s for the Kalman-Takens filter and 7.83 s for AUKF. Considering that both methods are using similar analysis filtering, this difference is due to the local scheme (based on d delays and N neighbor points) in the Kalman-Takens method. The values of delays d and neighbors N determine the number of local points used in the analysis and accordingly the size of the underlying vectors and matrices. AUKF, on the other hand, solves for all grid points altogether, which requires a larger amount of memory and time. In general, it is found that the Kalman-Takens is considerably less computationally demanding, i.e., $\sim$8 times faster for the entire experiment period, compared to the AUKF implementation for assimilating all observations into the system states.

9.5.3.2 Water Storage Update

In this section, we analyze the spatio-temporal increments derived by assimilating the GRACE TWS observations and explore their effects on the states. Figure 9.8 presents the average TWS time series after applying each filter, open-loop, and GRACE observations over Australia. Both filters largely decrease the misfits between the model states and the GRACE observations, which is expected since GRACE is used as a constraint. AUKF, however, has a larger impact on the states, especially where a significant TWS variation exists (e.g., 2006 and 2011–2012). The Kalman-Takens method, on the other hand, shows a smoother time series. Based on these results, we find that the Kalman-Takens approach is able to efficiently integrate observations into the model and correct missing trends as well as amplitudes and phases. Nevertheless, one can conclude that this method might not be able to efficiently extract spontaneous or high rate seasonal effects unless the training data has these variabilities/dynamics.

The correlation between the estimated TWS time series from the open-loop model run and the filters' estimates at each grid point within Australia and those of the GRACE TWS are presented in Fig. 9.9. The filters largely increase the correlation between model derived TWS and those of GRACE. The largest correlations (with 0.92 average) is obtained by AUKF suggesting that this method better reflects the GRACE TWS into the states. The average correlation between TWS of the Kalman-Takens and GRACE is 0.89 (0.03 less than AUKF), and when compared to only 0.52 obtained from the open-loop estimates, the efficiency of the method becomes visible.

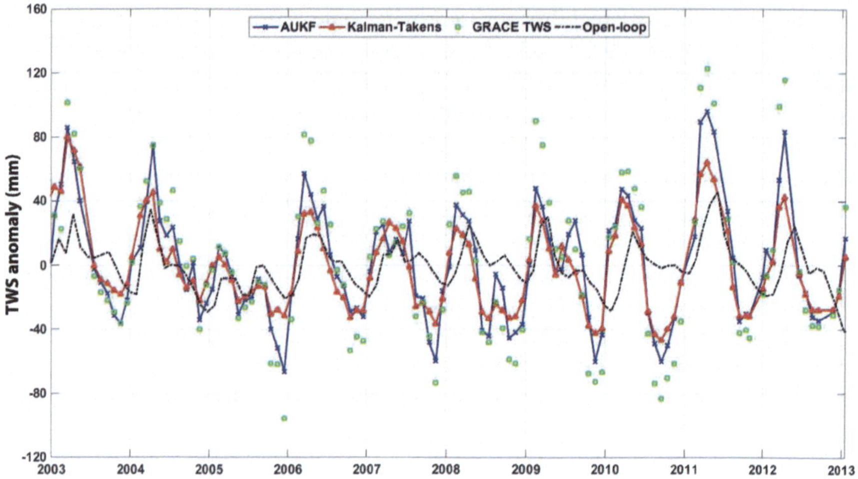

Fig. 9.8 Spatially averaged TWS time series of filters' estimates, GRACE TWS observations, and open-loop run within Australia

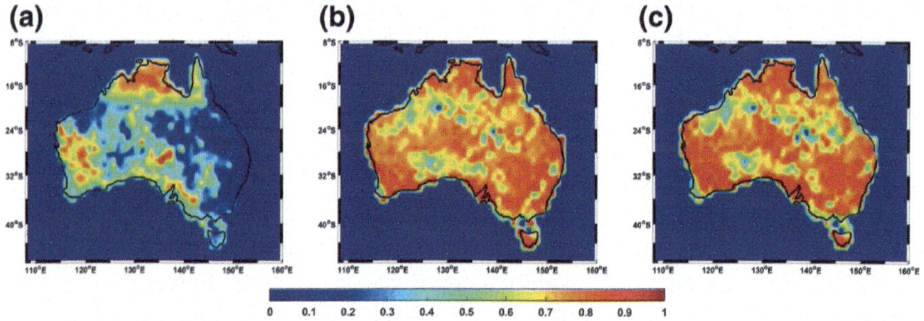

Fig. 9.9 Spatial correlations maps between GRACE TWS and open-loop run (**a**), AUKF estimates (**b**), and the Kalman-Takens filter (**c**)

Correlations of the open-loop TWS and GRACE are smaller over the mountainous area along the East coast compared to other parts of the country. This is due to difficulties of modeling hydrology in complex terrain areas (mountains). On the other hand, both assimilation methods show good performances by increased correlations with GRACE data. Over large parts of Australia, the performances of the Kalman-Takens filter and AUKF are found to be similar in terms of correlations with the GRACE TWS.

To further assess the capability of the filtering approaches for improving the model simulations, we test their ability in correcting the model variables for extreme and poorly known hydrological phenomena. To this end, the filters' TWS results are monitored between 2003 and 2012 over the Murray-Darling Basin. As shown by Schumacher et al. (2018), a long-term drought period (2001–2009), known as Millennium Drought (e.g., Ummenhofer et al. 2009; LeBlanc et al. 2012;

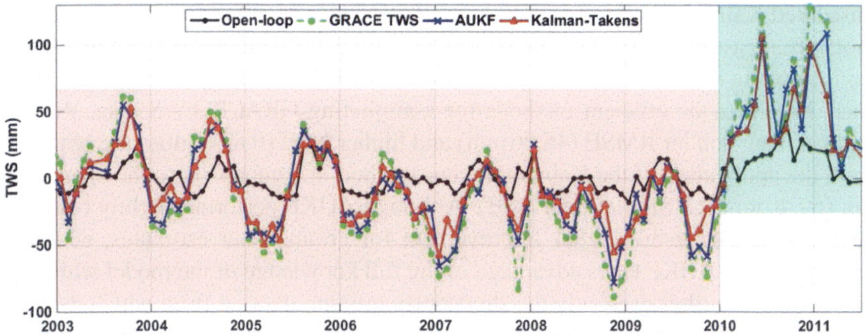

Fig. 9.10 Average TWS variations from the data assimilation filters, open-loop run, and GRACE TWS. The red shaded area shows the Millennium Drought and blue shaded area represent a strong ENSO effect

van Dijk et al. 2013), has remarkably affected TWS variations in the basin. This negative TWS trend has then been followed by an above average precipitation, mainly caused by El Niño Southern Oscillation (ENSO; see, e.g., Boening et al. 2012; Forootan et al. 2016) for the period of 2010–2012. Here, we investigate the capability of the open-loop, AUKF, and Kalman-Takens TWS estimates to capture these two extreme events. Figure 9.10 plots the average TWS time series of the above methods, as well as GRACE-derived TWS over the Murray-Darling Basin. As can be seen, while both Millennium drought (red shaded area) and ENSO effect (blue shaded area) are reflected in GRACE TWS time series, the open-loop run is unable to capture them, especially the drought effects. AUKF and the Kalman-Takens filter, on the other hand, successfully depict the negative trend between 2003 and 2010, followed by a positive anomaly after 2010. Except for few points such as 2004, 2007, and late 2009, the Kalman-Takens method presents a similar performance as AUKF in incorporating GRACE TWS data with states and reflecting extreme hydrological events.

9.6 Conclusions

The present study investigates the ability of the Kalman-Takens approach to reconstruct the nonlinear dynamics of a hydrological model. This is done to update observable state variables based on new observations when a physics-based model is not available. This implies that contrary to a standard data assimilation, the Kalman-Takens filter does not affect non-observed variables (e.g., water discharge in our case). In this work, we introduce a new setup for the Kalman-Takens filter to reconstruct additional states (e.g., soil moisture and groundwater) using the Gravity Recovery And Climate Experiment (GRACE) terrestrial water storage (TWS). The Kalman-Takens results are compared with a parametric forecasting approach of an adaptive unscented Kalman filtering (AUKF) as well as against in-situ groundwater and soil moisture measurements. The results prove a high capability of the Kalman-Takens for improving state estimates, largely comparable to the AUKF performance and as such, both provide efficient methods for assimilating GRACE TWS data. Results indicate that smaller RMSE (46.96 mm) and higher NSE (0.82) values are obtained from the application of the Kalman-Takens method in comparison to the open-loop run (69.40 mm RMSE and 0.58 NSE). Although AUKF performs slightly better in some cases, e.g., $\sim$3% higher improvement for groundwater estimates, which is expected since AUKF takes advantage of the full knowledge of the model while the non-parametric filter uses only the short noisy training data set from which to learn the dynamics, in all cases considered, the Kalman-Takens results are generally very close to those of AUKF. The data-driven approach also increases the NSE values between the estimated soil moisture variations and the OzNet in-situ measurements for all soil layers (11.83% on average) as compared to AUKF (13.77% on average). The proposed approach also reduces estimation complexities by using the local proxy model. The Kalman-Takens filter performs more efficient ($\sim$8 times faster) in terms of computational cost, which is very important to deal with a growing amount of data sets in high dimensional systems. This contribution, to the best of the authors' knowledge, is the first effort in using the data-driven approach in hydrological studies with complex state observation transition systems. Further research should be undertaken to investigate the Kalman-Takens filter in different hydrological applications and also to explore its capability in dealing with multiple satellite products.

Acknowledgements We would like to thank Tyrus Berry and Timothy Sauer for their valuable help in this study. M. Khaki is grateful for the research grant of Curtin International Postgraduate Research Scholarships (CIPRS)/ORD Scholarship provided by Curtin University (Australia). F. Hamilton is supported by National Science Foundation grant No. RTG/DMS-1246991. This work is a TIGeR publication. The GRACE data are acquired from the ITSG-Grace2014 gravity field model (Mayer-Gürr et al. 2014). In-situ groundwater and soil moisture measurements are obtained from the New South Wales Government (NSW; http://waterinfo.nsw.gov.au/pinneena/gw.shtml) and the OzNet network (http://www.oznet.org.au/), respectively. Meteorological forcing data are provided by Princeton University (http://hydrology.princeton.edu). Other data used in this study can be found at https://doi.org/10.6084/m9.figshare.5942548. A more detailed discussion of the results can be found in the supporting information (Huffman et al. 2007; Mu et al. 2011).

Chapter 10
Parametric and Non-parametric Data Assimilation Frameworks

Assimilation of multi-mission satellite products can be achieved using model-based and data-driven techniques. Kalman-Takens can perform better (than traditional data assimilation) for the cases the model is subject to error. In general, the performances of the data-driven Kalman-Takens approach are comparable to those of the standard SQRA.

—M. Khaki (This chapter is presented following Khaki et al. (2018g), "Assessing data assimilation frameworks for using multi-mission satellite products in a hydrological context")

10.1 Summary

With a growing number of available datasets especially from satellite remote sensing, there is a great opportunity to improve our knowledge of the state of the hydrological processes via data assimilation. Observations can be assimilated into numerical models using dynamics and data-driven approaches. The present study aims to assess these assimilation frameworks for integrating different sets of satellite measurements in a hydrological context. To this end, we implement a traditional data assimilation system based on the Square Root Analysis (SQRA) filtering scheme and the newly developed data-driven Kalman-Takens technique to update the water components of a hydrological model with the Gravity Recovery And Climate Experiment (GRACE) terrestrial water storage (TWS), and soil moisture products from the Advanced Microwave Scanning Radiometer - Earth Observing System (AMSR-E) and Soil Moisture and Ocean Salinity (SMOS) in a 5-day temporal scale. While SQRA relies on a physical model for forecasting, the Kalman-Takens only requires a trajectory of the system based on past data. We are particularly interested in testing both methods for assimilating different combination of the satellite data. In most of the cases, simultaneous assimilation of the satellite data by either standard SQRA or Kalman-Takens achieves the largest improvements in the hydrological state, in terms of the agreement with independent in-situ measurements. Furthermore, the Kalman-Takens approach performs comparably well to dynamical method at a fraction of the computational cost.

M. Khaki, *Satellite Remote Sensing in Hydrological Data Assimilation*,
https://doi.org/10.1007/978-3-030-37375-7_10

10.2 Introduction

The study of terrestrial water storage (TWS) and different water compartments, such as soil moisture, groundwater, and surface water storage, is essential because of their roles in the environment, hydroclimate impacts, and human life as a major fresh water resource. In this regard, hydrological models provide a unique opportunity to enhance our understandings of hydrological processes within land areas. The models have been used to analyze the spatiotemporal variations of hydrological components (e.g., Coumou and Rahmstorf 2012; Döll et al. 2003; Huntington 2006; van Dijk et al. 2013; Wooldridge and Kalma 2001). Nevertheless, there are factors such as inaccurate inputs and forcing fields, data deficiencies (e.g., limited ground-based observations), and imperfect modeling that impose a degree of uncertainties in models' simulations (van Dijk et al. 2011; Vrugt et al. 2013). High resolution (spatially and temporally) satellite remotely sensed observations of different water compartments can be assimilated to improve models performances (Khaki et al. 2017b; Schumacher et al. 2016). Accordingly, various approaches have been put forward to efficient incorporation of observations into the models (e.g., Bishop et al. 2001; Dreano et al. 2015; Evensen 2004; Kalnay 2003; Sauer 2004; Tippett et al. 2003).

Data assimilation provides a framework to integrate models simulations with new observations. When a physics-based model is available, data assimilation techniques constrain the model state with available observations in order to bring its outputs closer to the data according to their uncertainties (Bertino et al. 2003; Hoteit et al. 2012). This approach has been widely implemented in hydrological studies (e.g., Giustarini et al. 2011; Khaki et al. 2018b,c; Neal et al. 2009; Reichle et al. 2002; Seo et al. 2003; Tangdamrongsub et al. 2018; Vrugt et al. 2005; Weerts et al. 2006). In other cases, where the physical processes of the studied system are not available or perfectly understood, data-driven (or non-parametric) approaches may provide reliable alternatives (e.g., Dreano et al. 2015; Hamilton et al. 2016; Lguensat et al. 2017; Sauer 2004; Tandeo et al. 2015). Both dynamical and data-driven modeling approaches have their own advantageous and disadvantageous. Traditionally, data assimilation systems were implemented based on a physical model, which can lead to a better redistribution of increments between state variables but generally requires intensive computations in realistic applications (Tandeo et al. 2015). A data-driven model, on the other hand, only relies on data and their associated errors with no or limited knowledge of physical processes but computationally can be significantly less demanding.

The main aim of this contribution is to assess the performance of these frameworks for assimilating different combinations of multiple satellite remote sensing products within a hydrological context. For this purpose, we use an ensemble-based sequential technique, the Square Root Analysis (SQRA) filtering scheme (Evensen 2004) from dynamical, a modified version of the recently developed data-driven approach, Kalman-Takens filter (Hamilton et al. 2016) from the data-driven approach. Khaki et al. (2017b) recently studied the performance of various standard data assimilation schemes and showed that SQRA is highly capable of assimilating TWS data into a hydrological model (see also Schumacher et al. 2016). The method has also been found to outperform other existing filters, e.g., addressing the sampling error in covariance matrix, especially for the small-size ensembles and an efficient resampling process (see, e.g., Hoteit et al. 2015; Khaki et al. 2017b; Nerger 2004; Whitaker and Hamill 2002).

In addition to SQRA filter, a modified version of the recently developed data-driven approach, Kalman-Takens filter (Hamilton et al. 2016), is applied. Takens method has been used in various studies for non-parametric time series predictions (see, e.g., Packard et al. 1980; Sauer 2004; Sauer et al. 1991; Takens 1981). Hamilton et al. (2016) used this method and developed a new model-free filter for data assimilation when the physical model is not available. The Kalman-Takens method relies only on observations and a trajectory of the model to build a data-driven surrogate of the model dynamics, which is required to forecast the system state at a fraction of the computational time. The idea of using the model trajectory has also been used in Tandeo et al. (2015) and Lguensat et al. (2017) to simulate the dynamics of complex systems. All these studies have shown that the data-driven approach can perform well, sometimes comparable to a standard data assimilation.

Here, for the first time, the application of SQRA and Kalman-Takens are investigated for assimilating various observation sets including terrestrial water storage (TWS) derived from the Gravity Recovery And Climate Experiment (GRACE), soil moisture products from the Advanced Microwave Scanning Radiometer - Earth Observing System (AMSR-E) and Soil Moisture and Ocean Salinity (SMOS) into a hydrological model, and their combination. Several studies suggest that assimilating these products can successfully constrain the mass balance of hydrological models (e.g., Eicker et al. 2014; Khaki et al. 2017b, c; Reager et al. 2015; Thomas et al. 2014; Zaitchik et al. 2008). In addition, various studies validate AMSR-E (e.g.,Draper et al. 2009; Mladenova et al. 2011) and SMOS (e.g., Jing et al. 2018; Peischl et al. 2012) soil moisture products globally and suggest their capability for accurately reflecting soil moisture variations of the surface top soil layer. It has also been shown by Renzullo et al. (2014) and Tian et al. (2017) that assimilation of these products can effectively improve model soil moisture simulations over Australia. Different scenarios are tested here to achieve the best estimates of the water storage components. This involves using SQRA and the Kalman-Takens filters for integrating TWS and soil moisture observations separately and simultaneously and comparing their impact on different water compartments. Two different domains of Murray-Darling and Mississippi basins are selected for testing subject to the availability of in-situ measurements for evaluation of the results.

10.3 Materials

10.3.1 Data Assimilation (Forecast Step)

10.3.1.1 W3RA

Here we use $1^\circ \times 1^\circ$ grid-distributed biophysical model of the World-Wide Water Resources Assessment (W3RA) for the period from January 2003 to December 2012. W3RA is based on the Australian Water Resources Assessment system (AWRA) model, which is provided by the Commonwealth Scientific and Industrial Research Organisation (CSIRO) to monitor, represent and forecast Australian terrestrial water cycles (http://www.wenfo.org/wald/data-software/). Forcing fields of minimum and maximum temperature, downwelling short-wave radiation, and precipitation from Princeton University are used in this study (Sheffield et al. 2006, http://hydrology.princeton.edu). These $1^\circ \times 1^\circ$ daily forcing sets are only used to run the model. It is important to note that the inclusion of rain-gauge based product (e.g., Global Precipitation Climatology Centre; GPCC, Schneider et al. 2008) can potentially improve the model's performance (see, e.g., Massari et al. 2018). This study, however, uses Princeton Global Meteorological Forcing dataset (Sheffield et al. 2006, PRIN;), which has been shown, e.g., by Awange et al. (2016) to exhibit similar patterns of correlation to GPCC in some regions. The model parameters include effective soil parameters, water holding capacity and soil evaporation, relating greenness and groundwater recession, and saturated area to catchment characteristics (van Dijk et al. 2013). The water balance of a top soil, shallow soil and deep soil compartment, as well as groundwater and surface water dynamics, are simulated at grid resolution, in which each cell is modeled independently of its neighbors (van Dijk 2010). Water (e.g., from rainfall) redistribute between these compartments starting from the top soil layer and entering the deep root layer through capillary rise from groundwater. It then leaves the column through soil evaporation, extraction by shallow- and deep-rooted vegetation, or drainage into the groundwater store (Renzullo et al. 2014). Model state in the present study includes the W3RA water storages in the top, shallow and deep soil moisture, vegetation, snow, surface, and groundwater. Figure 10.1 summarizes the data and methodology (as it will discussed in Sect. 10.4) used in this study.

10.3.2 Data Assimilation (Analysis Step)

10.3.2.1 GRACE TWS

The GRACE spherical harmonic coefficients with their full error information are acquired from the ITSG-Grace2014 gravity field model (Mayer-Gürr et al. 2014). Here, we used Stokes' coefficients up to degree and order 90 (approximate spatial resolution of $\sim$300 by 300 km at the equator) covering 2003 to 2013. The following

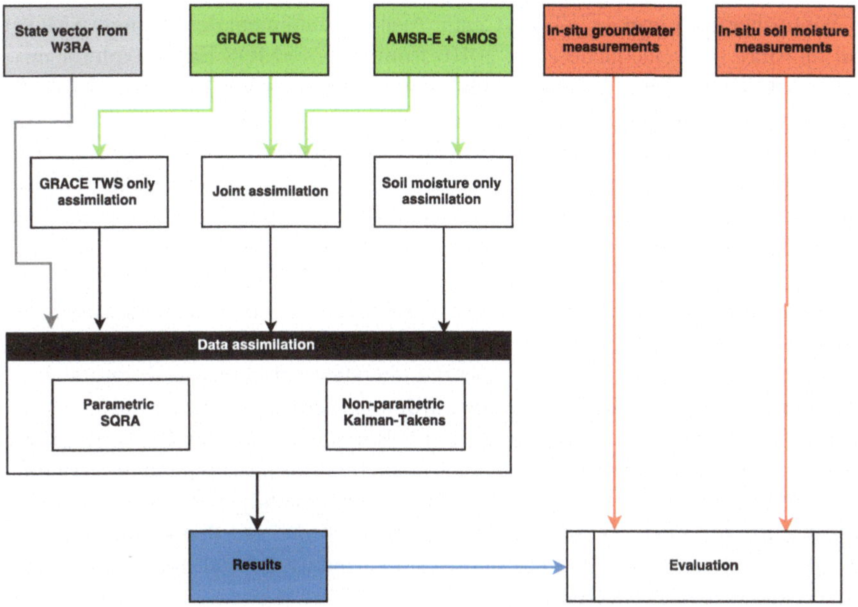

Fig. 10.1 A schematic illustration of the implemented data assimilation frameworks and data used

steps have been taken before converting the spherical harmonics to TWS. Degrees 1 and 2 are replaced with improved estimates since the GRACE-estimates are not very reliable (Cheng and Tapley 2004; Swenson et al. 2008). The L2 gravity fields are then converted into 5-day $3° \times 3°$ TWS fields (suggested by Khaki et al. 2017a, for data assimilation purposes) following Wahr et al. (1998). Note that colored/correlated noise in products is reduced by the Kernel Fourier Integration (KeFIn) filter proposed by Khaki et al. (2018a), which also accounts for signal attenuations and leakage effects caused by smoothing. The KeFIn filter works through a two-step post-processing algorithm. The first step mitigates the measurement noise and the aliasing of unmodelled high-frequency mass variations, and the second step contains an efficient kernel to decrease the leakage errors (see also Khaki et al. 2018a).

10.3.2.2 Soil Moisture

We use AMSR-E to derive soil moisture products. AMSR-E measures surface brightness temperature at twelve channels. This is highly correlated to surface soil moisture content (0–2 cm depth) and has been used to produce global data products of surface soil moisture content using satellite-based radiometer instruments (Njoku et al. 2003). Daily measurements of surface soil moisture from descending passes (see, e.g., De Jeu and Owe 2003; Su et al. 2013) with a spatial resolution of $0.25° \times 0.25°$ covering the period between 2003 and 2011 from the gridded Level-3 land surface product (Njoku 2004) are rescaled to a 5-day $1° \times 1°$ for the present study.

We further use Level 3 CATDS (Centre Aval de Traitement des Donnees SMOS) soil moisture data (Jacquette et al. 2010) from ESA's SMOS Earth Explorer mission. SMOS Microwave Imaging Radiometer using Aperture Synthesis (MIRAS) radiometer measures microwave emissions from Earth's surface to map land soil moisture ($\sim$0–5 cm depth). Here we use ascending passes of the satellite subject to their higher agreement to in-situ measurements (see, e.g., Draper et al. 2009; Jackson and Bindlish 2012). The soil moisture data temporal and spatial resolutions are three days and about 50 km, respectively. Similar to AMSR-E, SMOS data are rescaled to a 5-day (2011–2013) $1^\circ \times 1^\circ$ scale.

An important step is required to prepare soil moisture products for data assimilation and to remove the bias between the model simulations and observations. These measurements are mainly used to constrain the state variability, and not its absolute values. Several studies have applied different methods to rescale soil moisture measurements (see, e.g., Kumar et al. 2012; Reichle and Koster 2004). Here, we use cumulative distribution function (CDF) matching for rescaling the observations (Drusch et al. 2005; Reichle and Koster 2004). CDF matching relies on the assumption that the difference between observed soil moisture and that of the model is stationary and guarantees that the statistical distribution of both time series is the same (Draper et al. 2009; Renzullo et al. 2014).

10.3.3 Validation Dataset

10.3.3.1 Water Fluxes

For the sake of result assessment, water flux observations are also acquired. These include monthly precipitation data from TRMM-3B43 products in $0.25^\circ \times 0.25^\circ$ spatial resolution (Huffman et al. 2007; TRMM 2011), MOD16 evaporation data with eight days temporal resolution and one km spatial resolution from the University of Montana's Numerical Terradynamic Simulation group (Mu et al. 2011), and water discharge data from the Global Runoff Data Centre (GRDC) and United States Geological Survey (USGS), and the Australian Bureau of Meteorology under the Water Regulations (2008). All these products are rescaled into $0.25^\circ \times 0.25^\circ$ and to the same temporal resolution of data assimilation observations.

10.3.3.2 In-Situ Data

In-situ groundwater and soil moisture measurements are used to examine the results. Groundwater measurements are acquired from USGS for the Mississippi Basin and from New South Wales Government (NSW) for the Murray-Darling Basin. Specific yields are required to convert well-water levels to groundwater storage variations, which are unknown. Thus, following Strassberg et al. (2007), we use an average (0.15) of specific yields range from 0.1 to 0.3 as suggested by Gutentag et al. (1984)

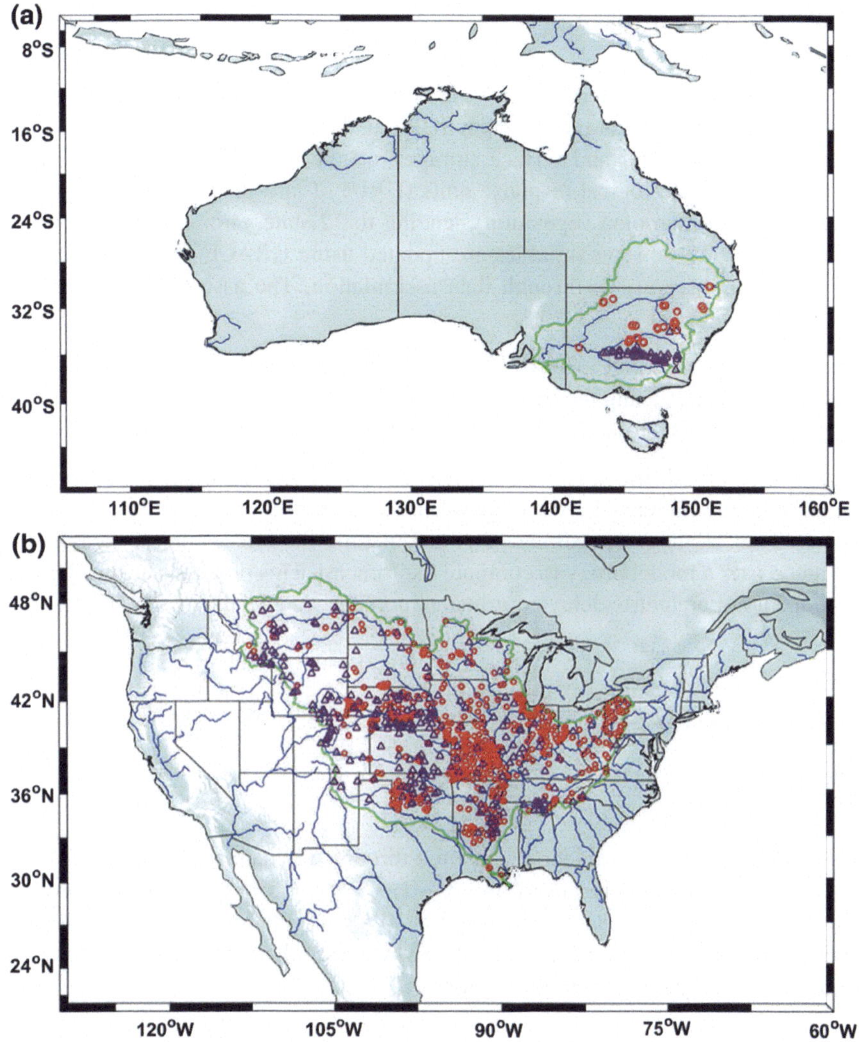

Fig. 10.2 Locations of Murray-Darling (**a**) and Mississippi (**b**) basins. A distribution of groundwater (circle) and soil moisture (triangle) in-situ stations are also displayed

over the Mississippi basin, and 0.13 specific yield from the range between 0.115 and 0.2 as suggested by the Australian Bureau of Meteorology (BOM) and Seoane et al. (2013) for the Murray-Darling basin. In-situ soil moisture data are obtained from the International Soil Moisture Network and the moisture-monitoring network over the Mississippi and Murray-Darling basins, respectively. The distribution of gauge stations over the study areas is presented in Fig. 10.2.

10.4 Data Assimilation

The model state includes top, shallow and deep soil moisture, vegetation, snow, surface, and groundwater storages for both SQRA and the Kalman-Takens approaches. Except for groundwater and surface storages, all the other components are simulated with two hydrological response units (HRU) of tall (deep-rooted vegetation) and short (shallow-rooted vegetation), leading to 12 state variables ($5 \times 2 + 2$) at each grid cell. These state variables are updated using GRACE TWS and satellite soil moisture observations through data assimilation. The assimilation procedure includes two step,

- *Forecast step.* $\mathbf{x}_{t-1}$ and its error covariance evolve through the time (t), the next assimilation step, using the dynamical model ($\mathcal{M}$).
- *Analysis step.* The forecast state ($\mathbf{x}_t^f$) is updated by the observation $\mathbf{y}_t$.

Here, both selected filters, i.e., SQRA and Kalman-Takens, use the same analysis step. The main difference between the two methods is that while a dynamics-driven model advances the state estimate forward in time for forecasting, a data-driven technique uses a model proxy to compute the forecast. This process can be achieved using the non-parametric delay-coordinate approach (see details in Sect. 10.4.1.2).

10.4.1 Forecast Step

10.4.1.1 SQRA

The model state ($\mathbf{x}_{t-1}$) is integrated in time through a dynamical model (Eq. 10.1). Given an ensemble member from time $t-1$, $\mathbf{x} = \{\mathbf{x_i}, \quad \mathbf{i} = \mathbf{1}, \ldots, \mathbf{n}\}$ the forecast state in SQRA can be calculated using Eq. 10.1. Observations at the assimilation time (t) are represented by $\{\mathbf{y}_t\}_{t=0}^{T} \in \mathbb{R}^{n_y}$, which are related to the state through a dynamical state-space system of the form,

$$\begin{cases} \mathbf{x_t}^f = \mathcal{M}_{t-1}(\mathbf{x}_{t-1}) + \boldsymbol{\nu}_t, & (10.1) \\ \mathbf{y}_t = \mathbf{H}_t \mathbf{x_t}^f + \mathbf{w}_t, & (10.2) \end{cases}$$

where 'f' stands for forecast (and 'a' in the following for analysis) and $\mathbf{x}^f$ refers to the forecast state at time t. $\mathcal{M}(.)$ is the model operator with the noise processes of $\boldsymbol{\nu} = \{\boldsymbol{\nu}_t\}_t$ drawn from $N(0, \mathbf{Q})$ with covariance matrix $\mathbf{Q}$, and $\mathbf{H}$ is the design matrix with $\mathbf{w} = \{\mathbf{w}_t\}_t$ drawn from $N(0, \mathbf{R})$ with covariance matrix R. Accordingly, the forecast state ($\mathbf{P}_t^f$) in SQRA can be derived by,

$$\mathbf{P}_t^f = \mathcal{M}_{t-1}\mathbf{P}_{t-1}\mathcal{M}_{t-1}^T + \mathbf{Q}_t. \quad (10.3)$$

10.4.1.2 The Kalman-Takens

The Kalman-Takens filter, initially proposed by Hamilton et al. (2016), is applied here after a few modification. As mentioned, the main different between this filter and SQRA is forecasting step while both methods use similar analysis scheme. The Kalman-Takens filter replaces model equations $\mathcal{M}$ with a local proxy $\tilde{\mathbf{f}}$ based on data. The method considers delay-coordinate vector (to replace the dynamical model for advancing the state forward in time. This delay-coordinate can be built using $[\mathbf{x^o}_t, \mathbf{x^o}_{t-1}, \ldots, \mathbf{x^o}_{t-d}]$, where $\mathbf{x^o}$ is the training data for reconstructing the system and d indicates the number of temporal delays.

In the original form of the method, it relies on observable $\mathbf{y}_t$ to create the delay-coordinate vector. Here, instead, we use a model trajectory to create the delay-coordinate vector. This is motivated by the fact that we are interested in updating the different water storage components while GRACE produces the summation of these compartments. We, therefore, assume that a trajectory generated by the model is readily available. In the present study the water storage components from W3RA, i.e., the open-loop top, shallow and deep soil moisture, vegetation, snow, surface, and groundwater are used to create the delay-coordinate vector.

Using the N nearest neighbors within a set of training data based on a given Euclidean distance, the delay-coordinate vectors at $t+1$, $\mathbf{x^o}^1_{t+1}, \mathbf{x^o}^2_{t+1}, \ldots, \mathbf{x^o}^N_{t+1}$, can be used to construct the local model for predicting $\mathbf{x}_{t+1}$. To this end, a locally constant model following Hamilton et al. (2016) is used (see also Hamilton et al. 2017). This model in its most basic form can be assumed as an average of the nearest neighbors, e.g.,

$$\tilde{f}(\mathbf{x}_t) = \left[\frac{\mathbf{x^o}^1_{t+1}, \mathbf{x^o}^2_{t+1}, \ldots, \mathbf{x^o}^N_{t+1}}{N}, \mathbf{x^o}_t, \ldots, \mathbf{x^o}_{t-d+1} \right]. \tag{10.4}$$

Once the local proxy $\tilde{f}$ is generated, the forecasting step can be carried out to estimate $\mathbf{x}^f$. Afterwards, the analysis step of SQRA is applied to reach $\mathbf{x}^a$ (see Sect. 10.4.2). Note that different values for the number of neighbors N and delays d were considered and their results are compared against in-situ measurement.

10.4.2 Analysis Step

The analysis step in SQRA involves first updating the forecast ensemble-mean ($\bar{\mathbf{x}}^f = \frac{1}{N}\sum_{i=1}^{N} \mathbf{x_i}^f$) as,

$$\bar{\mathbf{x}}^a = \bar{\mathbf{x}}^f + \mathbf{K}(\mathbf{y} - \mathbf{H}\bar{\mathbf{x}}^f), \tag{10.5}$$

with Kalman Gain (**K**)

$$\mathbf{K} = \mathbf{P}^f(\mathbf{H})^T(\mathbf{H}\mathbf{P}^f(\mathbf{H})^T + \mathbf{R})^{-1}, \tag{10.6}$$

and

$$\mathbf{P}^f = \frac{1}{N-1}\sum_{i=1}^{N}(\mathbf{x_i}^f - \bar{\mathbf{x}}^f)(\mathbf{x_i}^f - \bar{\mathbf{x}}^f)^T, \tag{10.7}$$

where $\bar{\mathbf{x}}^a$ is the analysis state, and the error covariance associated with observations ($\mathbf{y}$) is denoted by **R**. For each satellite observation set, a different **R** is used. Full error information of the L2 potential coefficients for each month are provided for GRACE data (cf. Sect. 10.3.2.1). These products are then converted from the GRACE coefficients to TWS errors following Schumacher et al. (2016). Regarding soil moisture observations, various error values are tested to monitor their impacts on data assimilation by comparing the results with independent measurements (cf. Sect. 10.3.3.2). This allows us to obtain optimum error values for soil moisture part of observation error covariance. Accordingly, **R** is assumed to be diagonal with an error standard deviation of 0.04 ($m^3\ m^{-3}$) for SMOS (suggested by Leroux et al. 2016) and 0.05 ($m^3\ m^{-3}$) for AMSR-E (suggested by De Jeu et al. 2008). It is worth mentioning that using spatially varying error information rather than the constant error assumption can lead to different results and potentially optimal data assimilation performance. However, providing such an information is difficult and out of the scope of our study, so we use the estimated optimal constant errors. Furthermore, we assume that GRACE data are uncorrelated from both SMOS and AMSR-E observations. An ensemble of anomalies, representing the deviation of the analysis ensemble members from the ensemble mean ($\bar{\mathbf{x}}^a$) is then sampled by,

$$\mathbf{A}^a = \mathbf{A}^f\mathbf{V}\sqrt{\mathbf{I} - \boldsymbol{\Sigma}^T\boldsymbol{\Sigma}}\boldsymbol{\Theta}^T, \tag{10.8}$$

where $\mathbf{A}^f = [\mathbf{A}_1{}^f \ldots \mathbf{A}_N{}^f]$ is the ensemble of forecast anomalies ($\mathbf{A}_i{}^f = \mathbf{x_i}^f - \bar{\mathbf{x}}^f$), $\boldsymbol{\Sigma}$ and $\mathbf{V}$ are obtained from the singular value decomposition (SVD) of $\mathbf{A}^f$ ($\mathbf{A}^f = \mathbf{U}\boldsymbol{\Sigma}\mathbf{V}^T$), and $\boldsymbol{\Theta}$ is a random orthogonal matrix for redistributing the ensemble variance (Evensen 2007; Hoteit et al. 2002). These perturbations are then added to the analysis state to form a new ensemble by,

$$\mathbf{x_i}^a = \bar{\mathbf{x}}^a + \mathbf{A}_i{}^a, \quad i = 1, \cdots, N. \tag{10.9}$$

$\mathbf{x_i}^a$ is then used to start the next forecasting cycle by integrating the $\mathbf{x_i}^a$ with the dynamical model to compute the next $\mathbf{x_i}^f$ (cf. Evensen 2004, 2007; Khaki et al. 2017b).

10.4.3 Filter Implementation

In order to generate the initial ensemble, we perturb the forcing fields according to their error characteristics. This is done using a Gaussian multiplicative error of 30% for precipitation, an additive Gaussian error of 50 Wm^{-2} for the shortwave radiation, and a Gaussian additive error of 2 °C for temperature (Jones et al. 2007; Renzullo et al. 2014). Monte Carlo sampling of multivariate normal distributions with errors representing the standard deviations of the forcing sets are then used to produce the ensemble (see details in Khaki et al. 2017b; Renzullo et al. 2014). The produced ensemble of perturbations of 72 members (suggested by Khaki et al. 2017b) are then integrated with model between 2000 and 2003 to generate an ensemble at the beginning of the assimilation period.

To mitigate for the standard issues related to the rank deficiency and the underestimation of the error covariance matrix of ensemble-based Kalman filters, which are due to the limited number of ensemble members and ensemble spread collapse (Anderson 2001; Houtekamer and Mitchell 2001), ensemble inflation and the Local Analysis (LA) scheme (Evensen 2003; Ott et al. 2004) applied. The ensemble inflation method uses a small coefficient to separately inflate prior ensemble deviation from the ensemble-mean and increases their variations (Anderson et al. 2007). Here, we use a constant factor ($S = 1.12$; Anderson 2001; Khaki et al. 2017b) to inflate the ensemble perturbations as,

$$\mathbf{x}'^f = S(\mathbf{x}^f - \bar{\mathbf{x}}^f) + \bar{\mathbf{x}}^f, \tag{10.10}$$

with X'^f representing the new forecast state, which contains the inflated ensemble perturbation. LA, on the other hand, spatially limits the impact of given measurements in the analysis step to the points located within a certain distance. In using LA, at each horizontal grid point (m, n), with m and n representing geographic latitude and longitude directions, respectively, the selected measurements close to the grid point contribute to the SQRA filtering process. This means that only particular state variables close to the point (m, n) within an assumed distance and corresponding observations at the same locations are used in the assimilation process (see details in Khaki et al. 2017b).

Figure 10.1 presents a summary of the data integration framework for the dynamics- and data-driven approaches. Different experimental scenarios in terms of methodology and assimilated observations are examined. Table 10.1 outlines the conducted experiments, indicating, in particular, the assimilated observations types and the model used for each case.

Table 10.1 A summary of the applied data assimilation scenarios. Note that all water storages includes top soil, shallow soil, deep soil water, snow, vegetation, surface, and groundwater storages

Assimilation case	Filtering technique	Observation type	State vector	Updated states
Case 1	SQRA	GRACE TWS	All water storages	Storages summation
Case 2	SQRA	AMSR-E + SMOS	Only soil storages (top, shallow, deep)	Scaling top soil layer (by field capacity value)
Case 3	SQRA	Joint observations	All water storages	Storages summation by observed TWS + Scaling top soil layer by observed soil measurements
Case 4	Kalman-Takens	GRACE TWS	All water storages	Storages summation
Case 5	Kalman-Takens	AMSR-E + SMOS	Only soil storages (top, shallow, deep)	Scaling top soil layer (by field capacity value)
Case 6	Kalman-Takens	Joint observations	All water storages	Storages summation by observed TWS + Scaling top soil layer by observed soil measurements

10.5 Results

In this section, we first analyze the results of different data assimilation methods and scenarios on the forecast estimates. This allows examining how each case incorporates different observations and how these effects are reflected in forecast state variables. Note that this is not a result validation process and the purpose of this analysis is to show the capability of different scenarios for forecast improving based on assimilated observation. We later evaluate the final results by comparing them against the reference fields. It is also worth mentioning that different scenarios are considered regarding the number of neighbors N (i.e., 2–40) and also the number of delays d (i.e., 1–25) in the Kalman-Takens implementation. It is found that increasing the number of neighbors can improve the approximation of training data for a particular point to a certain extent (due to the existing spatial correlations). However, selecting N too large can cause a rapid growth of errors, which is related to the effect of over-smoothing the training step. This is different for delays d, where much larger

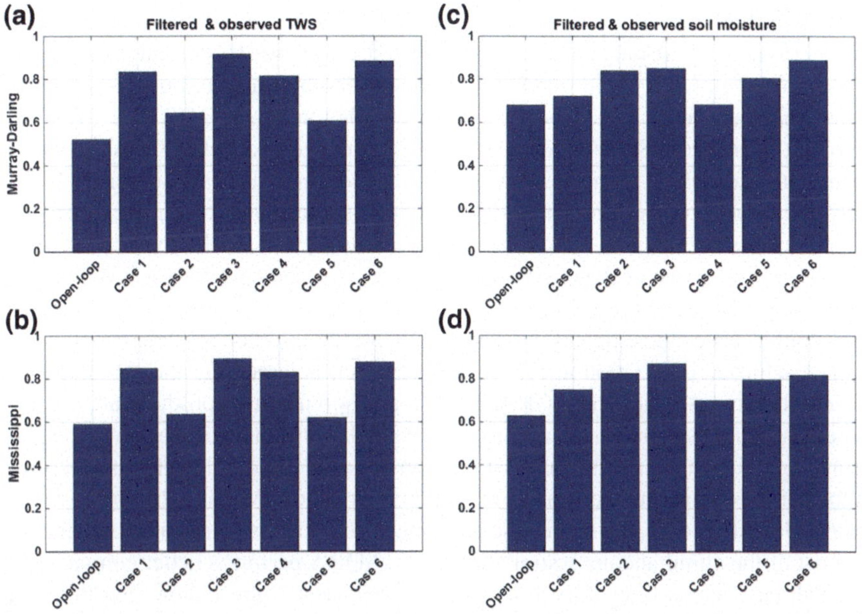

Fig. 10.3 Average correlations between observable variables and assimilated data sets for each case and open-loop at forecast steps. (**a**) and (**b**) indicate the correlations between estimated and observed TWS over Murray-Darling and Mississippi basins, respectively. The correlations between estimated top layer soil moisture and observations (SMOS+AMSR-E) are displayed in (**c**) for Murray-Darling basin and (**d**) for Mississippi basin

errors are present for smaller values that underestimate temporal variabilities in the data. Accordingly, we set $N = 14$ and $d = 11$ as they lead to the best assimilation performances for the entire experiment (see more details in Khaki et al. 2018f).

Figure 10.3a, b plot correlations between the estimated TWS by each filtering method and GRACE TWS over Murray-Darling and Mississippi basins, respectively. Correlations between the filters estimates and observed soil moistures (from satellites) are also depicted respectively in Fig. 10.3c, d for the Murray-Darling and Mississippi basins. Note that the correlation values are calculated for all grid points within the basins (at 95% confidence interval) and their averages at forecast steps for each case is presented in Fig. 10.3.

The minimum correlation values are found for the open-loop run while all the other cases demonstrate higher correlations. Comparable performances are achieved by SQRA and Kalman-Takens methods. This is clear from the close correlations for cases 1 and 4, cases 2 and 5, and cases 3 and 6, regardless of whether GRACE TWS only, soil moisture measurements only, or both of them are assimilated. Based on Fig. 10.3, one can see that both SQRA and Kalman-Takens that assimilate GRACE TWS and satellite soil moisture data simultaneously, i.e., case 3 and case 6, exhibit the highest correlations over the Murray-Darling and Mississippi basins. This can be seen for both sets of observations, i.e., GRACE TWS and soil moisture measurements. In cases where only one data is assimilated, e.g., cases 1, 2, 4, and 5, the largest correlation is generally achieved between the observables and assimilated observations. For example, as it is expected, a larger correlation between GRACE TWS and TWS estimates from SQRA and Kalman-Takens are achieved when GRACE data is assimilated compared to the cases when satellite soil moisture is assimilated. Similarly, the correlation between the estimated and observed soil moisture fields are the largest for cases 2 and 5 over both basins. Interestingly, the results show that assimilating even only one of the observation data sets, e.g., either GRACE TWS or soil moisture products, can also improve the correlations for non-observable variables. This demonstrates the efficient impacts of data assimilation on all state variables.

The achieved correlation improvement, however, is largest for the simultaneous assimilation cases, where both GRACE TWS and soil moisture products are assimilated. This suggests that simultaneous data assimilation can lead to better forecasts. From Fig. 10.3, the simultaneous assimilation in cases 3 and 6, lead to larger correlations between the filters estimates of soil moisture and TWS, and the observations over both basins compared to the case only one observation is assimilated. In general, in most of the simultaneous assimilation cases, SQRA performs better compared to the Kalman-Takens filter. Nevertheless, the correlation values show that this is a marginal superiority for TWS correlations while in soil moisture correlation over the Murray-Darling the Kalman-Takens filter reaches larger correlation values. To better analyze the impact of data assimilation, results of these two simultaneous assimilation cases over Murray-Darling and Mississippi basins are plotted in Fig. 10.4. Both cases successfully reduce the misfits between the estimates and GRACE TWS as well as soil moisture observations for both basins. Major improvements can also be seen compared to open-loop time series. This figure along with Fig. 10.3 illustrate that assimilating both observation sets can better balance the effects of observations between all state variables. It is particularly of interest to see that the computationally less demanding Kalman-Takens performs closely to the dynamical method, and even better in some cases.

To better show how each method can reduce the misfits between observations and state variables, two extreme events including an above average precipitation, mainly caused by El Niño Southern Oscillation (ENSO; see, e.g., Boening et al. 2012; Forootan et al. 2016) for the period of 2010–2012 over the Murray-Darling basin and the El Niño events in 2010 over the Mississippi basin (e.g., Munoz and Dee 2004) are selected. This experiment is undertaken to monitor each case performance for reflecting the above events in the system. Average TWS estimates from each case are compared with GRACE TWS in Fig. 10.5, where the first row shows precipitation and GRACE TWS time series while the second row demonstrates differences between assimilated observations and filter estimates. It can be seen that least errors are calculated for simultaneous assimilation using SQRA and to a lesser degree simultaneous assimilation by the Kalman-Takens method. This shows that both methods perform well in reducing the discrepancy between model and observations in such extreme anomalies. GRACE data assimilation using SQRA and Kalman-Takens appear to be more successful to capture these events that satellite soil moisture only assimilation.

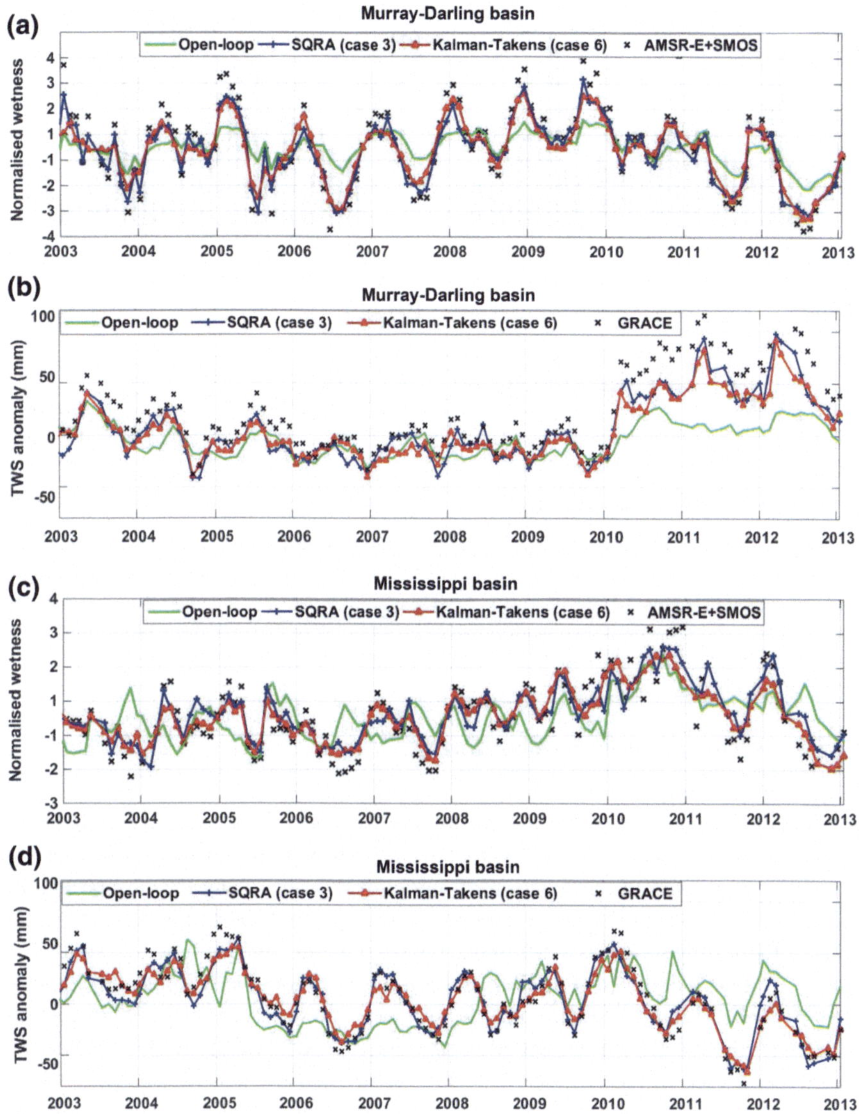

Fig. 10.4 **a** and **b** show soil moisture and TWS variation time series of simultaneous data assimilations using SQRA and Kalman-Takens over the Murray-Darling, respectively. **c** and **d** are the same as (**a**) and (**b**) but for the Mississippi basins. The figure also contains average time series of open-loop and observations

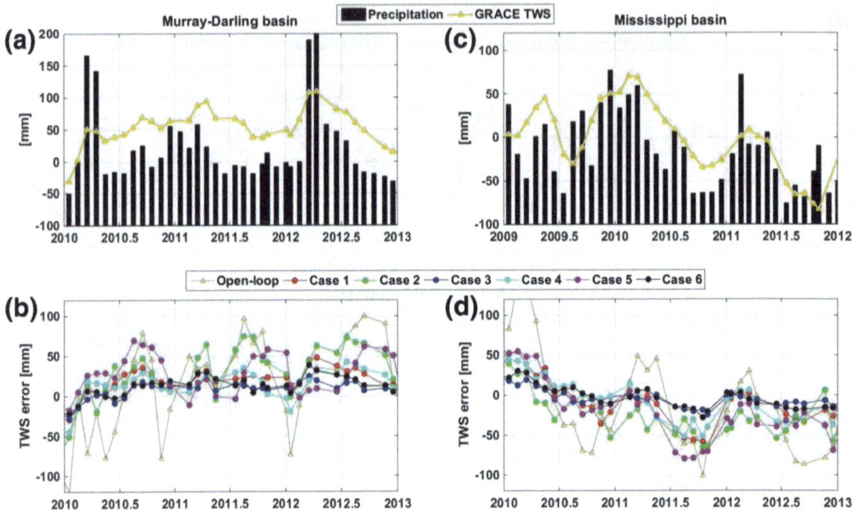

Fig. 10.5 **a** and **b** represent the average rainfall and GRACE TWS variations, respectively, over the Murray-Darling basin. The average rainfall and GRACE TWS variations over the Mississippi basin are shown in (**c**) and (**d**), respectively. Note that rainfall bar plots are shifted (−100 mm) for a better presentation. Second row: the differences between GRACE TWS and TWS estimated by each data assimilation case, as well as the open-loop run for the corresponding basins

10.5.1 Groundwater Evaluation

To assess the results of each data assimilation scenario, independent groundwater in-situ measurements are used. Estimated groundwater in-situ measurements are spatially interpolated to the location of model grid points using the nearest neighbor (the closest four grid values) to compare with groundwater time series by each method. Error time series, as a difference between in-situ and estimated groundwater values, are then calculated. For every station, we compute the Root-Mean-Squared Error (RMSE), standard deviation (STD) of error time series and also the correlation between in-situ measurements and filters results. Figure 10.6 displays the results corresponding all assimilation cases over the Murray-Darling and Mississippi basins. One can see that the simultaneous data assimilation using both filtering schemes perform closely and better than other cases. The least RMSE values are achieved from these cases compared to other scenarios. After these, assimilating only GRACE TWS using SQRA, and the Kalman-Takens filter, obtain smaller RMSE and STD values. This figure further demonstrates the capability of Kalman-Takens for assimilating multiple observation data sets, leading to comparable results to the traditional data assimilation system. Detailed results of all tested cases are presented in Table 10.2. Note that a significance test for the correlation coefficients is applied using t-distribution. The estimated t-value and the distribution at 0.05 significant

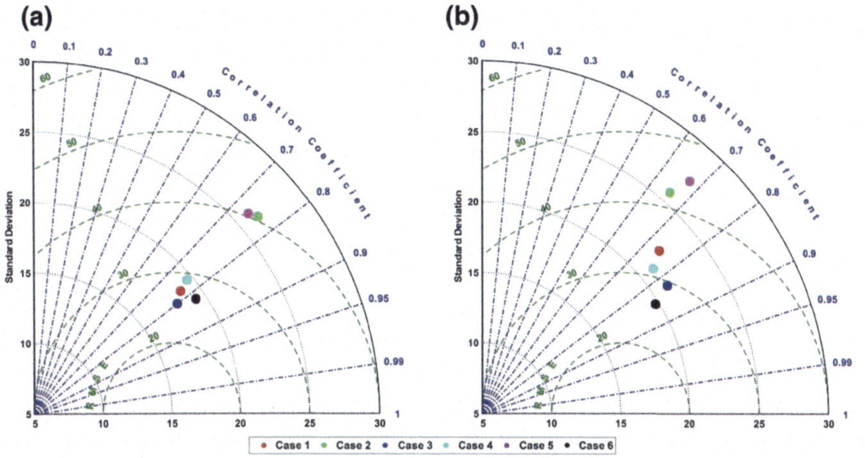

Fig. 10.6 Comparison between different data assimilation cases over the Murray-Darling (**a**) and Mississippi (**b**) basins. Groundwater estimates by filters are compared with in-situ measurements to calculate RMSE, STD, and correlation

Table 10.2 Summary of statistical values derived from implemented methods using the groundwater in-situ measurements. For each method the RMSE average and its range (±XX) at the 95% confidence interval is presented. The improvements in the analysis state RMSE estimates are calculated using the in-situ measurements in comparison to the forecast states and open-loop run

Method	Murray-Darling basin		Mississippi basin		RMSE Reduction (%)	
	RMSE (mm)	Correlation	RMSE (mm)	Correlation	Murray-Darling	Mississippi
Case 1	26.90 ± 6.32	0.78	33.54 ± 7.56	0.73	36.42	27.59
Case 2	40.72 ± 7.29	0.75	42.08 ± 8.24	0.65	3.76	9.15
Case 3	24.85 ± 5.74	0.80	29.51 ± 6.18	0.82	41.27	36.29
Case 4	28.68 ± 7.18	0.76	31.72 ± 7.06	0.77	32.21	31.52
Case 5	40.09 ± 8.92	0.74	44.29 ± 8.15	0.68	5.25	4.38
Case 6	25.78 ± 5.46	0.82	27.11 ± 6.25	0.84	39.07	40.47

level are used to calculate p-value. The correlations with p-values that lie under 5% are assumed to be significant.

Results in Table 10.2 demonstrates improved estimates after assimilation for all the cases in comparison to the open-loop, 25% RMSE reduction and 28% correlations (on average). The best performance is achieved from case 3 (simultaneous assimilation using dynamics method) for the Murray-Darling basin and from case 6 (simultaneous assimilation using Kalman-Takens) for the Mississippi basin. In most of the cases, more RMSE reductions are obtained over the Mississippi basin, especially using Kalman-Takens. The better performance of Kalman-Takens in cases 4 and 6 in comparison to the cases 1 and 3 within the Mississippi basin could be attributed to model errors that can degrade the performance of the parametric approach that relies on the model algorithms. GRACE TWS suggests larger effects on RMSE reduction than satellite soil moisture products. Simultaneous assimilation using either SQRA or Kalman-Takens results in the least RMSEs. Over Murray-Darling, assimilation of GRACE TWS only leads to better results in comparison to assimilating only soil moisture measurements. This, however, is different for the Mississippi basin, where assimilating only soil moisture observations in case 2 provides better results. On the other hand, Kalman-Takens leads slightly to better results when assimilating GRACE TWS.

Overall, based on Table 10.2, simultaneous data assimilation gives the best groundwater estimates with larger correlations and less RMSE with respect to the in-situ groundwater measurements. The Kalman-Takens results are not only close to those of SQRA but also in some cases show larger improvements. More importantly, the Kalman-Takens method is found to be less demanding computationally, i.e., ~6 times faster for the study period, compared to SQRA. Knowing that both methods exploit similar analysis scheme, the main reason for such superiority refers to faster forecasting in the Kalman-Takens filter, which is based on a local approximation (using the proxy model) and requires much less computation than a physics-based model.

10.5.2 Soil Moisture Evaluation

We further examine the assimilation results by comparing the soil moisture estimates with independent in-situ measurements. Here, we only investigate the correlation between the estimate and in-situ data because converting the assimilation outputs (as column water storage measured in mm) into volumetric units similar to the in-situ soil moisture measurements is likely to introduce a bias (Renzullo et al. 2014). Estimated soil moisture at the model top layer is compared with 0–8 cm measurements over the Murray-Darling basin and 0–10 cm over the Mississippi basin. We also use 0–30 cm and 0–50 cm measurements over the Murray-Darling and Mississippi basins, respectively, to examine the summation of the model top, shallow and a portion of deep-root soil layers. Lastly, 0–90 cm (for Murray-Darling) and 0–100 cm (for Mississippi) soil measurements are compared with the summation of the model top, shallow, and deep soil moisture layers. Similar to groundwater assessment, estimated soil moisture time series are spatially interpolated at the locations of the in-situ measurements using the nearest neighbor. The correlation is then calculated between estimated and in-situ time series and the results are demonstrated in Fig. 10.7.

It is clear from Fig. 10.7 that assimilating observations, especially GRACE TWS, mainly affect deep soil moisture layers and improve their estimates. The least improvement can be seen for the model top layer. Improvements with respect to the open-loop are achieved in all scenarios. These improvements, however, are different for each filtering method. Overall, assimilating only soil moisture measurements (as in cases 2 and 5) achieves better results in comparison to GRACE only assimilation

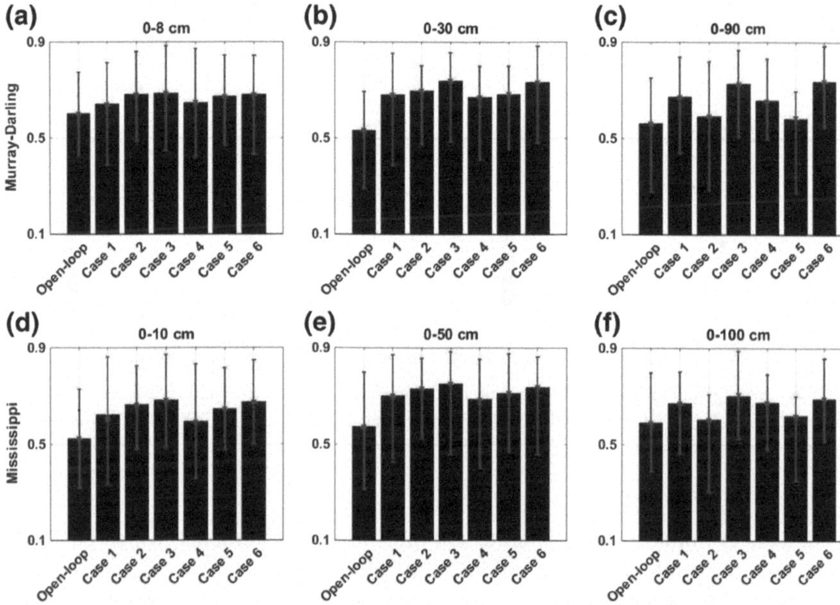

Fig. 10.7 Average correlations between soil moisture estimated by each applied case and the open-loop run with in-situ measurements at different layers including 0–8 cm (**a**), 0–30 cm (**b**), and 0–90 cm (**c**) over the Murray-Darling basin, and 0–10 cm (**a**), 0–50 cm (**b**), and 0–100 cm (**c**) over the Mississippi basin

(as in cases 1 and 4) over top layers. Simultaneous data assimilation using either SQRA or Kalman-Takens achieves the largest correlations to the in-situ measurements for all layers. This demonstrates the benefit of assimilating multiple data sets. Again, comparable results are obtained from both filtering schemes.

10.5.3 Water Fluxes Assessment

Comparison between estimated water storage changes, $\Delta\mathbf{s}$, and water fluxes, namely precipitation $\mathbf{p}$, evaporation $\mathbf{e}$, discharge $\mathbf{q}$, is assumed here. These components are related to each other in reality through the water balance equation (i.e., $\Delta\mathbf{s} = \mathbf{p} - \mathbf{e} - \mathbf{q}$). The correlation between the estimated $\Delta\mathbf{s}$ from all assimilation cases and each flux observation is calculated over the Murray-Darling and Mississippi basins. The average correlation values are presented in Fig. 10.8. Larger correlations are obtained for assimilation cases compared to the open-loop run results. Smaller improvements are achieved from the assimilation of only soil moisture measurements in comparison to the GRACE, as well as simultaneous data assimilation. Similar to the previous results, it can be concluded that GRACE TWS has larger impacts on state estimates

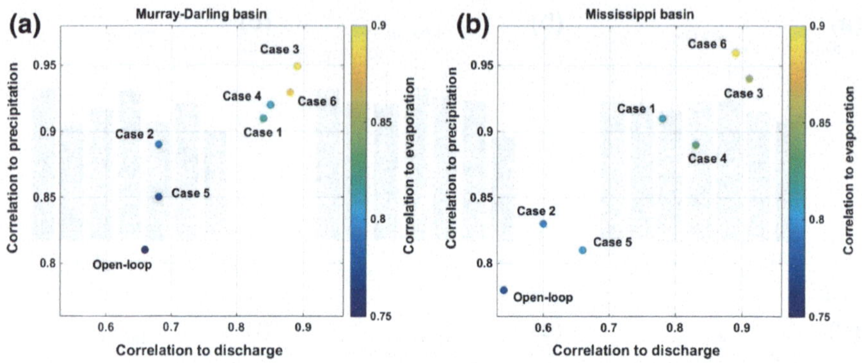

Fig. 10.8 Average correlations between water storage changes, $\Delta\mathbf{s}$, estimated by each applied case and the open-loop run with water flux observations over the Murray-Darling (**a**) and Mississippi (**b**) basin

during data assimilation than satellite soil moisture measurements, which basically update only the model top layer soil moisture component.

Between flux observations, it is found that, in general, larger correlations are achieved between $\Delta\mathbf{s}$ and $\mathbf{p}$, which is due to the larger influences of rainfall on water storage variations over the basins. SQRA reaches higher correlation values to $\mathbf{q}$ over both basins. In terms of $\mathbf{p}$ and $\mathbf{e}$, on the other hand, the Kalman-Takens filter obtains larger correlations over the Mississippi basin. It can also be seen that larger correlation of $\Delta\mathbf{s}$ to $\mathbf{p}$, generally leads to larger correlation to $\mathbf{e}$ in different cases (e.g., simultaneous assimilation using SQRA and Kalman-Takens). From Fig. 10.8, it is also clear that GRACE only data assimilation has better influences on the Murray-Darling basin, close to the simultaneous assimilation results. These results confirm previous outcomes that the Kalman-Takens filter performs well during assimilation comparable to the standard data assimilation using SQRA.

10.6 Discussion

The results of Sect. 10.5 suggest that in all cases, assimilation improves groundwater estimates in comparison to the open-loop (∼38% RMSE reduction). Simultaneous data assimilations, i.e., simultaneous assimilations of observations using dynamical method (case 3) and the Kalman-Takens (case 6) lead to the largest RMSE reductions of 41.27% with 39.07%, respectively. This is in agreement with the founding of previous literature (see, e.g., Montzka et al. 2012; Renzullo et al. 2014; Tian et al. 2017; Zobitz et al. 2014), which suggested that better results can be achieved by assimilating multi-satellite products when properly accounting for the measurement errors. Larger impacts on results are found for assimilating GRACE compared to satellite soil moisture observations. This, in particular, is evident by monitoring data assimilation results against in-situ soil moisture networks with the Murray-Darling and Mississippi basins. More pronounced improvements (12% on average) are obtained in the deep soil moisture layers, where GRACE TWS has the larger impacts on state estimates. Approximately 31% improvements in groundwater estimations are obtained from GRACE TWS only (in cases 1 and 4) as compared to soil moisture assimilation in cases 2 and 5 regardless of the filtering method. A similar impact was also suggested by Khaki et al. (2017a). Overall, close performances are observed from the dynamical and data-driven approaches. Interestingly, the Kalman-Takens outperforms SQRA filter in some cases, e.g., 2.23% more RMSE reduction over the Mississippi basin. Hamilton et al. (2016) explained that in cases where the model is subjected to larger errors, the Kalman-Takens could provide better forecasts. We further find that the Kalman-Takens is much less computationally demanding (~6 times faster) compared to the standard SQRA implementation, which can be very important especially in cases with high spatio-temporal resolutions.

10.7 Conclusion

Assimilation of multi-mission satellite products can be achieved using model-based and data-driven techniques. We assimilate the Gravity Recovery And Climate Experiment (GRACE) terrestrial water storage (TWS) and soil moisture products from the Advanced Microwave Scanning Radiometer - Earth Observing System (AMSR-E) and Soil Moisture and Ocean Salinity (SMOS) using the Square Root Analysis (SQRA) and data-driven Kalman-Takens techniques to assess their performances. Independent groundwater and soil moisture in-situ measurements are used to examine the data assimilation results over the Murray-Darling and Mississippi basins. Our results indicate that in most of the cases, simultaneously assimilation of observations using either SQRA or Kalman-Takens provides the best results with respect to in-situ measurements. These variants can also better distribute the effects of observations between all state compartments such as different soil layers and groundwater. This is shown by the better agreement between assimilation results corresponding to cases 3 and 6 and both groundwater and soil moisture in-situ measurements. More improvements in both water components estimates are obtained within Mississippi basin, particularly using Kalman-Takens. This could be attributed to the larger model errors, which have larger impacts on the parametric method that uses model dynamics. It can be concluded that the Kalman-Takens can perform better for the cases the model is subject to error. In general, the performances of the data-driven Kalman-Takens approach are comparable to those of the standard SQRA. This study suggests that the data-driven filtering technique can be a capable alternative for the traditional data assimilation.

Part VII
Hydrologic Applications

Data assimilation can be useful to study hydrological processes over different regions by improving models' performances. The approach allows for introducing missing trend and climatic related extreme events into hydrologic water compartments. It can also practically allow us to downscale relatively coarse resolution satellite data such as GRACE TWS by merging them into the models, and further to separate its aggregated water storage information into different water compartments based on the physical processes implemented in the model equations. This makes it possible to separately analyze various water compartments, e.g., soil moisture and groundwater, using the improved estimates. Yet, the application of data assimilation should be investigated over various scales and parts of the world to assess its ability for these improvements. More importantly, every region has unique characteristics in terms of hydrological processes, climate variabilities, human impacts on the environment, and geographic properties, which make the data assimilation a challenging task. Its success, thus, depends on taking into an account various factors such as treating un-modelled (or poorly modelled) phenomena and considering hydroclimate connectivity while running the process. This generally requires the use of additional (non-assimilated) observations such precipitation and El Niño/Southern Oscillation (ENSO) products (for climate part), surface runoff and satellite altimetry data (for surface water), and water use datasets for farming, industry, and human consumption (for human factors).

Some applications of data assimilation are assumed here. These include investigating surface and sub-surface water storage variations within Bangladesh, Iran, and various basins of South America. This is done to demonstrate the capability of satellite data integration for using non-regional models to study water storage changes in various areas, which is facilitated by data assimilation. The global distribution of these regions, as well as their unique hydrologic properties, allows us to better explore the effectiveness of the approach. Besides, the relationship between sub-surface water storage changes and both climate (e.g., precipitation) and anthropogenic (e.g., farming) impacts are investigated. Both these effects have remarkable impacts on water resources on all regions. As an instance, Iran has suffered from the dry climatic condition and huge groundwater abstraction, the climate has largely impacted South America, and both anthropogenic and climate extreme events have affected Bangladesh. In addition, surface water storage is an important factor over these areas, which has undeniable impacts on total water storage variations. However, this component has not usually modelled properly in the hydrologic system, and therefore, requires a special concentration during data assimilation. The enhanced knowledge of water storage changes in each area is crucial to better understand hydrological processes and can be used for water management and agriculture objectives.

Chapter 11
Groundwater Depletion Over Iran

Data assimilation successfully correct for the open-loop simulation variations, e.g., in terms of trends and multi-year variations, especially for groundwater storage. Based on the improved estimates, we find that groundwater trends in a large part of the country's (Iran) central, western and southern areas are negative representing a significant water availability issue.

—M. Khaki (This chapter is presented following Khaki et al. (2018c), "Determining water storage depletion within Iran by assimilating GRACE data into the W3RA hydrological model")

11.1 Summary

Groundwater depletion, due to both unsustainable water use and a decrease in precipitation, has been reported in many parts of Iran. In order to analyze these changes during the recent decade, in this study, we assimilate Terrestrial Water Storage (TWS) data from the Gravity Recovery And Climate Experiment (GRACE) into the World-Wide Water Resources Assessment (W3RA) model. This assimilation improves model derived water storage simulations by introducing missing trends and correcting the amplitude and phase of seasonal water storage variations. The Ensemble Square-Root Filter (EnSRF) technique is applied, which showed stable performance in propagating errors during the assimilation period (2002–2012). Our focus is on sub-surface water storage changes including groundwater and soil moisture variations within six major drainage divisions covering the whole Iran including its eastern part (East), Caspian Sea, Centre, Sarakhs, Persian Gulf and Oman Sea, and Lake Urmia. Results indicate an average of -8.9 mm/year groundwater reduction within Iran during the period 2002 to 2012. A similar decrease is also observed in soil moisture storage especially after 2005. We further apply the canonical correlation analysis (CCA) technique to relate sub-surface water storage changes to climate (e.g., precipitation) and anthropogenic (e.g., farming) impacts. Results indicate an

M. Khaki, *Satellite Remote Sensing in Hydrological Data Assimilation*,
https://doi.org/10.1007/978-3-030-37375-7_11

average correlation of 0.81 between rainfall and groundwater variations and also a large impact of anthropogenic activities (mainly for irrigations) on Iran's water storage depletions.

11.2 Introduction

Water scarcity has become a serious issue in the Islamic Republic of Iran (abbreviated here as Iran) in recent years (e.g., Amery and Wolf 2000; Wolf and Newton 2007; Trigo et al. 2010; Madani 2014; Michel 2017). With the increased extraction of groundwater, its level has been reported to fall significantly (see, e.g., Sarraf et al. 2005; Motagh et al. 2008; Mohammadi-Ghaleni and Ebrahimi 2011; Van Camp et al. 2012; Afshar et al. 2016). There have been studies that investigate surface and groundwater changes in Iran during the last decade (2003 onward) mainly using Terrestrial Water Storage (TWS) data from the Gravity Recovery And Climate Experiment (GRACE, Tapley et al. 2004). For example, Voss et al. (2013) reported $\sim$143.6 km^3 reduction of freshwater from 2003 to 2009 over the north-central area of the Middle East, which largely covers the Tigris-Euphrates Basin. Forootan et al. (2014a) applied a statistical inversion to separate GRACE TWS using hydrological model outputs and altimetry data as a priori information, and found a decrease in water storage with an average linear rate of ~ -15 mm/year between 2002 and 2011. A large negative trend (2003–2012) in TWS was observed by Joodaki et al. (2014) using GRACE TWS data over the western Iran and eastern Iraq.

Estimating sub-surface water storages is very important since they support the life in semi-arid areas like Iran. Fatolazadeh et al. (2016) used the wavelet approach to improve estimates of groundwater storage variations from GRACE and found a remarkable decrease in groundwater in 2008, 2010 and particularly in 2011. Forootan et al. (2017) compared changes in water storage and hydrological water fluxes in Iran using GRACE and climate reanalysis data. Their results indicated that the decline of TWS in the Urmia and Tigris-Euphrates basins are greater than the decrease in the monthly accumulated total water fluxes. Therefore, it was concluded that the anthropogenic contribution on surface and groundwater flow is significant, and results in the storage decline within Iran.

These studies have proved the effectiveness of GRACE to enhance the understanding of water storage changes within the country. However, they do not provide a full understanding of spatially distributed water resources changes in different water compartments in Iran. GRACE TWS measures the summation of all water masses in the surface and sub-surface compartment of the terrestrial water storage (vegetation, snow, surface waters, soil, groundwater, etc.). Therefore, GRACE TWS must be separated into different storage compartments, which has been achieved to date through a forward modeling or an inversion framework as is described in Forootan et al. (2014a) and the literature mentioned before.

To complement previous attempts, the aims of this study are to (i) update hydrological model simulations of sub-surface water storage changes (including water stored in the soil and groundwater storage) within Iran using GRACE data assimilation, and (ii) investigate climate and anthropogenic impacts on the estimated sub-surface water storages in (i). This study is the first data assimilation attempt to integrate GRACE TWS into the World-Wide Water Resources Assessment (W3RA; van Dijk 2010) hydrological model over Iran. This methodology has been implemented in studies to constrain the mass balance of hydrological models over different river basins (e.g., Zaitchik et al. 2008; van Dijk et al. 2014; Eicker et al. 2014; Reager et al. 2015; Girotto et al. 2016; Schumacher et al. 2016). The main rationale in following this approach is that one relies on the physical processes, implemented in the model equations, to separate GRACE TWS into water compartments (see similar arguments, e.g., in Bertino et al. 2003). Thus, by generating ensemble members for a model derived water storage simulation, we will compute a priori estimates of mass redistribution in the country. Then, by assimilating GRACE data, while considering their uncertainty, we update (correct) these model estimations. A similar concept has also been followed in studies in hydrology, climate, and oceanography (see, e.g., Garner et al. 1999; Bennett 2002; Kalnay 2003; Schunk et al. 2004; Lahoz 2007; Khaki et al. 2017b, c). In addition, by applying data assimilation, we will likely be able to reliably separate GRACE TWS data into different water compartments since both model and observation errors are considered. Considering that the spatial resolution of models is usually better than GRACE data, through the assimilation procedure, GRACE observations are downscaled, and therefore, higher resolution estimations of water storages will be available within the country (see also Schumacher et al. 2016; Khaki et al. 2018b).

Once improved model simulations are obtained, by assimilating GRACE TWS, relationships between the model-derived groundwater and soil moisture storages and climatic variables within Iran are investigated. To investigate the impacts of climate on the regional water storage estimates, precipitation from satellite remote sensing, temperature, and vegetation changes through the Normalized Difference Vegetation Index (NDVI) are used. Furthermore, anthropogenic effects are explored using the changes in water use for farming, industry, and human consumption. To this end, Canonical Correlation Analysis (CCA) is applied to provide an insight into the relations between model-derived water storages and both climatic and anthropogenic impacts by extracting spatio-temporal correlations between these inter-related data sets. For a better spatial analysis of water storage and to reduce the uncertainty of estimations, the study area is divided into six major areas: the eastern part of Iran (indicated by East), Caspian Sea, Centre, Sarakhs, Persian Gulf and Oman Sea, and Lake Urmia (Fig. 11.1).

11.3 Study Area and Data

11.3.1 Iran

Located in an arid and semi-arid region, Iran experiences strong regional differences in climate (Fig. 11.1). Subtropical conditions are dominant over the northern part, but 90% of the country has limited rainfall with extremely hot summers in the central and southern coastal regions (Golian et al. 2015). Much of the west to northwest of Iran is located in high plateaus and mountain ranges associated with strong temperature differences between winter and summer. By contrast, the centre to southern parts are warm (cf. Fig. 11.1) for most of the year with mild winters and hot summers. Annual rainfall, the main source of freshwater in Iran, varies from 50 mm in the deserts to 2275 mm in the northern part of the country (FAO 2009). Only a fraction of the country receives enough rainfall for agriculture. A growing use of irrigation for agricultural productions (Ardakani 2009) and the increasing population (from ~55 million in 1990 to ~80 million in 2015 Karamouzian and Haghdoost 2015), make water availability an important issue across the country (Michel 2017).

11.3.2 W3RA Hydrological Model

The present study uses the globally distributed World-Wide Water Resources Assessment system (W3RA) model, run at 1°×1°. W3RA, based on the Australian Water

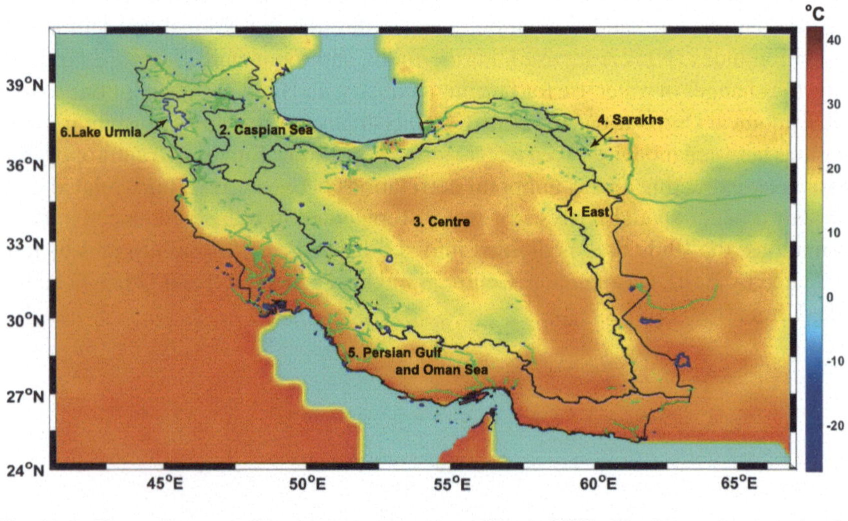

Fig. 11.1 The study area and its average temperature (Harris 2008). The figure also contains the locations of 6 major catchments separated by black solid lines

Resources Assessment system (AWRA) model (version 0.5) developed in 2008 by the Commonwealth Scientific and Industrial Research Organization (CSIRO) is a daily grid-distributed biophysical model that simulates landscape water stored in the vegetation and soil systems (see details in van Dijk 2010). The model represent and forecast terrestrial water cycles (van Dijk 2010; Renzullo et al. 2014). W3RA does not consider anthropogenic effects (e.g., irrigation). Therefore, by assimilating GRACE TWS, which integrates both natural and anthropogenic signals (e.g., Schumacher et al. 2018), we hope to constrain the model's water storage simulations and introduce the missing variations. Meteorological forcing data that is used here are minimum and maximum temperature, down-welling short-wave radiation, and precipitation from the Princeton University (Sheffield et al. 2006). The model contains effective soil parameters, water holding capacity and soil evaporation, relating greenness and groundwater recession, and saturated area to catchment characteristics parameters (van Dijk et al. 2013). This one-dimensional grid-based water balance model represents the water balance of the soil, groundwater and surface water stores in which each cell is modeled independently of its neighbors (van Dijk 2010; Renzullo et al. 2014). The model state, which is used for data assimilation (2002–2012), is composed of W3RA storages of the top, shallow root and deep root soil layers, and groundwater storage in an one-dimensional system (vertical variability).

11.3.2.1 Satellite-Derived Observations

We use monthly GRACE level 2 (L2) gravitational Stokes' coefficients truncated up to spherical harmonic degree and order 90 along with their full error information from 2002 to 2012 provided by the ITSG-Grace2016 gravity field model (Mayer-Gürr et al. 2014). The monthly full error information of the Stokes' coefficients is used to construct an observation error covariance matrix for the GRACE TWS fields to be used for data assimilation (Schumacher et al. 2016). Degree 1 of Stokes' coefficients are replaced with those estimated by Swenson et al. (2008) to account for the movement of the Earth's center of mass. Degree 2 and order 0 (C20) coefficients are replaced by those from Satellite Laser Ranging solutions due to unquantified large uncertainties in this term (e.g., Cheng and Tapley 2004; Chen et al. 2007). Afterward, following Wahr et al. (1998), the L2 gravity fields is converted to gridded TWS fields with a $1^\circ \times 1^\circ$ spatial resolution.

Correlated noise in data due to anisotropic spatial sampling, instrumental noise (K-band ranging system and GPS), and temporal aliasing caused by the incomplete reduction of short-term mass variations (Forootan et al. 2014b) can be reduced by smoothing filters (e.g., Kusche et al. 2009). The application of smoothing, however, causes a spatial leakage problem that can be problematic given that strong water resources of Tigris River and the Persian Gulf Basin can affect GRACE signals, as leakage-in errors, over the northwest and south of Iran, respectively. To tackle these errors, we use a Kernel Fourier Integration (KeFIn) filter, proposed by Khaki et al. (2018a), which defines an efficient averaging kernel to improve GRACE TWS within

Iran. The KeFIn filtering method accounts for signal attenuations and leakage effects caused by smoothing in a two step filtering scheme (see more details in Khaki et al. 2018a). Lastly, in order to reach absolute TWS estimates (similar to W3RA), the mean TWS for the study period is taken from W3RA and added to the GRACE TWS anomalies time series.

Furthermore, since W3RA does not simulate lake dynamics, one needs to account for the existing surface water storage over the Lake Urmia before assimilation of the GRACE TWS data. Water level height datasets from satellite radar altimetry of Jason-1 (260 cycles from 2002 to 2008) and Jason-2 (165 cycles from 2008 to 2012) are used to separate groundwater and surface water storage from GRACE TWS (more details in Sect. 11.4.1.2). We use the ExtR post-processing technique (Khaki et al. 2014, 2015) to retrack the data and improve water level measurements, which are erroneous within inland waters. Filtered surface heights are then used to create time series for virtual gauge stations over the Lake Urmia. These time series are subsequently used to remove the contribution of surface water storage changes from GRACE TWS data before implementing the proposed data assimilation (see also the procedure in Forootan et al. 2014a).

Satellite-derived precipitation data of TRMM-3B43 products (TRMM 2011) from the Tropical Rainfall Measuring Mission Project (TRMM; version 7) is used to study rainfall variations. We convert the gridded precipitation products provided with a $0.25° \times 0.25°$ spatial scale to $1° \times 1°$ for the period between 2002 and 2012. In addition, we use Version 4 gridded daily Normalized Difference Vegetation Index (NDVI) derived from the NOAA Climate Data Record (CDR) between 2002 and 2012 to further investigate climatic impacts. This dataset is produced by the NASA Goddard Space Flight Center (GSFC) and the University of Maryland with a $0.05° \times 0.05°$ spatial resolution. The datasets are rescaled to a $1° \times 1°$ spatial resolution. A summary of the data sets and links to download the data are provided in Table 11.1.

11.3.2.2 Temperature

Monthly average temperature data for the temporal period of 2003 to 2012 is acquired from Climatic Research Unit (CRU; Harris 2008), which is used in CCA as a climate indicator. This data is provided using more than 4000 weather stations distributed around the world. For the sake of consistency with other data sets, the collected $0.5° \times 0.5°$ spatial scale data is converted to $1° \times 1°$.

11.3.3 In-Situ Data

We use in-situ groundwater level data collected from 562 observation wells distributed over the six drainage divisions of East, Caspian Sea, Centre, Sarakhs, Persian Gulf and Oman Sea, and Lake Urmia water (cf. Fig. 11.1) to compare them with our results. Datasets are provided by the Iran Water Resources Management Company

Table 11.1 A summary of the datasets used in this study

Description	Platform	Data access
Terrestrial water storage (TWS)	GRACE	https://www.tugraz.at/institute/ifg/downloads/gravity-field-models/itsg-grace2014/
Precipitation	TRMM-3B43	https://disc.gsfc.nasa.gov/datacollection/TRMM_3B43_7.html
Normalized Difference Vegetation Index (NDVI)	NASA-GSFC	ftp://eclipse.ncdc.noaa.gov/pub/cdr/avhrr-land/ndvi/
Hydrological model	W3RA	http://www.wenfo.org/wald/data-software/
Temperature	Harris (2008)	https://crudata.uea.ac.uk/cru/data/hrg/
Groundwater in-situ measurements	IWRMC	http://www.wrm.ir/
Average water consumption	IWRMC	http://www.wrm.ir/
Discharge data	IWRMC	http://www.wrm.ir/
Number of groundwater bore holes	IWRMC	http://www.wrm.ir/
Altimetry-derived level height	Jason-1	http://podaac.jpl.nasa.gov
Altimetry-derived level height	Jason-2	http://avisoftp.cnes.fr/

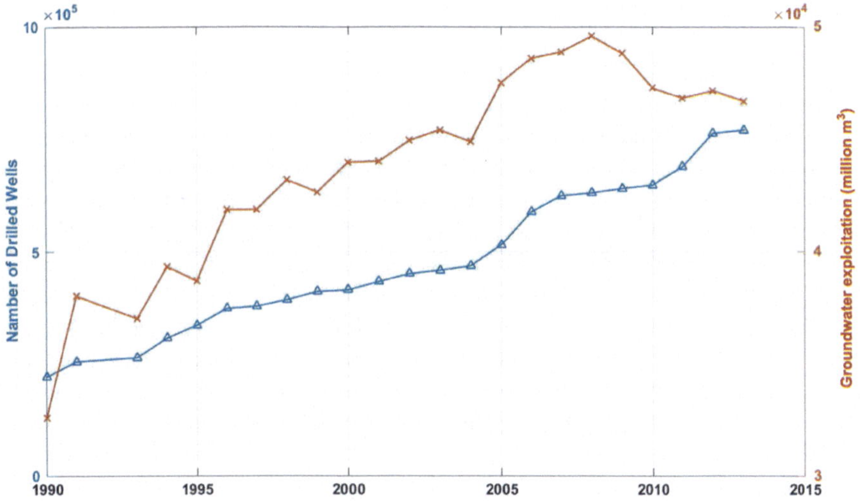

Fig. 11.2 Groundwater depletion and the number of drilled wells in Iran from IWRMC

(IWRMC) and are categorized based on Iran's six largest provinces on a yearly temporal scale presenting groundwater storage changes for an entire aquifer (Forootan et al. 2014a). Figure 11.2 shows an annual increase in groundwater extraction and the number of drilled wells for the entire country derived from IWRMC data sets. The IWRMC volumetric groundwater change measurements are converted to equivalent water height using the area of each aquifer. The area-averaged time series of groundwater changes for each aquifer is then generated and used for evaluating the results. The modified in-situ groundwater time series are compared separately to the average assimilation results for the same aquifer. River water discharge, the number of bore holes, and average water use for farming, industry, and urban use provided by IWRMC are also used in the CCA process (see Sect. 11.4.2). Details of all the applied datasets, as well as the model are presented in Table 11.1.

11.4 Method

11.4.1 Data Assimilation

11.4.1.1 EnSRF Filtering

In order to assimilate GRACE data into the W3RA model, we use the Ensemble Square-Root Filter (EnSRF) following Whitaker and Hamill (2002). EnSRF is an extended version of traditional Ensemble Kalman Filter (EnKF) that does not require the observations to be perturbed by introducing a new sampling scheme. Here, EnSRF is selected to avoid sampling errors that can be reflected in the background covariance matrix especially in using a limited number of ensembles (Whitaker and Hamill 2002). Khaki et al. (2017b) showed that this method is highly capable of assimilating GRACE TWS data into a hydrological model amongst the most commonly used filters. EnSRF adopts a similar analysis step to the EnKF in the sense that the analysis perturbations are computed from the forecast perturbations by updating each ensemble perturbation with a Kalman-like update step. In the present study X consists of six different water storages including top soil, shallow soil, and deep soil water, vegetation, snow, and groundwater storages. Previous studies, e.g., Forootan et al. (2014a) and Tourian et al. (2015), have investigated the surface water variations, specifically, in the Lake Urmia Basin and the Caspian Sea as the major source of surface water storage changes in Iran. Therefore, here, we only focus on the estimation of sub-surface compartments including groundwater and soil moisture. The modified GRACE TWS data (see Sect. 11.3.2.1 for details) is then used to update the above water compartments excluding surface storage.

The forecast model state, the integrated model state by a dynamical model for N times (N is the ensemble number), is represented by $X^f = [{X_1}^f \ldots {X_N}^f]$, where ${X_i}^f$ $(i = 1 \ldots N)$ is the ith ensemble (hereafter 'f' refers to forecast and 'a' represents analysis). The corresponding model state forecast error covariance of P^f is defined by:

$$P^f = \frac{1}{N-1}\sum_{i=1}^{N}(X_i{}^f - \bar{X}^f)(X_i{}^f - \bar{X}^f)^T, \tag{11.1}$$

$$\bar{X}^f = \frac{1}{N}\sum_{i=1}^{N}(X_i). \tag{11.2}$$

The update stage in EnSRF contains two steps. First, it updates the ensemble-mean following,

$$\bar{X}^a = \bar{X}^f + K(y - H\bar{X}^f), \quad i = 1 \ldots N, \tag{11.3}$$

$$K = P^f(H)^T(HP^f(H)^T + R)^{-1}, \tag{11.4}$$

where K is the Kalman gain, y is the observation vector. The transition matrix and the observation covariance matrix are indicated by H and R, respectively. Next, EnSRF updates the forecast ensemble of anomalies $A^f = [A_1{}^f \ldots A_N{}^f]$ into the analysis ensemble deviation A^a. A^f as the deviation of model state ensembles from the ensemble mean is derived from,

$$A_i{}^f = X_i{}^f - \bar{X}^f. \tag{11.5}$$

EnSRF exploits the serial formulation of the Kalman filter analysis step in which the observations are assimilated each at a time to compute the analysis perturbations that exactly match the Kalman filter covariance (Hoteit et al. 2008) using the modified gain ($\tilde{K}$) with,

$$A^a = (I - \tilde{K}H)A_i{}^f, \tag{11.6}$$

$$\alpha = \left(1 + \sqrt{\frac{R}{HP^fH^T + R}}\right)^{-1}, \tag{11.7}$$

where I is an identity matrix. More details on EnSRF can be found in Whitaker and Hamill (2002) and Tippett et al. (2003).

11.4.1.2 Assimilating GRACE TWS into W3RA

Monthly gridded GRACE TWS data are assimilated into W3RA to update the model states, a summation of model vertical water compartments (here soil moisture, vegetation biomass, snow, and groundwater). Note that no parameter adjustment is considered here and the observations are only used to constrain the system states. The monthly increment (i.e., the difference between the monthly averaged GRACE TWS and simulated TWS) can be added to each day of the current month, which guarantees that the update of the monthly mean is identical to the monthly mean of the daily updates. In practice, the differences between the predictions and the updated states are added as offsets to the state vectors at the last day of each month to generate the ensembles for the next month assimilation step. We use Monte Carlo sampling of multivariate normal distributions with the errors representing the standard deviations of the forcing sets (precipitation, temperature, and radiation) to generate an initial ensemble (Renzullo et al. 2014). The perturbed meteorological forcing datasets, then, are integrated forward with the model from 2000 to 2002 providing 72 sets of state vectors (as suggested by Oke et al. 2008) at the beginning of the study period.

An application of small ensemble size is problematic in ensemble data assimilation systems, as it can lead to filter divergent or inaccurate estimation (Tippett et al. 2003). Therefore, we apply ensemble inflation that uses a small coefficient factor (here 1.12; Anderson 2001) to inflate prior ensemble deviation from the ensemble-mean and increases their variations (Anderson et al. 2007). Furthermore, the Local Analysis (LA) scheme (Evensen 2003; Ott et al. 2004) is applied for localization. LA improves the assimilation procedure by restricting the observations used for the covariance matrix computation to a spatially limited area (Khaki et al. 2017a). As a result, only those measurements located within a certain distance from a grid point have an impact on the updated states (Evensen 2003; Ott et al. 2004). Different localization lengths are tested and their results are assessed against in-situ groundwater measurements (Sect. 11.3.3) to reach the best case scenario (i.e., 5° half-width used in this study).

As mentioned, it is necessary to remove surface water storages from GRACE TWS data over Lake Urmia before data assimilation. For this purpose, following Forootan et al. (2014a) who undertook water analysis over the same area, we use satellite altimetry time series over the lake to derive surface water storage. The Global Land Data Assimilation System (GLDAS) outputs of total column soil moisture, snow water equivalent, and vegetation biomass water storage as well as water level variations from altimetry are used to estimate temporal and spatial patterns of surface water storage using Independent Component Analysis (ICA). The extracted patterns are then adjusted to GRACE TWS products using a least squares adjustment (LSA) procedure (see details in Forootan et al. 2014a). The GRACE data after removing surface water storage is used for the data assimilation process over Lake Urmia.

11.4.2 Canonical Correlation Analysis (CCA)

The present study applies Canonical Correlation Analysis (CCA) to find the linear connection of two sets of multidimensional variables of predictor (x_c) and criterion (y_c) values. CCA is selected here rather than simple correlation analysis due to its ability in establishing the relationships between multiple intercorrelated variables. CCA extracts canonical coefficients that represent common processes between two or more variables (Chang et al. 2013) using an eigenvector decomposition that yields linear weights, known as canonical coefficients, which describe maximum correlations between variables (see details in Steiger and Browne 1984). The combination of variables with the first canonical coefficient for each set has the highest possible multiple correlations with the variables in the other set. CCA extracts canonical coefficients u and v such that $X_c = {x_c}^T u$ and $Y_c = {y_C}^T v$ (X_c and Y_c are canonical variates) possess a maximum correlation coefficient (Chang et al. 2013) using the following function,

$$\begin{aligned} R &= \frac{E[X_c Y_c]}{sqrt(E[X_c^2]E[Y_c^2])} \\ &= \frac{E[u^T x_c {y_c}^T v]}{sqrt(E[u^T x_c {x_c}^T u]E[v^T y_c {y_c}^T v])} \\ &= \frac{u^T C_{x_c, y_c} v}{sqrt(u^T C_{x_c, x_c} u v^T C_{y_c, y_c} v])}, \end{aligned} \tag{11.8}$$

where C_{x_c, x_c} and C_{y_c, y_c} are covariance matrices of x_c and y_c, respectively and the objective in above function is to maximize the correlation R. We use an eigenvalue decomposition procedure to find the linear weights producing canonical coefficients, which imply maximum possible correlations (see details in Steiger and Browne 1984). There are different canonical coefficients within each set leading to different uncorrelated coefficients. Nevertheless, the combination of variables with the first canonical coefficient for each set has the highest possible multiple correlations with the variables in the other set.

Two scenarios are considered for prediction: (i) the predictor (x_c) contains time series of both groundwater used for farming, industry, and human consumption from IWRMC and climate-related variables of precipitation, NDVI, and temperature (provided by Harris 2008), and (ii) the predictor (x_c) includes only climate-related variables of precipitation, NDVI, and temperature. This is done to explore the impact of each scenario on water variations. The criterion (y_c) in both scenarios contains water storage (groundwater and soil moisture) and discharge (from IWRMC) variations. By applying CCA, we establish the best combinations between two sets of variables in two different cases. By comparing the results of these two scenarios, we can investigate how water use and climate variabilities impact water storage changes within Iran. Nevertheless, there are other effective components (e.g., large-scale ocean-atmosphere phenomenon, evaporation, and droughts) on the water storage, which is difficult to include all of them in the process. This CCA scheme, however, could provide an insight on the connection between the above components. Table 11.2 summarizes the experiments undertaken in this study. The corresponding research objectives and related sections that contain each experiment's results are also listed in the table.

Table 11.2 The undertaken experiments and corresponding research objectives. The result section associated to each experiment is also presented

Experiment	Research objective	Result section
Simulated assimilation	To assess the impacts of GRACE observations on different water storage	Section 11.5.1
Evaluation procedure	To examine the validity of results against independent observations	Section 11.5.2
Water storage analysis	To analyze spatio-temporal variations of groundwater and soil moisture	Section 11.5.3
Climatic impacts using PCA	To investigate the impacts of climate indicators (e.g., precipitation) on water storage	Section 11.5.4
CCA	To establish the relations between water storages and human-as well as climate-related variables	Section 11.5.5

11.5 Results and Discussion

11.5.1 Simulated Assimilation

In the following, we analyze the effect of various scenarios of observations on the assimilation. As mentioned earlier, GRACE TWS observations are used to update the sum of soil moisture, vegetation, snow, and groundwater compartments at each grid cell. Thus, it is important to investigate the distribution of the increments between these compartments, especially soil moisture and groundwater storage while the influence of the remaining storages (i.e., vegetation and snow) is negligible. In particular, we are interested in monitoring the impacts of trends in observations time series on different water components. Schumacher et al. (2018) showed that assimilating GRACE TWS data can improve model simulation of seasonality and trend of TWS, as well as individual water storage components. This point is important because the largest part of GRACE TWS trends caused by groundwater variations that originate from both natural and human-induced (e.g. water use) changes while soil moisture variations generally follow climate pattern. Simulation experiments are undertaken to monitor how observations' variations, and particularly their trends are reflected in soil moisture and groundwater estimates during assimilation.

To illustrate how GRACE data assimilation can improve model states, we perform a synthetic study, in which arbitrary errors (uncertainty with different magnitudes) are assigned to different model derived water storage states. We evaluate whether these states accurately receive increments from GRACE TWS. To this end, we introduce different uncertainties to model states and test how these are transferred to the assimilation forecast steps (cf. Eqs. (11.3), (11.4)). Figure 11.3 shows the relationship between selected uncertainties of water states and their corresponding weights in the (synthetic) assimilation. Based on this setup, six different scenarios are considered to explore the impact of weights as the ratio of the assigned increment derived for

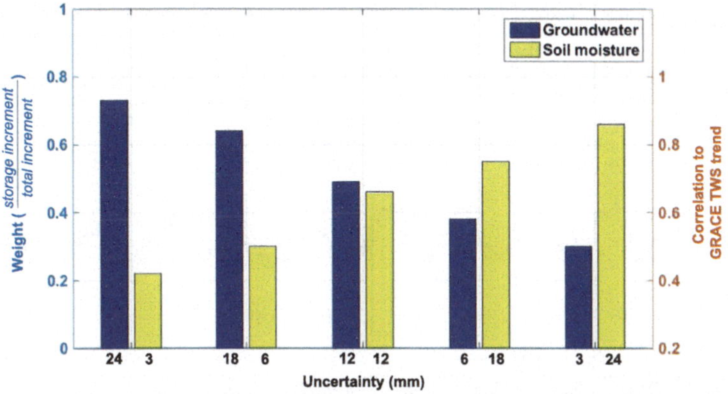

Fig. 11.3 Relationships between groundwater and soil moisture state variable uncertainties and corresponding weights during data assimilation

each storage state to the summation of all states. The results presented in Fig. 11.3 indicate an average influence of assimilating GRACE TWS data into W3RA over Iran between 2003 and 2013. In general, as theoretically expected, higher weight (i.e., larger increment) is assigned to a variable with a smaller uncertainty. In other words, by assimilating GRACE TWS, the model's water states with larger uncertainty receive larger increments, and this is reverse for states with smaller uncertainty. These results approve the recent results of Schumacher et al. (2018), who assimilate GRACE TWS data into WGHM model over Australia. Figure 11.3 also shows that the average correlations between the individual estimated storage in each scenario and the assimilated GRACE TWS. The correlations are calculated after removing seasonal effects on time series to focus on trends. It can be seen that larger correlations to the GRACE TWS trends are obtained for a compartment with larger uncertainty and correspondingly with a larger increment. This means that the assimilation process transfers the observation trends into the more uncertain storage, which receives the larger corrections.

Another synthetic experiment is also implemented, where, different observation sets are assimilated into W3RA but this time without manipulating their uncertainties. The aim is to investigate whether the distribution of increments of different water states changes when the TWS observations change. Here, four different synthetic observation scenarios are considered, which include two versions of the WaterGAP Global Hydrology Model (WGHM; more details on Döll et al. 2003; Müller et al. 2014) TWS estimates with and without water abstractions, GRACE-derived TWS, and GRACE TWS minus WGHM soil moisture that roughly gives groundwater observations. The spatially averaged time series of the TWS observations (for the first three cases) over Iran are displayed in Fig. 11.4a. The difference between the WGHM TWS observations with and without water use clearly show the anthropogenic impacts as a distinct negative trend in WGHM with water abstraction impact. A similar trend can also be seen in GRACE TWS. Assimilation of these

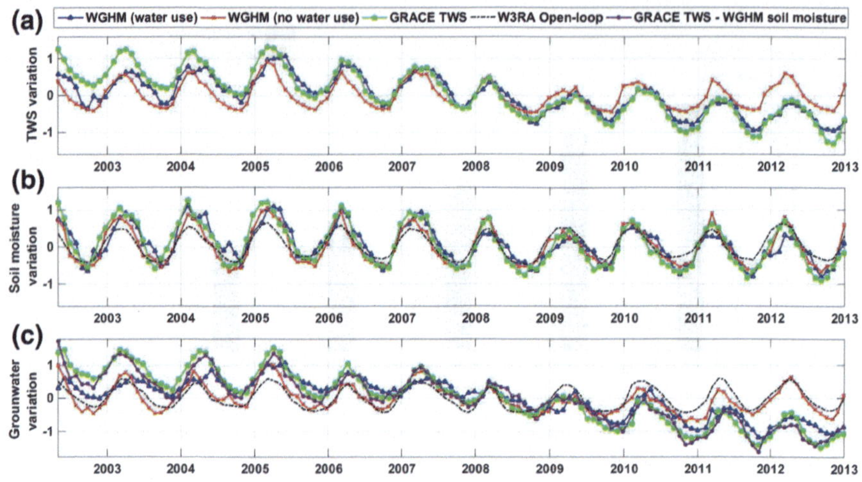

Fig. 11.4 **a** Simulated average TWS observations using WGHM with and without human use, and W3RA open-loop plus GRACE trend. Average soil moisture (**b**) and groundwater (**c**) estimates from data assimilation based on simulated observations in (**a**)

observations can show how water storages, for example their trends, are distributed between soil moisture and groundwater estimates. Assimilating WGHM TWS without water use, which does not show any significant trends, might better estimate soil moisture. This is due to the fact that the main source of TWS's negative trends is groundwater exploitation, while soil moisture variations generally are related to climatic (e.g., precipitation) variations. Hence, comparing the soil moisture results of assimilating GRACE TWS and WGHM TWS with water use with those of WGHM TWS without water use can help to assess the performance of data assimilation in updating soil moisture. Furthermore, while the first three observation sets (i.e., WGHM with and without water use and GRACE-derived TWS) are used to update the summation of all compartments, the last case (GRACE TWS minus WGHM soil moisture) is used to update only the groundwater simulations. The main rationale for updating only groundwater in the last experiment is to compare its results with the other scenarios, which can help to investigate how accurate groundwater corrections are applied from TWS increments in the other cases, where different compartments are available.

The results of the data assimilation variants are shown in Fig. 11.4b, c and updated groundwater estimates from assimilating GRACE TWS minus WGHM soil moisture is plotted in Fig. 11.4c. The assimilation results for soil moisture (Fig. 11.4b) and groundwater (Fig. 11.4c) show that the negative TWS trends are largely reflected only in groundwater time series. The average correlation between the above TWS observations and corresponding groundwater estimates is 0.92, 42% (on average) larger than for the open-loop run, which indicates the suitability of data assimilation for constraining system states. For the entire area, there is a stronger agreement between the soil moisture from assimilation compared to the open-loop run, e.g., 22% (on average) for the GRACE TWS case and 28% (on average) for the WGHM

TWS with water use case. Lower correlations are obtained for assimilating WGHM TWS without water use in comparison to other data assimilation scenarios (see also Fig. 11.4b). Furthermore, groundwater variations from the assimilated GRACE TWS are largely correlated to the groundwater estimates from assimilating only groundwater observations (GRACE TWS minus WGHM soil moisture). TWS observations of WGHM without water use have the least effect on groundwater variations.

It can be concluded from Fig. 11.4 that the data assimilation process successfully distributes TWS increments between soil moisture and groundwater storages. These results indicate that the largest part of increments during data assimilation is assigned to groundwater. The larger impact on groundwater, based on Fig. 11.3, suggests that the groundwater estimation of W3RA is more uncertain than its soil moisture and as a result it receive larger updates. This is even more clear in Fig. 11.5, where groundwater and soil moisture estimates by ensemble members between 2004 and 2008 are shown. This time period is selected because it includes an episode with strongly negative groundwater trend after 2005 (see also Fig. 11.4c), where ensemble spreads show a different pattern, e.g., larger spreads. The propagated groundwater ensemble members are more dispersed than those of soil moisture, which causes larger ensemble deviations from its mean and consequently larger uncertainty for the states (cf. Eqs. (11.1), (11.2)). This can be due to the point that the W3RA model has a simplified simulation of groundwater dynamics for unconfined groundwater and does not simulate confined groundwater dynamics or anthropogenic groundwater extraction (Tregoning et al. 2012). The larger corrections applied to groundwater is also realistic considering the fact that a majority of water depletion in Iran occurs in

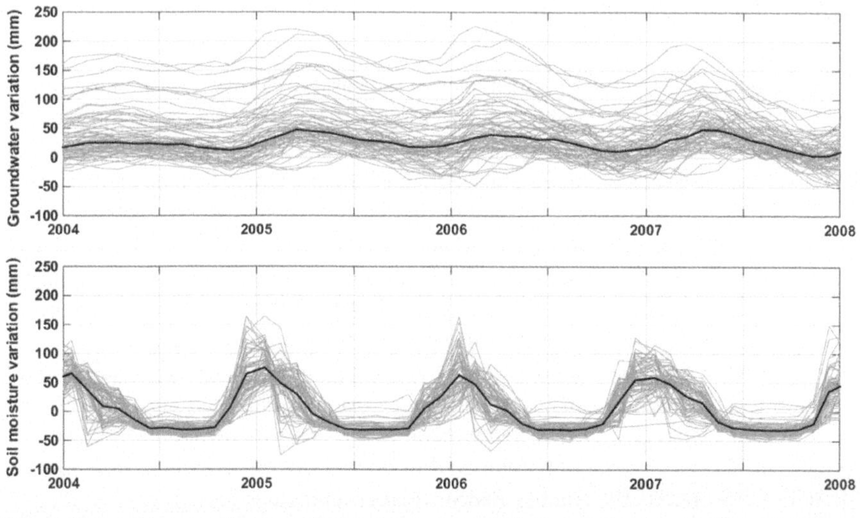

Fig. 11.5 Average groundwater and soil moisture ensemble spreads between 2004 and 2008 over Iran. Gray lines indicate ensemble members and the black solid line present ensemble mean. Larger ensemble propagation is evident compared to that of soil moisture that represents larger uncertainties in the former water storage compartment

groundwater due to large extractions for irrigation. The applied irrigation water is likely to locally increase total soil column water storage, which may contribute to a smaller decline in soil water content (Michel 2017).

Even though the results indicate good performance of GRACE data assimilation, one might still expect artefacts from the TWS increments on the state estimates. The absence of groundwater abstractions and anthropogenic impacts in most hydrological models, especially where the rate of this extraction is high, can cause a misinterpretation of a negative TWS trend captured by GRACE in the system states. As shown by Girotto et al. (2017), the assimilation of GRACE TWS can successfully introduce the negative trends in the modeled TWS and groundwater, however, this can also introduce unrealistic decline in other components, e.g., soil moisture and evapotranspiration. This effect can be exacerbated when groundwater extraction is large and occurs over an extended period. The model dynamical range of groundwater may not be sufficient to accommodate the assimilated values (Zaitchik et al. 2008; Li et al. 2015). Despite these, merging GRACE TWS data with high resolution models is the most efficient existing approach to analyze groundwater changes over wide areas, which in most cases results in an improvement in the estimates (Li et al. 2015; Girotto et al. 2017). Here, we addressed this challenge by conducting a synthetic experiment, as well as by independently assessing groundwater and soil moisture from assimilation. However, more investigations are needed to be extended and the impacts of various data assimilation scenarios on each individual water compartments need to be tested. These investigations are, however, out of the scope of this study.

11.5.2 Result Evaluation

In this section, we assess the performance of data assimilation using in-situ groundwater measurements. To examine the validity of data assimilation results, in-situ groundwater measurements of the six major drainage regions in the area including the East, Caspian Sea, Centre, Sarakhs, Persian Gulf and Oman Sea, and Lake Urmia (cf. Fig. 11.1) are used. For each basin in Fig. 11.1, we calculate the spatial average time series of groundwater storages with and without data assimilation and compare them with the IWRMC in-situ and WGHM groundwater variation. We first analyze the performance of two assimilation cases of GRACE TWS and GRACE TWS minus WGHM soil moisture data assimilation experiments for improving groundwater estimates. Figure 11.6 shows the average root-mean-square error (RMSE) and standard deviation (STD) calculated using groundwater from assimilation cases and in-situ measurements. Both cases perform comparably in terms of RMSE and STD with an average of 38% error reduction compared to open-loop. Nevertheless, assimilating GRACE TWS obtains the smaller RMSE than groundwater only data assimilation. This further confirms the effectiveness of the applied data assimilation for distribution TWS increments, especially for groundwater storage. Based on this assessment, hereafter only the results for GRACE TWS data assimilation are presented.

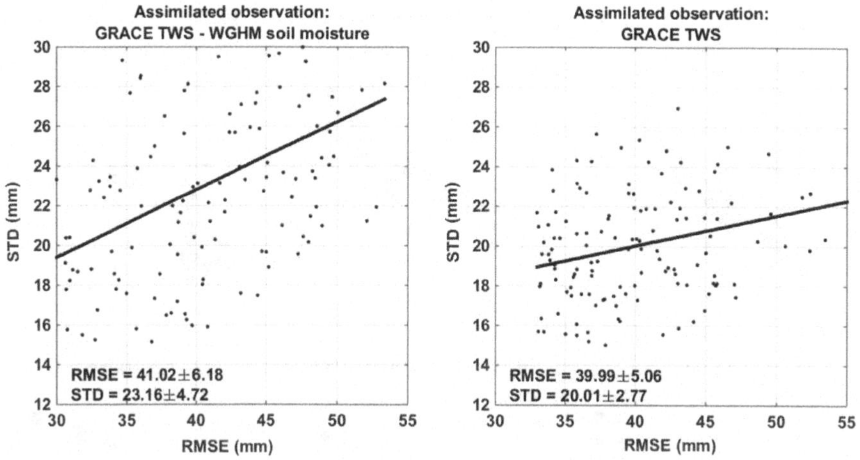

Fig. 11.6 Average groundwater RMSE and STD from assimilating GRACE TWS and GRACE TWS minus soil moisture

The results for groundwater examination from data assimilation, WGHM, and the open-loop run for each drainage division are illustrated in Fig. 11.7, which show that the strongest agreement between groundwater estimates and in-situ measurements occur in the assimilation results. In most of the cases, WGHM performs better than the open-loop. For a better assessment of data assimilation results, additional agreement statistics using RMSE and correlation analysis are calculated and reported in Table 11.3. Significance at $p < 0.05$ was calculated using the Student's t-test with consideration of temporal autocorrelation through effective sample size.

The computed time series for each region is compared to IWRMC data for the corresponding region in order to estimate the reported statistics in Table 11.3. Generally, the assimilation results are largely correlated with the in-situ data (0.85 on average) after data assimilation, with an improvement of 35% over open-loop results. The largest improvements in terms of correlation increase and RMSE reduction with respect to the in-situ measurements are achieved over Lake Urmia, Sarakhs, and to a lesser degree Persian Gulf and Oman Sea. Table 11.3 shows considerable groundwater decline in most of the regions especially within the Persian Gulf and Oman Sea and Lake Urmia (both mostly located in the western areas). The largest negative groundwater trend is exhibited for Lake Urmia while the lowest trend is found for the Caspian Sea division in the north, which could be attributed to a large amount of precipitation in the latter region.

We further examine the soil moisture estimates from data assimilation. In the absence of reliable in-situ soil moisture measurements over the study area, we use satellite-derived and independent model soil moisture products. Soil moisture observations from the Advanced Microwave Scanning Radiometer—Earth Observing System (AMSR-E) and Soil Moisture and Ocean Salinity (SMOS) are compared to the assimilated top layer soil moisture estimates. The motivation behind this comparison

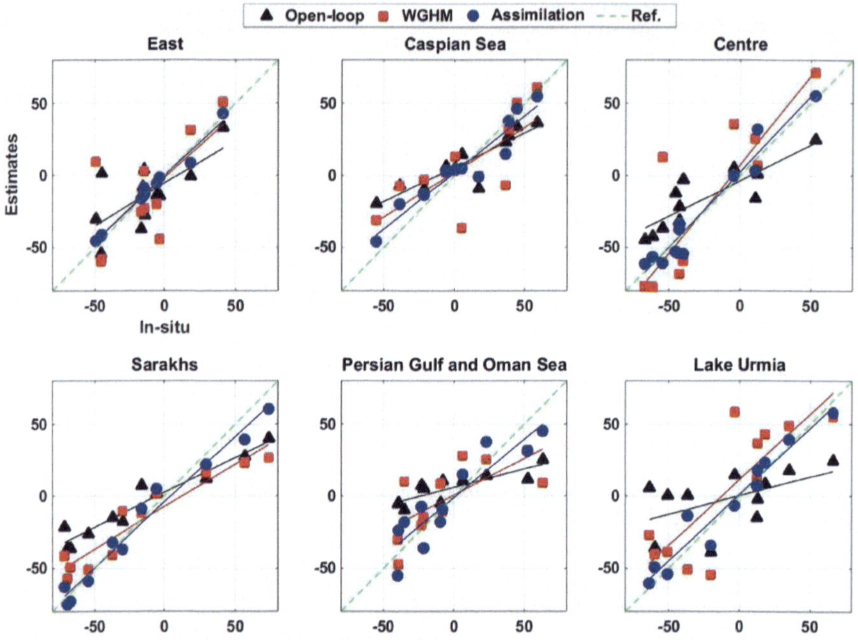

Fig. 11.7 Comparison between in-situ groundwater measurements and those estimated by open-loop run, data assimilation, and WGHM over different catchments (units are mm)

Table 11.3 Statistics of groundwater variations and its errors with respect to the in-situ observations. For each region the RMSE average and its range (±XX) at the 95% confidence interval is presented. Improvements in data assimilation results are calculated for each catchment in relation to the water storages from the model without implementing data assimilation

Region	Groundwater trend (mm/year)	Assessment with In-situ				Improvement (%)
		Open-loop		Assimilation		
		Correlation	RMSE (mm)	Correlation	RMSE (mm)	
East	−3.8	0.57	60 ± 8.66	0.84	38 ± 4.64	36.29
Caspian sea	−2.1	0.64	64 ± 9.19	0.73	46 ± 5.13	28.13
Centre	−6.7	0.63	55 ± 7.84	0.65	41 ± 5.01	26.55
Sarakhs	−5.4	0.61	52 ± 7.58	0.82	32 ± 4.26	38.64
Persian Gulf and Oman sea	−9.3	0.56	79 ± 9.07	0.75	49 ± 5.17	37.81
Lake Urmia	−11.8	0.52	69 ± 8.28	0.81	40 ± 4.25	41.90

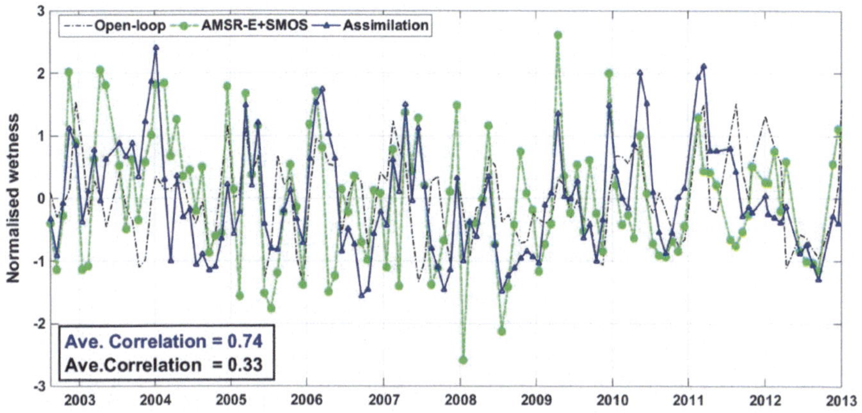

Fig. 11.8 Comparison between the average estimated top layer soil moisture with and without (open-loop) data assimilation and soil moisture observations from satellite remote sensing (AMSR-E+SMOS). Correlations between the satellite measurements and both open-loop and assimilation estimates are also repotted in the figure

is based on the fact that SMOS and AMSR-E measurements are largely correlated, respectively, to surface 0–5 cm and 0–2 cm soil moisture content (Njoku et al. 2003). Figure 11.8 shows the average time series of the above comparison within the study period. It can be seen that the assimilation top layer soil moisture is better matched (41% improvement in correlation) to the satellite measurements in comparison to the open-loop estimates. This shows a successful impact of GRACE TWS data assimilation on the model top layer.

Total soil moisture estimates from data assimilation, i.e., summation of soil moisture at top, shallow- and deep-root layers, are compared with soil moisture estimates of WGHM, the Global Land Data Assimilation System (GLDAS, Rodell et al. 2004), and soil moisture provided by van Dijk et al. (2014), who combined different data (e.g., GRACE) and model outputs (indicated here as W3). The results are displayed in Fig. 11.9. In all cases, data assimilation leads to a better agreement to other products with an average 25% improvement. The largest correlation, as well as the greatest improvement, are found for soil moisture after assimilation of WGHM. There is also a considerable correlation between the results and W3.

11.5.3 Water Storage Analysis

Based on the improved soil moisture and groundwater estimates, spatio-temporal variations of both compartments are analyzed in this section. The variation of groundwater storages within Iran before and after data assimilation are illustrated in Fig. 11.10. The blue graph in Fig. 11.10 represents the average groundwater variations of all grid points after data assimilation. This graph clearly shows a negative

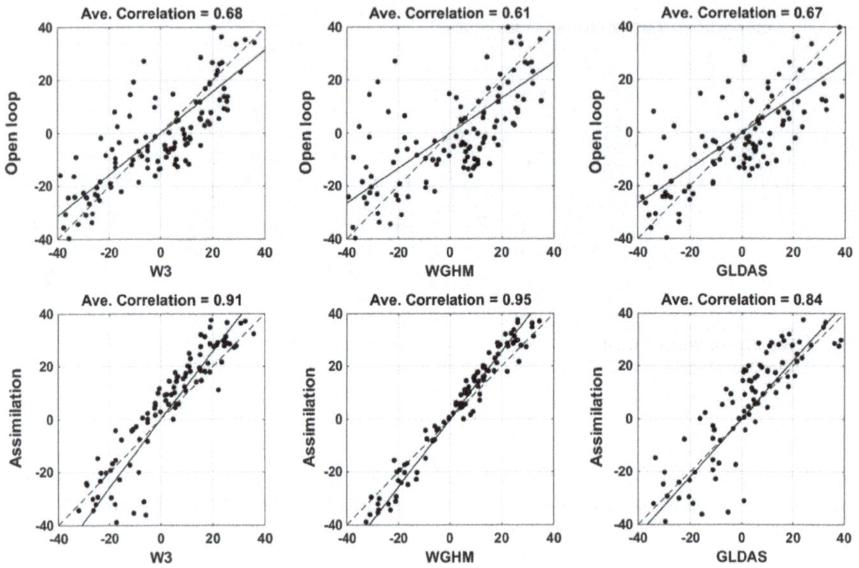

Fig. 11.9 Comparison between the average soil moisture estimates from open-loop and data assimilation, and soil moisture products of W3, WGHM, and GLDAS (units are mm)

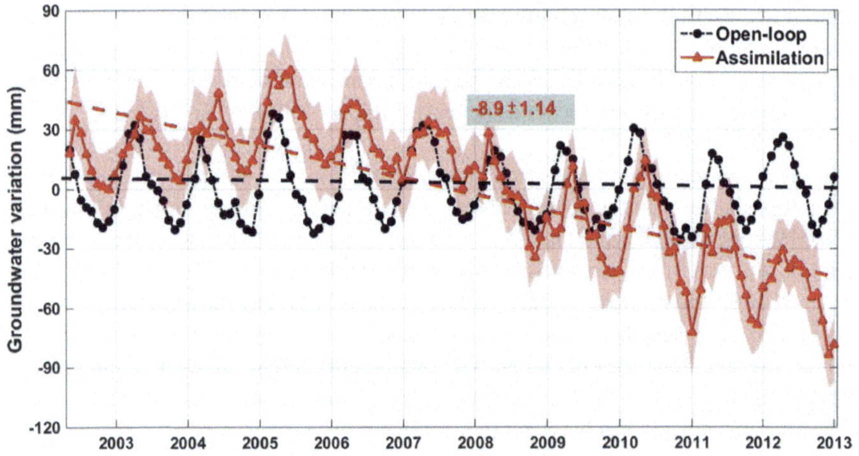

Fig. 11.10 Average groundwater variations within Iran from open-loop and data assimilation results and corresponding 95% confidence intervals (shaded blue). Trend lines for time series are also displayed by dashed lines. Note that the open-loop time series slop is not reported because no significant trend is observed

trend between 2002 and 2013 with an average −8.9 mm/year groundwater depletion for the entire country. However, such a trend is not present in the open-loop time series. GRACE TWS data assimilation constrains groundwater estimates and introduces this negative trend into the state as it exists in GRACE TWS observations (cf. Fig. 11.4). It is evident that the W3RA without data assimilation is not able to provide reliable long-term changes of groundwater, e.g., trend and multi-year variations. Therefore, data assimilation is vital for reliable interpretation of ground water beyond the annual cycle. However, without additional information the data assimilation results cannot differentiate between natural and anthropogenic causes. Apart from the trends, Fig. 11.10 also shows a multi-year cycle, e.g., positive trend between 2002 and 2005 and a stronger negative trend for the later years 2006 to 2013. Again, this trend is not visible in the open-loop simulations.

Furthermore, we separately analyze water compartments for each of Iran's major drainage regions. The soil moisture and groundwater average time series from W3RA before and after assimilating GRACE TWS for each of the divisions are shown in Figs. 11.11 and 11.12, respectively. Larger soil moisture variations (in terms of amplitude) exist for the data assimilation results compared to open-loop results in Fig. 11.11. In particular, this is evident for the Persian Gulf and Oman Sea and Caspian Sea. This could be due to a larger amount of annual precipitation over these areas. Declines in soil water content can be seen in Sarakhs, especially between 2005

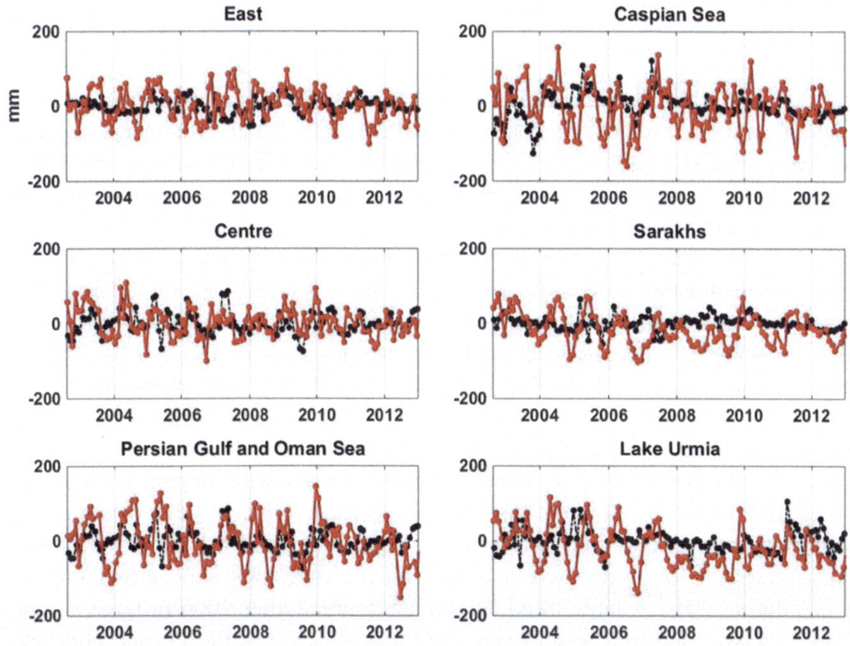

Fig. 11.11 Average time series of soil moisture variations over different catchments with (blue) and without (black) data assimilation

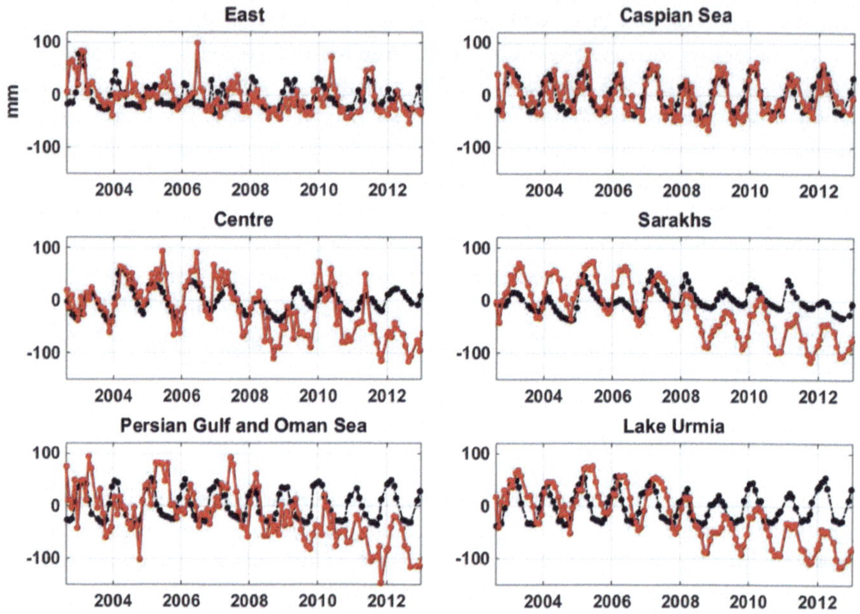

Fig. 11.12 Average time series of groundwater variations over different catchments with (blue) and without (black) data assimilation. The correlations of time series with the in-situ measurements, as well as the trends of assimilation results are reported in Table 11.2

and 2009, and Lake Urmia. In most of the regions, increases (e.g., large positive variations) are observed during 2004 and 2010. Overall, better agreements between open-loop and assimilation time series are found over East and Centre regions, where a semi-arid climate condition is dominant. GRACE data assimilation has the least impact on soil moisture estimates within these areas.

Figure 11.12 depicts groundwater variations for each individual drainage division. Similar to soil moisture analysis (cf. Fig. 11.11), data assimilation results demonstrate larger magnitudes than open-loop results. Except for the Caspian Sea, all the regions show a considerable decline in groundwater estimates during the study period. In particular, this is clear in Lake Urmia, Sarakhs, and Centre, especially after 2007. These trends are absent in the open-loop time series and derive from GRACE TWS after implementing data assimilation, which confirm the results shown in Fig. 11.10. Larger groundwater declines are found in regions over the western parts of the country (e.g., the Persian Gulf and Oman Sea and Lake Urmia). In most of the cases, groundwater rise is observed as a positive trend between 2004 and 2005. These increases are then followed by consistent declines despite some short-term increases such as during 2010. A large trend decline is observed after 2006 in Lake Urmia, Centre, Sarakhs, and to a lesser degree in Caspian Sea. For the Persian Gulf and Oman Sea, Sarakhs, and Center, the groundwater negative trend is remarkable after 2008. Despite a small negative trend in East for the study period, the groundwater variations have the smallest amplitudes in this region compared to other areas.

Seasonal variations can clearly be seen in most of the regions while this pattern is dominant mostly in Caspian Sea. Figure 11.12 and the reported negative trends in Table 11.3 show that groundwater depletion is a major issue in most parts of Iran resulting in a remarkable dryness across the country.

11.5.4 Climatic Impacts

We further investigate the connection between climatic impacts and water storage variations. A comparison between groundwater and soil moisture variations and climate-related variables such as precipitation and NDVI can reveal such interactions these parameters. Figure 11.13 shows maps of temporal average precipitation, soil moisture, and groundwater maps during the study period. The first row in Fig. 11.13 represent the average applied increment to soil moisture and groundwater storages, the second row indicates variations (average of time series at each grid point) of

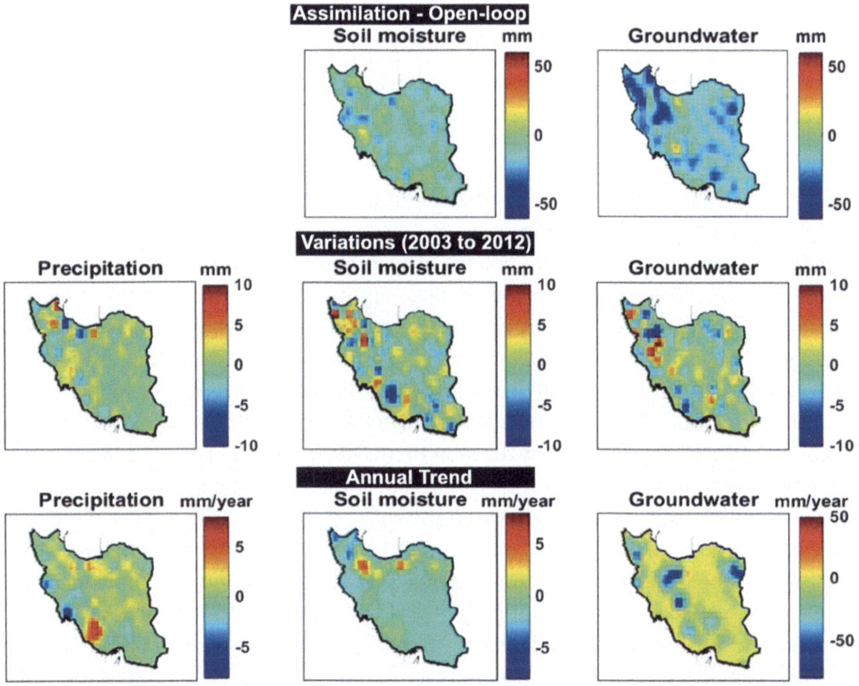

Fig. 11.13 First row: temporally averaged increments applied to soil moisture and groundwater storages. Second row: variation of precipitation, soil moisture, and groundwater (after data assimilation) estimated as the average of each time series at each grid point. Third row: gridded trend of time series precipitation, soil moisture, and groundwater (after data assimilation) time series

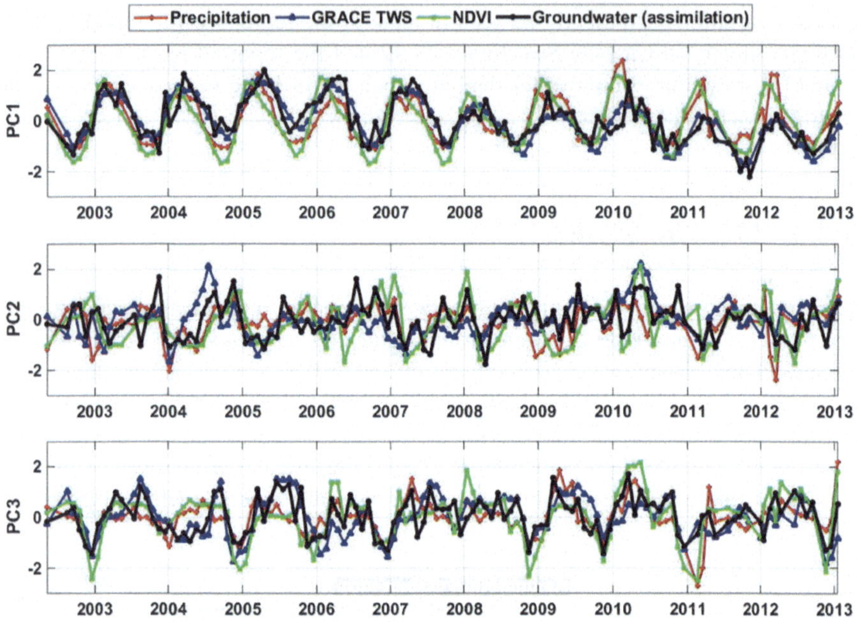

Fig. 11.14 The three first principal components of precipitation, GRACE TWS, NDVI, and groundwater

precipitation, soil moisture, and groundwater, and trends for each variable at each grid point are depicted in the third row.

Figure 11.13 shows the spatial pattern of increments, i.e., the difference between assimilation results and open-loop estimates, applied to the system states. It can be seen that the largest increments are applied to groundwater storage as can be expected from Figs. 11.3 and 11.4. These corrections are mostly focused on the northwest to south and the eastern part of Iran. In soil moisture, the increments can be found across the country, again, with larger concentrations in the western areas. The effect of data assimilation clearly can be seen by the increments illustrated in Fig. 11.13. The spatial pattern of precipitation, soil moisture, and groundwater variations in Fig. 11.13 show larger variations over the north toward northwest parts, where the Alborz mountain range cover a large portion of the areas. A similar pattern can also be seen in western parts, where the Zagros mountain range is located. Overall, the soil moisture map more closely reflects the precipitation patterns compared to groundwater variations, which can be attributed to impacts from water uses. Contrary to precipitation and soil moisture, negative groundwater variations are found over different regions, especially the north-western and southern parts. There are very limited variations in terms of amplitude changes for precipitation, soil moisture, and groundwater within the centre, eastern, and partially south-eastern parts of Iran. Trend maps (last row in Fig. 11.13) illustrate spatial patterns for each component. Both precipitation and soil moisture show increasing trends in the north and to a lesser degree in the south. Groundwater trends are generally negative in all regions, but more strongly in the west, where Lake Urmia is located. A significant groundwater depletion can be observed in the central parts extended to the north, where Tehran, Iran's capital city is located. Large groundwater extractions in Tehran during the study period can be the main reason for this while in other areas, an excessive irrigation is a potential candidate for the observed depletion. It can be seen that there is an agreement between the applied increment by data assimilation, especially for groundwater, and the negative observed trends. Again it can be concluded that without using assimilation, these negative trends are not captured.

To better quantify the spatio-temporal variations of water storage and climate variabilities, principal component analysis (PCA Lorenz 1956) is applied on precipitation, NDVI, GRACE TWS, and groundwater time series. This allows us to monitor the relationship between the estimated groundwater and GRACE TWS, as well as their connection to climatic impacts through precipitation and NDVI. The first three extracted principal components (PC1, PC2, and PC3) of each component are plotted in Fig. 11.14. There is good agreement between the time series for all three cases, in particular for seasonal variations. All time series in PC1 show a clear annual variation. Negative trends, especially after 2009 are only captured by PC1 of GRACE TWS and groundwater. Stronger agreements between precipitation and NDVI PCs can be found. This can be attributed to vegetation growth response to rainfall and soil moisture. The assimilated groundwater storage variations largely follow the GRACE TWS variation patterns, both in terms of variability and trend, mainly due to the application of GRACE data assimilation. Both of these variables are strongly correlated with rainfall time series in PC2 and PC3 with an average correlation of 0.86. Various strong anomalies are occur in the time series, e.g., in 2005 and 2010. Increases in the time series occur in PC1 for all variables between 2004 and 2006 and during 2010 and 2012. PC2 shows similar rises in 2008 and 2010 followed by a strong decrease. PC3 shows an increase in 2009 and 2010 in the precipitation, GRACE TWS, and groundwater which explains the corresponding increase in water storages (cf. Figs. 11.10 and 11.12). Some negative anomalies are found in PC3 in 2003, 2005, and 2011 and also in 2006 and 2013. The other variables generally demonstrate the same variation pattern as precipitation, which shows a strong connection between water storage variations and climatic changes. Water storage variations in Iran, however, are also affected by non-climate factors (e.g., anthropogenic impacts), which are likely the cause of the observed negative trends in PC1 for GRACE TWS and groundwater.

The corresponding empirical orthogonal functions (EOF1, EOF2, and EOF3) extracted by applying PCA on precipitation, NDVI, GRACE TWS, and groundwater from data assimilation are shown in Fig. 11.15. Overall, the mode 1 represents a strong annual signal (as would be expected), mode 2 shows some deviations from the annual signal (e.g. inter-annual variations) in the same regions as for mode 1. Mode 3 to some extent shows inter-annual variations but importantly shows some extreme values. The spatial patterns of NDVI, GRACE TWS, and groundwater are largely correlated to rainfall pattern, especially in EOF1 and EOF2. Larger spatial variations exist over the northern and western parts of Iran, which seem to cause

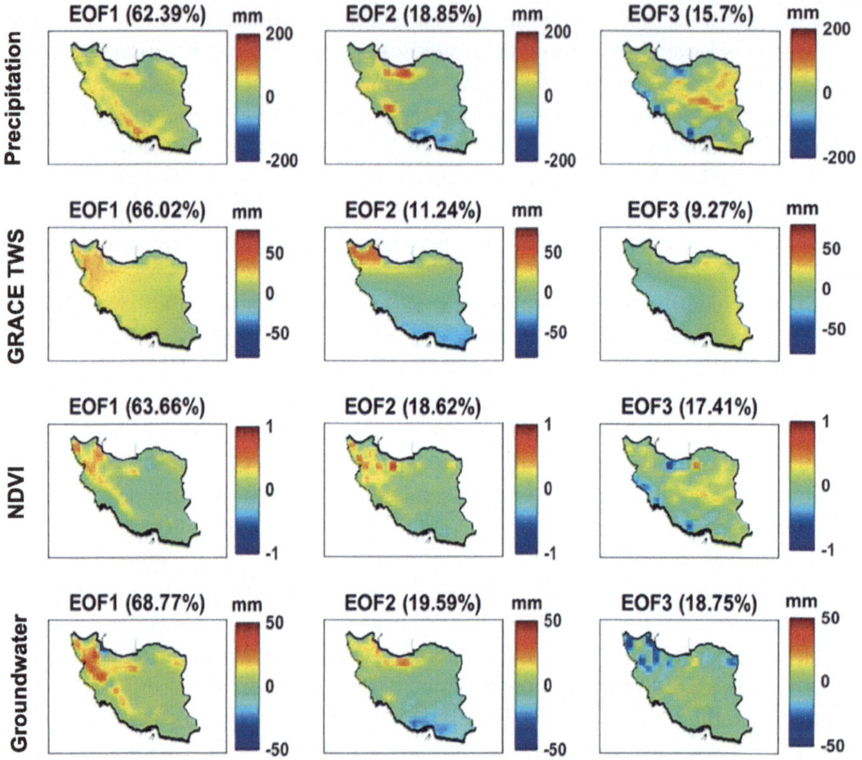

Fig. 11.15 The empirical orthogonal functions (EOF1, EOF2, and EOF3) extracted from precipitation, GRACE TWS, NDVI, and groundwater

larger water storage and NDVI changes in the same areas. These are the parts with higher altitudes in which precipitation rates are generally high. GRACE TWS and groundwater EOF2 maps show strong positive signals over the north toward the northwest and partially in western areas. The rainfall EOF2 map, however, does not show a large signal over the north-western part but only over the northern and western parts, where the Alborz and Zagros mountain ranges are located. On the other hand, all variables show a negative signal in the south-eastern part. Positive signals over the eastern parts, with smaller amplitudes, compared to EOF1 and EOF2 for NDVI, GRACE TWS, and groundwater are displayed by EOF3 maps. Negative signals can be seen in EOF3 maps, especially for groundwater mostly over the northwestern areas, where Lake Urmia is located, as well as the northeast and Sarakh.

11.5.5 CCA Results

We further implement CCA on the estimated water compartments (from the data assimilation) on the one hand, and human- as well as climate-related variables on the other hand in two different scenarios, i.e., (i) the predictor contains time series of both groundwater used (e.g., for farming and industry) and climate-related variables (precipitation, NDVI, and temperature), and (ii) the predictor includes only climate-related variables of precipitation, NDVI, and temperature (cf. Sect. 11.4.2). By this, we can establish the relations between water storages and other factors. CCA is applied to the spatially averaged time series of all variables to estimate canonical coefficients. Canonical loadings are used to interpret the CCA results, which measure the simple linear correlation between an observed variable and the estimated canonical variates (Dattalo 2014). The interpretation is mostly based on examining the sign and the magnitude of the canonical coefficients assigned to each variable. Variables with larger coefficients contribute more to the variates and variables with opposite signs exhibit an inverse relationship with each other while those with the same sign exhibit a direct relationship. Detailed results of the CCA experiment for each scenario applied within Iran are presented in Table 11.4.

The table summarizes the contribution of each variable in CCA. Results indicate that scenario (i) leads to larger canonical correlation coefficients in comparison to scenario (ii). This means that variations in water storages are more correlated to variations of the combined human- and climate-related parameters. Note that CCA

Table 11.4 Average canonical correlation coefficients and variable loadings for the data inputs in CCA for each scenario

			Scenario (i)	Scenario (ii)
			Canonical coefficients	Canonical coefficients
Canonical correlation coefficient			0.972	0.841
Predictor variables		Precipitation	0.721	0.749
		NDVI	0.365	0.412
		Temperature	−0.591	−0.681
	Water use for	# Farming	−0.938	–
		# Industry	−0.758	–
		# Drink (Urban use)	−0.820	–
		Number of bore holes	−0.893	–
Criterion variables		Groundwater	0.938	0.705
		Soil moisture	0.633	0.617
		Water discharge	0.174	0.249

extract different sets of results (roots), thus, we only use the first root that is statistically significant (for a significant level of 0.05). It can be seen from Table 11.4 that the water use has strong negative correlations to water storage variations, especially groundwater, which has the largest loading. This means that water consumption for various uses (especially farming) is a very effective factor within the country that causes the greatest impact on groundwater (with 0.938 canonical correlation). Among climate variables, precipitation, and to a lesser degree temperature have also a considerable influence on water storage variations. Not surprisingly, an increase (or decrease) in rainfall directly leads to increase (or decrease) in water storages as indicated by the same signs. Table 11.4 suggests that variations in groundwater use and climate parameters in both scenarios have minimum impact on water discharge. This may be due to the fact that surface waters compose a relatively small amount of water availability across Iran in comparison to other storages such as groundwater.

It can also be inferred from Table 11.4 that removing the water use from scenario (i) results in smaller canonical correlation in (ii), which means a smaller agreement between variables in scenario (ii) and water storage changes, even though this removal causes ~3% and 5% increase in loadings of precipitation and temperature, respectively. Comparing the results of both scenarios implies the large anthropogenic impact (more than climate-related factors) on water storages variations, which makes it essential to include this impact along with climatic effects while one studies subsurface water storage variations in Iran. Figure 11.16 depicts scatter bi-plots and the linear trend which represents the correspondence between two sets of variables using

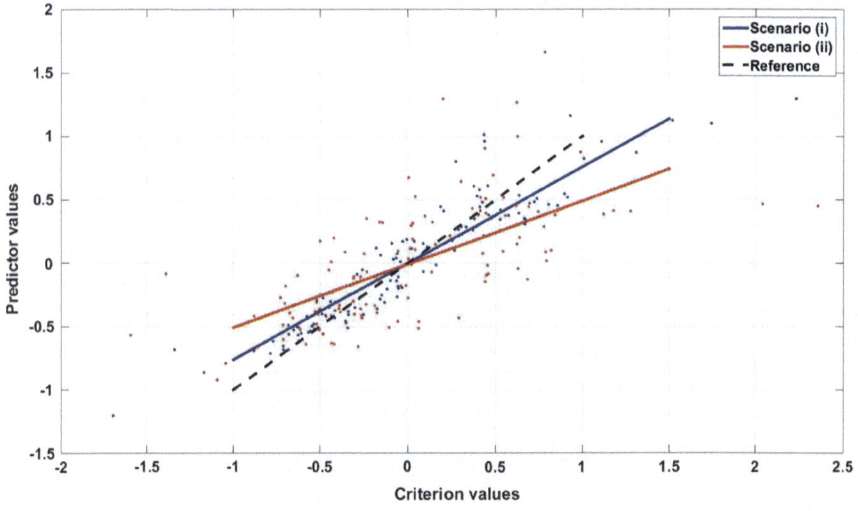

Fig. 11.16 Scatter bi-plots (circles) and the linear trend (solid lines) of average canonical coefficients from CCA for each scenario applied. The combination of the water storages and discharge data and their canonical coefficients are in the x-axis (as criterion variables), the y-axis represents the combination of the predictor variables. Black solid line represents the reference line

average canonical coefficients for each scenario. It can be seen that the distribution of the two datasets in scenario (i) has smaller deviations and is more symmetric (closer to the reference line than scenario (ii)), which leads to higher canonical correlation for the first scenario. Figure 11.16 shows that incorporating the water use results in a better agreement between the criterion, i.e., water storage variations and predicant. This stresses the necessity of considering the water use and anthropogenic impacts (e.g., irrigation) on water storages analyzes, which cannot be happen without inclusion of GRACE TWS into the process.

11.6 Conclusions

Sub-surface water storages are a major source of freshwater in Iran. With increased population and irrigated land, water availability has become a serious issue across the country. In the present study we assimilate GRACE TWS into W3RA to separately analyze different water compartments including groundwater, soil moisture, and surface water storages. The six major drainage divisions in the area including the eastern part of Iran (East), Caspian Sea, Centre, Sarakhs, Persian Gulf and Oman Sea, and Lake Urmia are considered to better understand water availability in the different regions. An analysis is undertaken to examine the effects of GRACE data assimilation on different water storage compartments. It is found that the implemented process can effectively distribute the TWS increments between groundwater and soil moisture storages. Although the results show improvements in both groundwater and soil moisture, the data assimilation still may have introduced some artefacts into the simulated groundwater dynamics due to the massive effects of groundwater extraction within the country, which requires an independent extensive study and more comprehensive analysis.

It is found that the application of GRACE TWS data assimilation can significantly improve the performance of W3RA. Data assimilation successfully correct for the open-loop simulation variations, e.g., in terms of trends and multi-year variations, especially for groundwater storage. Based on the improved estimates, we find that groundwater trends in a large part of the country's central, western and southern areas are negative representing a significant water availability issue. An average −8.9 mm/year water storages decline is observed during 2002 to 2012 with a larger rate since 2005 suggesting that Iran is becoming considerably dryer. Larger water store depletions are found to occur in the Persian Gulf and Oman Sea and Lake Urmia with lesser effects on soil moisture in these regions. In the Caspian Sea region, however, due to a large amount of precipitation, smaller groundwater and soil moisture trends are observed. In the Persian Gulf and Oman Sea, −9.3 mm/year (on average) groundwater trend is found, which is the second largest negative trend after that of Lake Urmia.

Furthermore, PCA is applied to investigate the relationship between the estimated groundwater and GRACE TWS, as well as their connection to climatic impacts in various parts of Iran. Larger water storage spatial variations are observed over the northern and western parts of Iran with higher altitudes in which precipitation rates are generally high. Contrary to rainfall maps, strong positive GRACE TWS and groundwater signals are found over the north toward the northwest and partially in western areas. In terms of temporal variations, water storage variables generally demonstrate the same variation pattern as precipitation, however, they are also affected by non-climate factors (e.g., anthropogenic impacts), which are likely the cause of the observed negative trends in GRACE TWS and groundwater time series. Therefore, CCA is applied to explore the relationship between water storages estimated by data assimilation and climatic, as well as anthropogenic indicators. The application of CCA reveals strong correlation (0.89 in average) suggesting that the groundwater use has a major impact on water storage variations.

Acknowledgements M. Khaki is grateful for the research grant of Curtin International Postgraduate Research Scholarships (CIPRS)/ORD Scholarship provided by Curtin University (Australia). This work is a TIGeR publication.

Chapter 12
Water Storage Variations Over Bangladesh

In the absence of any considerable decrease in precipitation over the region, a remarkable groundwater reduction is observed from the first and second Intrinsic Mode Functions (IMFs), which can be referred to human impacts. High spatio-temporal resolution remote sensing products along with the data assimilation methodology show a high capability for studying water storages in Bangladesh.

—M. Khaki (This chapter is presented following Khaki et al. (2018b), "A study of Bangladesh's sub-surface water storages using satellite products and data assimilation scheme")

12.1 Summary

Climate change can significantly influence terrestrial water changes around the world particularly in places that have been proven to be more vulnerable such as Bangladesh. In the past few decades, climate impacts, together with those of excessive human water use have changed the country's water availability structure. In this study, we use multi-mission remotely sensed measurements along with a hydrological model to separately analyze groundwater and soil moisture variations for the period 2003–2013, and their interactions with rainfall in Bangladesh. To improve the model's estimates of water storages, terrestrial water storage (TWS) data obtained from the Gravity Recovery And Climate Experiment (GRACE) satellite mission are assimilated into the World-Wide Water Resources Assessment (W3RA) model using the ensemble-based sequential technique of the Square Root Analysis (SQRA) filter. We investigate the capability of the data assimilation approach to use a non-regional hydrological model for a regional case study. Based on these estimates, we investigate relationships between the model derived sub-surface water storage changes and remotely sensed precipitations, as well as altimetry-derived river level variations in Bangladesh by applying the empirical mode decomposition (EMD) method. A larger correlation is found between river level heights and rainfalls (78% on average) in comparison to groundwater storage variations and rainfalls (57% on average).

M. Khaki, *Satellite Remote Sensing in Hydrological Data Assimilation*,
https://doi.org/10.1007/978-3-030-37375-7_12

The results indicate a significant decline in groundwater storage (~32% reduction) for Bangladesh between 2003 and 2013, which is equivalent to an average rate of 8.73 ± 2.45 mm/year.

12.2 Introduction

South Asia, and in particular Bangladesh, is amongst the most water vulnerable regions of the world exhibiting an increase in droughts and floods due to climate change (McCarthy et al. 2001). Groundwater is the main source of drinking and irrigation water (almost 90%) in the country Islam et al. 2013). Any considerable change in climate will, therefore, affect Bangladesh's available water, which is stored in different forms including aquifers, soils, surface waters as rivers, lakes, man-made reservoirs, wetlands and seasonally inundated areas (Papa et al. 2015). Understanding the interaction between precipitation (mainly provided during the Monsoon season) and water storage changes is important to relate climate variability to hydrology. An in-depth understanding of this interaction can be more difficult in Bangladesh due to the changing behavior of monsoonal precipitation (Wang et al. 2006) as well as the lack of knowledge on their influence on the hydrology of the region (Shahid 2010; Rafiuddin et al. 2010).

Groundwater accessibility has made Bangladesh an agro-based country with the main product being rice, making it one of the world's largest rice producer (Abdullah Aziz et al. 2015). The excessive groundwater usage during the last two decades has resulted in serious problems of both rapid falling of groundwater levels and the deterioration of its quality (Qureshi et al. 2015). Groundwater depletion has been reported by Shamsudduha et al. (2009) between 1985 and 2005 within different regions in Bangladesh such as north-central, northwestern, and southwestern parts of the country. This has also been shown by Shamsudduha et al. (2012) for the period of 2003 to 2007. Moreover, Sengupta et al. (2013) reported that groundwater in 63 (out of 64) districts of Bangladesh are seriously contaminated with arsenic, which is partly attributed to its depletion. A number of studies attribute the drop in groundwater level since 1972 to the rainfall decrease and increase in human water usage (see, e.g., Mainuddin 2002; Ahmed 2006; McBean et al. 2011; Dey et al. 2011; Adhikary et al. 2013). The Groundwater Monitoring Survey Report of Bangladesh Agricultural Development Corporation (BADC) and Institute of Water Modeling (IWM) showed a three-meter drop of groundwater levels in Dhaka (Sumon and Abul Kalam 2014). Knappett et al. (2016) claimed that an excess extraction caused the groundwater level to decline more than one meter near the Buriganga River, which passes in the southwest outskirts of Dhaka resulting in insufficient resources available for the rapidly growing population.

Soil water storage variation is another important factor that worsens the situation and affects agriculture. Furthermore, a considerable amount of surface water from rainfall is consumed by human and thus is not able to recharge the groundwater (e.g., Kanoua and Merkel 2015; Qureshi et al. 2015; Alimuzzaman 2017), which can aggravate the conditions mentioned above. Apart from efforts by these studies, a comprehensive study is missing to account for both groundwater and soil moisture variations and their connections to climate variability and change over the entire Bangladesh.

In this regard, hydrological models are important tools for simulating and predicting sub-surface water storages with high saptio-temporal resolutions (e.g., Wooldridge and Kalma 2001; Döll et al. 2003; van Dijk et al. 2013). However, imperfect modeling of complex water cycle processes, data deficiencies on both temporal and spatial resolutions (e.g., limited ground-based observations), and uncertainties of (unknown) empirical model parameters, inputs and forcing data cause some degrees of deficiencies in them (Vrugt et al. 2013; van Dijk et al. 2011, 2014). These limitations are addressed through data assimilation, which is a technique that incorporates additional observations into a dynamic model to improve its state estimations (Bertino et al. 2003; Hoteit et al. 2012). The technique has been widely applied and validated in the fields of oceanography, climate, and hydrological science (Garner et al. 1999; Elbern and Schmidt 2001; Bennett 2002; Moradkhani et al. 2005; van Dijk et al. 2014; Reager et al. 2015). Several studies indicate that terrestrial water storage (TWS) derived from the Gravity Recovery And Climate Experiment (GRACE) can play a significant role in better understanding surface and sub-surface processes related to water redistribution within the Earth system (e.g., Huntington 2006; Chen et al. 2007; Kusche et al. 2012; Forootan et al. 2014b; van Dijk et al. 2014). In particular, Shamsudduha et al. (2012) showed a high capability of GRACE measurements for studying water storage variations in the Bengal Basin. A growing number of studies have also assimilated GRACE TWS in order to constrain the mass balance of hydrological models (e.g., Zaitchik et al. 2008; Thomas et al. 2014; van Dijk et al. 2014; Eicker et al. 2014; Tangdamrongsub et al. 2015; Reager et al. 2015; Khaki et al. 2017c; Schumacher et al. 2018).

The present study aims at assimilating GRACE TWS into the World-Wide Water Resources Assessment (W3RA) hydrological model (van Dijk 2010) to analyze groundwater and soil moisture changes within Bangladesh. While the main focus is on groundwater and soil moisture, surface water as an important water source in Bangladesh is also studied since some surface water sources (e.g., lakes and rivers, except major ones) are not modeled in W3RA. Moreover, since GRACE TWS reflects the summation of all water compartments, for the first time, we use three different scenarios to account for surface water storage changes before data assimilation (see details in Sect. 12.4.1). The main reason for using the W3RA model to perform our investigations is to rely on the physical processes implemented in the model equations to consistently separate GRACE TWS (since both model and observation errors are considered) into different water compartments that includes groundwater and soil moisture. As hydrological models are usually better resolved than GRACE data during the assimilation procedure, observations are downscaled, and therefore, higher spatial resolution estimations of water storages will be available within the study region (see, e.g., Schumacher et al. 2016). Here, we use the ensemble-based sequential technique of the Square Root Analysis (SQRA) filtering scheme (Evensen 2004) to assimilate GRACE TWS into W3RA. SQRA is preferred over the traditional ensemble Kalman filter since it offers a higher computational speed, simplicity, and independence of observation perturbations. Besides, Khaki et al. (2017b) showed that this method is highly capable of assimilating GRACE TWS data into a hydrological model.

After data assimilation, we investigate the connections between the estimated groundwater and soil moisture storages (from improved model) and both surface water level variations and rainfall from multi-mission satellite remote sensing data over Bangladesh. Satellite radar altimetry products of Jason-1 and -2, and Envisat are used in this study to provide 19 virtual river gauge stations for the period 2003 to 2013 distributed across Bangladesh. Since satellite altimetry was initially designed for ocean studies (Fu and Cazenave 2013), its observations over inland water bodies must be carefully post processed (Birkett 1998; Calmant et al. 2008; Khaki et al. 2015). Therefore, the Extrema Retracking (ExtR) technique, proposed by Khaki et al. (2014), is applied to retrack satellite waveform data to improve range estimations and consequently derive better water level estimations.

Further, we apply the statistical method of empirical mode decomposition (EMD, Chen et al. 2007) to explore connections between the groundwater and surface water from the model, rainfall data from the Tropical Rainfall Measuring Mission (TRMM), and retracked surface water heights. EMD is an efficient approach to extract cyclic/semi-cyclic components and is preferred over the classical techniques such as the Fourier analysis (Chen et al. 2007; Pietrafesa et al. 2016).

12.3 Study Area and Data

12.3.1 Bangladesh

Bangladesh is located in the Bengal Basin, where the Ganges, Brahmaputra, and the Meghna rivers converge. The average temperature of the country ranges from 17 to 20.6 °C during winter and 26.9 to 31.1 °C during summer (Rajib et al. 2011). Thus, it is placed in the sub-tropical region with a humid, warm, and tropical climate, which is dominated by a subtropical monsoon originating over the Indian Ocean, which carry warm, moist, and unstable air (Ahmed 2006; Khandu et al. 2017). An average drought frequency in the country is reported to be equivalent to 2.5 years (Adnan 1993; Hossain 1990) when rainfall, as the most important water supply, drops by almost 46% (Dey et al. 2011). The annual precipitation ranges from less than 1500 to ~5000 mm and varies over different parts of the country, e.g., 1276 mm and 1337 mm in the central and western regions, respectively (see, e.g., Hasan et al. 2013; Islam et al. 2014).

12.3.2 W3RA Hydrological Model

The globally distributed 1° × 1° World-Wide Water Resources Assessment system (W3RA) model is used to simulate water storage over Bangladesh. W3RA is a daily grid distributed biophysical model developed in 2008 by the Commonwealth Scientific and Industrial Research Organisation (CSIRO). The model simulates water storage and flows to monitor, represent, and forecast terrestrial water storages (van Dijk 2010; Renzullo et al. 2014). The meteorological forcing data sets for the model include minimum and maximum temperature, downwelling short-wave radiation, and precipitation from Princeton University (see detail in Sheffield et al. 2006). Effective soil parameters, including water holding capacity, and soil evaporation, related greenness and groundwater recession, and saturated area to catchment characteristics are the model parameters (van Dijk et al. 2013). The model states used in this study include the top, shallow, and deep root soil layers, groundwater storage, and surface water storage in a one-dimensional system (vertical variability). More detailed information on W3RA can be found in van Dijk et al. (2013).

12.3.3 Remotely Sensed Observations

12.3.3.1 GRACE

The GRACE level 2 (L2) monthly Stokes' coefficients up to degree and order 90 and their full error information (2003–2013) are obtained from the ITSG-Grace2014 gravity field model (Mayer-Gürr et al. 2014) and used in the data assimilation process. The monthly full error information of the Stokes' coefficients is used to construct an observation error covariance matrix for the GRACE TWS fields (Eicker et al. 2014; Schumacher et al. 2016). Note that different GRACE products from various centers can lead to different results depending on their data processing strategies (Shamsudduha et al. 2017). However, for the sake of data assimilation, in addition to GRACE observations, we also need full error information associated with the observations. Schumacher et al. (2016) and Khaki et al. (2017a) show that it is important to consider GRACE full error covariance matrix to conduct data assimilation experiments. A more comprehensive analysis of different GRACE products has already been performed in a recently published paper of Schumacher et al. (2018). Their results indicate that while using the full covariance matrix in the data assimilation procedure, differences between the GRACE products do not significantly change to affect the final results. Therefore, we only use ITSG-Grace2014 data for which we are sure that the full covariance field is well representative of the GRACE data's error structure.

Degree 1 of Stokes' coefficients are replaced with those estimated by Swenson et al. (2008) to account for the change in the Earth's center of mass. Degree 2 and order 0 (C20) coefficients are replaced by those from Satellite Laser Ranging solutions due to unquantified large uncertainties in this term (e.g., Cheng and Tapley 2004; Chen et al. 2007). Colored/correlated noises in the L2 products are reduced using the DDK2 smoothing filter following Kusche et al. (2009). This smoothing causes some degree of signal attenuation (Klees et al. 2008) and moving anomalies from outside the region (e.g., Bay of Bengal) (Chen et al. 2007; Khaki et al. 2018a). To mitigate this issue, following Swenson and Wahr (2002), we apply a Lagrange multiplier filter to decrease leakage errors over the entire Bangladesh. This filter uses a basin averaging kernel method expanded in terms of spherical harmonics and subsequently combined with L2 potential coefficients to improve the GRACE estimates (see details in Swenson and Wahr 2002). The L2 gravity fields are then converted to $1° \times 1°$ TWS fields following Wahr et al. (1998). Note that the GRACE data provide changes in TWS while W3RA produces absolute TWS. Accordingly, the mean TWS for the study period is taken from W3RA and is added to the GRACE TWS change time series to obtain absolute values and make them comparable with model outputs (Zaitchik et al. 2008).

12.3.3.2 Satellite Radar Altimetry

Satellite radar altimetry data of Jason-1 and -2, i.e., 20-Hz sensor geographic data records (SGDR), and Envisat, i.e., 18-Hz SGDR products are applied in this study. The data includes 260 cycles of Jason-1 covering 2002–2008, 166 cycles of Jason-2 covering 2008–2013, and 113 cycles of Envisat covering 2002–2012. Jason-2 is a follow-on mission of Jason-1 with a similar temporal resolution of $\sim$9.915 days and the ground cross-track resolution of $\sim$280 km (over the equator), with the same characteristics as Topex/Poseidon altimetry mission (Benada 1997; Papa et al. 2010a). Jason-1 and-2 data are obtained from the Physical Oceanography Distributed Active Archive Center (PO.DAAC) and AVISO, respectively. Additionally, Envisat RA2 products with a 35 days repeat cycle (30 days for new orbit after October 2010) are derived from ESA (Table 12.1).

Altimeter ranges should be corrected for atmospheric impacts such as ionospheric, tropospheric, and electromagnetic effects (Benada 1997). We apply geophysical correction, including solid earth tide, pole tide, and dry tropospheric (Birkett 1995) to correct the ranges. The ExtR post-processing technique (Khaki et al. 2014) is applied on waveforms to retrack datasets and improve range measurements. The retracked altimetry data are then used to build virtual time series for 19 different points (Fig. 12.1) located on the satellite ground tracks and distributed throughout the study area. At each virtual point, several points belonging to the same satellite cycle are considered, and the median value of the retracked altimetry-based water levels is computed to address the hooking effect (Frappart et al. 2006). While a satellite is passing above a water body, its emitted signal is locked over a spatially limited part of the water, which can result in an error. The hooking effect, caused as result of off-nadir range measurements, leads to less accurate height estimations (Seyler et al. 2008; Boergens et al. 2016). Afterwards, time series of water level variations from Jason-1 and -2 are combined with those of Envisat products to produce monthly

Table 12.1 A summary of the datasets used in this study

Description	Platform	Detail	Data access
Terrestrial water storage (TWS)	GRACE	GRACE level 2 (L2)	https://www.tugraz.at/institute/ifg/downloads/gravity-field-models/itsg-grace2014/
Altimetry-derived level height	Jason-1	Pass numbers 90 and 231	http://podaac.jpl.nasa.gov
	Jason-2	Pass numbers 90 and 231	http://avisoftp.cnes.fr/
	Envisat	Pass numbers 337, 438, 795, 896, and 982	http://envisat.esa.int/dataproducts/ra2-mwr/
Precipitation	TRMM-3B42	Daily accumulated precipitation	http://disc2.gesdisc.eosdis.nasa.gov/data/TRMM_L3/TRMM_3B42_Daily.7
Hydrological model	W3RA	The Commonwealth Scientific and Industrial Research Organisation (CSIRO)	http://www.wenfo.org/wald/data-software/
Surface water storage		Satellite-derived surface water storage in the Ganges-Brahmaputra River Basin	Papa et al. (2015)
In-situ measurements	BWDB	http://www.ffwc.gov.bd/	

surface levels. Details of the datasets, model, and pass numbers of the altimetry missions used in this study are presented in Table 12.1.

12.3.3.3 Precipitation

We use precipitation data of the Tropical Rainfall Measuring Mission Project (TRMM-3B43 products; version 7, TRMM 2011; Huffman et al. 2012) to assess climate variability over Bangladesh. Incorporating more microwave sounding and imagery records as well as implementing better processing algorithms have caused a large improvement in this version of data (Huffman et al. 2012; Fleming and Awange 2013). The data sets, validated by Khandu et al. (2017) over the study region showed promising performance. The gridded ($0.25° \times 0.25°$) precipitation products (2003 to 2013) are converted to $1° \times 1°$ and used to investigate their connection to water storage changes.

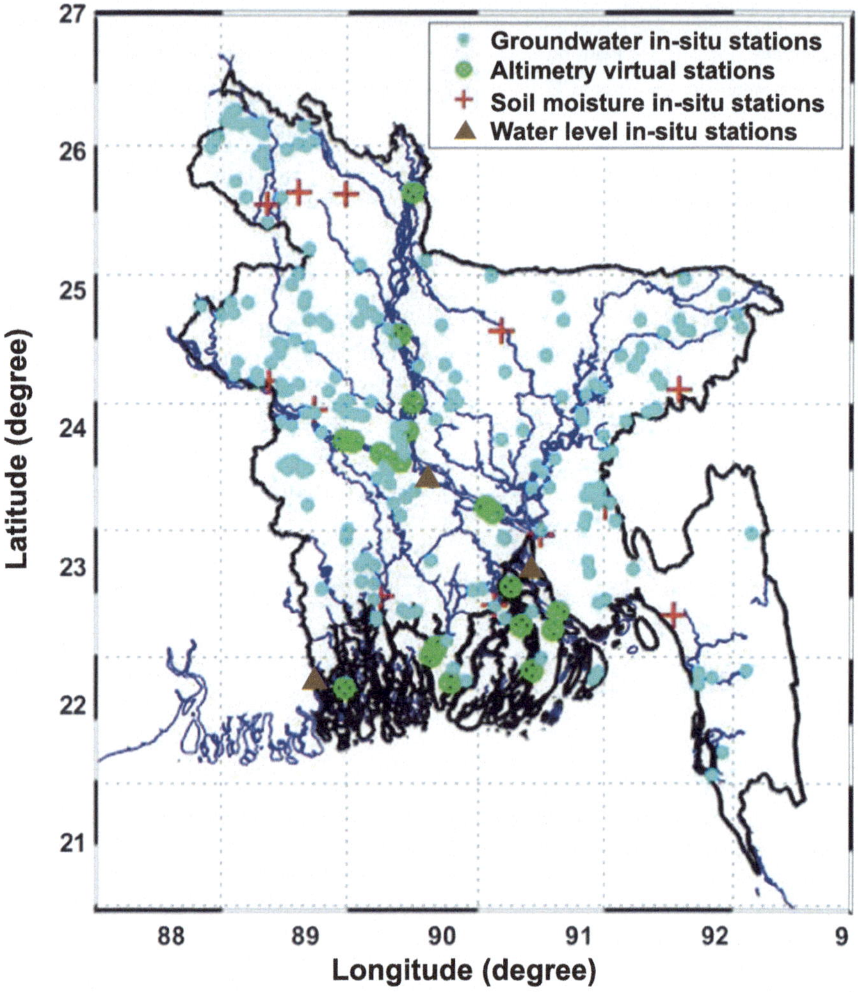

Fig. 12.1 The study area is represented by black solid line. The figure also contains the locations of virtual stations for satellite altimetry time series and various in-situ stations

12.3.4 Surface Storage Data

For the objective of data assimilation, considering that many surface water sources (in different forms, e.g., lakes and rivers except few major ones) are not modeled in W3RA, surface water storages should be removed from GRACE TWS data. To this end, we use satellite-derived surface water data in the Ganges–Brahmaputra River Basin (as the main source of surface water in Bangladesh) provided by Papa et al. (2015). The data is based on a multi-satellite approach that combines surface water extent from the Global Inundation Extent from Multi-Satellite (GIEMS, Papa et al. 2006, 2010b; Prigent et al. 2012) and level height variations of water bodies from Envisat radar altimetry to estimate surface water storage (Frappart et al. 2012) covering the period from 2003 to 2007. Since study period is 2003 to 2013, canonical correlation analysis is applied to extend the data from 2007 to 2013. Satellite derived river height fluctuations of Sect. 12.3.3.2 that are distributed across the study area are used in the process of extending the surface water storage of Papa et al. (2015). More details on CCA are provided in Sect. 12.4.4.

12.3.5 In-Situ Measurements

To evaluate the performance of data assimilation, 198 groundwater stations and 12 soil moisture stations (see Fig. 12.1) are used. These data are provided by the Bangladesh Water Development Board (BWDB) and Institute of Water Modelling (IWM) in The Asian Development Bank (2011). Figure 12.2 shows the sample products of different groundwater stations, as well as soil moisture variations measured at various depths. Specific yields ranging from 0.01 to 0.2 (Shamsudduha et al. 2011; BWDB 1994) are used to convert well-water levels to variations in groundwater storage. Details of the datasets used in this study are outlined in Table 12.1.

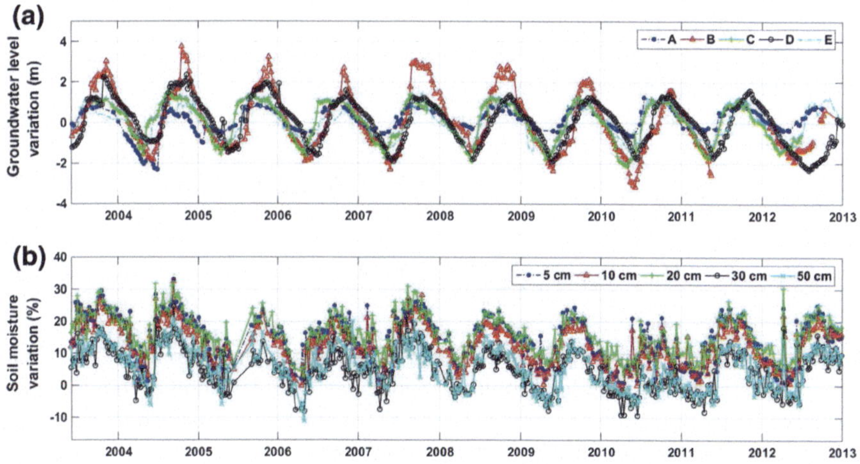

Fig. 12.2 **a** In-situ groundwater level variations of various stations (A: 21.790° latitude and 91.870° longitude, B: 24.450° latitude and 88.598° longitude, C: 26.190° latitude and 88.750° longitude, D: 25.590° latitude and 88.900° longitude, E: 22.110° latitude and 92.100° longitude). **b** Soil moisture variations at deferent depths belong to Rajshahi in-situ station (24.3670° latitude and 88.7000° longitude)

12.4 Method

12.4.1 Data Assimilation

12.4.1.1 Filtering Method

The square root analysis (SQRA) scheme for the ensemble Kalman filter (EnKF), presented in Evensen (2004) is used to assimilate the GRACE TWS into W3RA. SQRA, which is a deterministic form of ensemble-based Kalman filters, uses a statistical sample of state estimates and unlike traditional Kalman filtering method, does not need an observation perturbation (Burgers et al. 1998; Sakov and Oke 2008; Khaki et al. 2017b). Instead, by introducing a new sampling scheme, SQRA uses unperturbed observations without imposing any additional approximations like uncorrelated measurement errors (Evensen 2004). The update stage in SQRA includes two steps starting with updating the ensemble-mean as,

$$\bar{X}^a = \bar{X}^f + K(y - H\bar{X}^f), \quad i = 1 \ldots N, \tag{12.1}$$

$$K = P^f (H)^T (H P^f (H)^T + R)^{-1}, \tag{12.2}$$

where 'f' stands for forecast, 'a' for analysis, and N is the ensemble number. In Eq. 12.1, $\bar{X}^a$ is the mean analysis state, K represent the Kalman gain, and y is the observation vector. The transition and observation covariance matrices are indicated by H and R, respectively. $\bar{X}^f$, which represents the forecast ensemble mean, and the model state forecast error covariance (P^f) are derived by,

$$\bar{X}^f = \frac{1}{N}\sum_{i=1}^{N}(X_i), \tag{12.3}$$

$$P^f = \frac{1}{N-1}\sum_{i=1}^{N}(X_i{}^f - \bar{X}^f)(X_i{}^f - \bar{X}^f)^T. \tag{12.4}$$

The model state contains N different vectors of the model state variables. Note that $A^f = [A_1{}^f \ldots A_N{}^f]$ is the ensemble of anomalies, the deviation of model state ensembles from the ensemble mean ($A_i{}^f = X_i{}^f - \bar{X}^f$). In the second update step, SQRA computes the ensemble perturbations through.

$$A^a = A^f V \sqrt{I - \Sigma^T \Sigma \Theta^T}, \tag{12.5}$$

where Σ and V are result from singular value decomposition of A^f ($A^f = U \Sigma V^T$). Γ refers to the singular value decomposition and Θ is a random orthogonal matrix (e.g., the right singular vectors from a singular value decomposition of a random $N \times N$ matrix) for ensemble redistribution of the variance reduction (cf. Evensen 2004, 2007; Khaki et al. 2017b).

12.4.1.2 Assimilation of GRACE Data

To assimilate GRACE TWS into the model, we use a summation of model's vertical water compartments (e.g., soil moisture, groundwater, and surface water) at 13 grid points. This summation is then updated by the GRACE TWS at the same location at every assimilation step (whenever a new observation is available). Initial ensemble members are generated by perturbing the meteorological forcing fields following Renzullo et al. (2014). In this regard, the three most important forcing variables; precipitation, temperature, and radiation are perturbed using Monte Carlo sampling of multivariate normal distribution (with the errors representing the standard deviations) to produce an ensemble (with 72 members as suggested by Oke et al. 2008). The perturbed meteorological forcing datasets are then integrated forward with the model from 2000 to 2003 to provide a set of state vectors at the beginning of the study period.

Two widely used tuning techniques of ensemble inflation and localization are applied to enhance the assimilation performance especially when a limited ensemble size is assumed. Ensemble inflation uses a small coefficient (i.e., 1.12 in our study; Anderson 2001) to inflate prior ensemble deviation from the ensemble-mean to increase their variations and alleviate the inbreeding problem (Anderson et al. 2007). For localization, the Local Analysis (LA) scheme (Evensen 2003; Ott et al. 2004; Khaki et al. 2017a) is applied. LA restricts the impact of a given measurement in the update step to the points located within a certain distance (3° following Khaki et al. 2017a) from the measurement location. We also implement three different cases to deal with surface water storage during data assimilation.

- Case 1: Assimilating the GRACE TWS data after removing surface storages into the model states except for the surface water compartment.
- Case 2: Adding surface water storage to model surface water compartment and using the GRACE TWS to update the summations of all water compartments.
- Case 3: Assimilating the GRACE TWS to update the summations of all water compartments (including surface water compartment).

In Sect. 12.5.1, the results of all the case scenarios are compared with each other and evaluated against in-situ groundwater measurements.

12.4.2 Empirical Mode Decomposition (EMD)

The empirical mode decomposition (EMD) proposed by Chen et al. (2007) is used for analyzing multivariate datasets of this study. EMD establishes different frequencies and trends within time series, which are called Intrinsic Mode Functions (IMFs), by considering local oscillations (Rilling et al. 2003; Flandrin et al. 2004). The idea is that a signal is composed of fast oscillations superimposed by slow oscillations (Flandrin et al. 2004). Thus, EMD decomposes any complicated data set into a finite and often small number of IMFs hidden in the observations (Huang et al. 1998). We apply EMD on all available time series of this study to extract different frequencies and also to find local trends for a better understanding of their interrelationships.

12.4.3 Retracking Scheme

In this study, we use the retracking method to improve altimetry estimations of river height variations. The retracking process is essential since complex waveform patterns are usually observed over rivers. To this end, Extrema Retracking (ExtR) post-processing technique (Khaki et al. 2014, 2015) is used. The ExtR is a three step filter that starts by applying a moving average filter to reduce the random noise of the waveforms. It then identifies extremum points of the filtered waveforms, and finally, extracts the main leading edge amongst the established extremum points. The method is applied to process different types of waveforms and improve level estimations as demonstrated in Khaki et al. (2014, 2015). The filter is employed here to retrack satellite radar altimetry data and then removing surface storage from TWS (see Sect. 12.4.4). Figure 12.12 shows river level fluctuations for different parts of Bangladesh (Fig. 12.3a) and the entire area of the country (Fig. 12.3b).

12.4.4 Canonical Correlation Analysis (CCA)

Canonical correlation analysis (CCA) seeks to find the linear relationship between two sets of multidimensional variables x and y. The process extracts canonical coefficients u and v such that $X = x^T u$ and $Y = y^T v$ (X and Y are canonical variates) possess a maximum correlation coefficient (Chang et al. 2013) using the following function,

$$\begin{aligned} P &= \frac{E[XY]}{sqrt(E[X^2]E[Y^2])} \\ &= \frac{E[u^T xy^T v]}{sqrt(E[u^T xx^T u]E[v^T yy^T v])} \\ &= \frac{u^T C_{xy} v}{sqrt(u^T C_{xx} u v^T C_{yy} v])}, \end{aligned} \tag{12.6}$$

where C_{xx} and C_{yy} are covariance matrices of x and y respectively and the objective in above function is to maximize the correlation P. Once the coefficients are calculated, they can be used to find the projection of x and y onto u and v as canonical variates with maximum correlation. Here, x contains the vectors of surface water storages from Papa et al. (2015) at each grid point and y includes river heights variations from satellite radar altimetry in a same temporal scale as the former data (2003 to 2007). After performing canonical correlation analysis, the computed canonical coefficient of u and v, and a new set of variables y (from 2007 to 2013) are used to estimate the canonical variate of x. The combination of surface water storages (x) using the extracted u from the first part has the maximum correlation to the altimetry-derived river heights variability. Hence, this coefficient vector can be used to transform river heights into surface waters at each grid point.

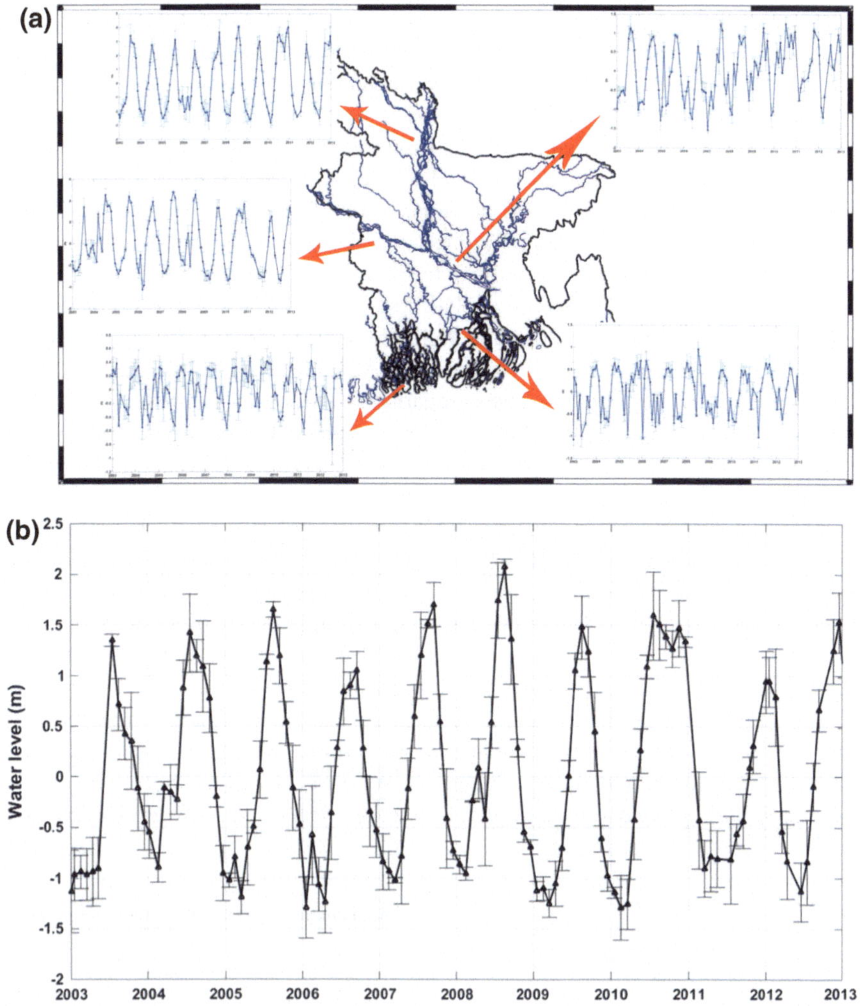

Fig. 12.3 Average river height variation time series from satellite radar altimetry for different parts (**a**) and for the entire area (**b**) of Bangladesh. The average error for each measurement is presented as error bars

12.5 Results

12.5.1 Data Assimilation

Before discussing groundwater and soil moisture variations within Bangladesh, the effect of data assimilation on terrestrial water storage time series and its capability to improve model simulations are investigated. Figure 12.4 shows average TWS time series within Bangladesh before (model-free run) and after data assimilation. The figure also contains GRACE TWS time series. It can be seen that data assimilation largely reduces misfits between model-free run and observations.

To assess whether data assimilation (e.g., in Fig. 12.4) can result in better water storage estimates, in-situ groundwater and soil moisture measurements are used for validation. Time series of groundwater and soil moisture anomalies are generated for each station. Groundwater and soil moisture results from all the three assimilation cases (cf. Sect. 12.4.1.2) are spatially interpolated using the nearest neighbour (the closest four data values) to the location of the in-situ measurements. This is also done for outputs of the WaterGAP Global Hydrology Model (WGHM; more details on Döll et al. 2003; Müller et al. 2014), as well as estimated storages by van Dijk et al. (2014), indicated here by W3, who merged GRACE observations with an ensemble of hydrological model outputs. The comparison between these products and data assimilation results allows us to better investigate any achieved improvements. For this purpose, the RMSE and correlations between in-situ and estimated time series are calculated.

Table 12.2 summarizes the average RMSE and correlation for each of the three data assimilation case. From Table 12.2, it can be seen that the groundwater results are more correlated to in-situ measurements after the application of each assimilation case (0.81 on average), 0.39 larger than model simulations without applying data assimilation (model-free run). An average RMSE improvement of 51.16% (at 0.95 confidence level) in case 1 shows a significant influence of the data assimilation scheme, approximately 4.44% and 39.11% larger than cases 2 and 3, respectively. It is also evident from Table 12.2 that data assimilation results, especially cases 1 and 2 outperform groundwater estimates of WGHM. Note that the provided W3 does not include groundwater and therefore we use it for soil moisture comparison

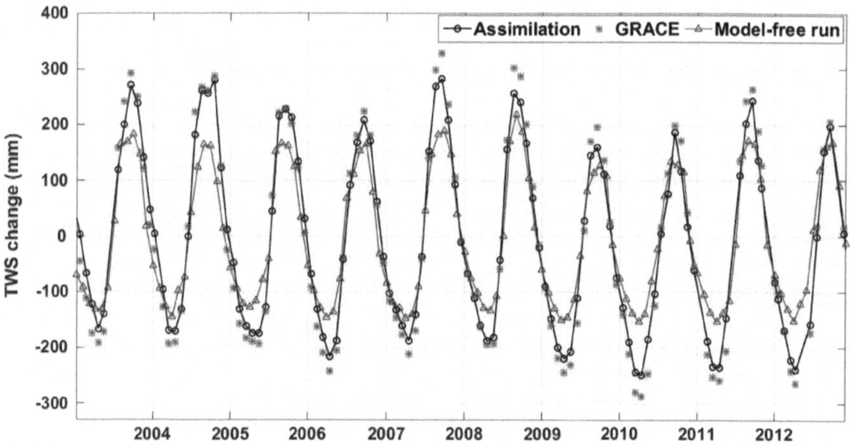

Fig. 12.4 Average TWS change time series from data assimilation (case 1), model-free run, and GRACE

Table 12.2 Statistics of groundwater errors. For each case, the RMSE average and its range (±XX) at the 95% confidence interval is presented. Improvements in data assimilation results are calculated with respect to the groundwater storages from the model without implementing data assimilation

			Improvement (%)	
Assimilation scenario	Correlation	RMSE (mm)	Correlation	RMSE (mm)
Case 1 [Removed surface storages from GRACE TWS]	0.86	35 ± 5.65	51.16	57.36
Case 2 [Added surface storages to W3RA surface water]	0.82	39 ± 5.18	48.78	52.92
Case 3 [No surface storage correction applied]	0.75	68 ± 7.72	44.02	18.25
WGHM	0.79	57 ± 5.37	46.83	30.89
Mode-free run	0.42	83 ± 9.29	–	–

Table 12.3 Average correlations improvements (at 95% confidence interval) between in-situ and soil moisture estimates with respect to model-free run

Filter	0–10 cm	0–20 cm	0–50 cm
Case 1	10.42	19.27	25.25
Case 2	11.10	17.88	24.48
Case 3	5.25	8.34	12.91
WGHM	–	–	17.51
W3	–	–	15.23

only. Table 12.2 emphasizes that model groundwater estimations can be successfully improved with respect to the in-situ measurements if they are fine tuned by GRACE data through the assimilation especially for cases 1 and 2.

Furthermore, correlation analysis is carried out between in-situ soil moisture measurements at various depths and data assimilation results from different scenarios, as well as soil moisture estimates of WGHM and W3 (Table 12.3). In-situ measurements at different depths are compared with different layers from data assimilation results. For this purpose, in-situ soil moisture time series of 0–10 cm, 0–30 cm, and 0–50 cm depths are compared to the model top, shallow, and deep soil moisture layers. While the model soil moisture of top layer corresponds to the thickness between 5 and 10 cm, the model shallow and deep-root soil layers broadly represent 10–21 cm and 3–6 m soil thicknesses (see also Renzullo et al. 2014; Tian et al. 2017). Here, we compare W3RA's top layer estimations with in-situ of 0–10 cm, top layer plus shallow-root simulations with in-situ of 0–20 cm, and summation of the top, shallow, and a portion of deep-root soil layers with 0–50 cm in-situ measurements. Note that WGHM and W3 outputs are provided at a single aggregated layer and correspondingly are compared with in-situ soil time series at the depth 0–50 cm. Table 12.3 shows that the highest correlation improvements, 18.31% (on average) for all layers and 25.25% for the deep layer. Case 2 also represents considerable improvements slightly smaller than case 1, still 11.57% larger than case 3, 6.97% larger than WGHM, and 9.25% larger than W3. Both Tables 12.2 and 12.3 demonstrate a high capability of data assimilation in improving model simulations of different compartments. These tables also indicate a better performance of the implemented data assimilation, specifically cases 1 and 2, compared to WGHM and W3.

To better analyze the differences between each assimilation case, we compare their RMSE during 2007. In 2007, a major flooding (following ENSO rains) occurred across South Asia affecting Bangladesh (Gaiha et al. 2010). This phenomenon can help us to monitor performances of each case in such an extreme situation and their ability to distribute observed TWS between all water compartments. Groundwater estimates from each case and in-situ measurements are used to calculate RMSE for each assimilation case (Fig. 12.5), where the least errors are estimated by cases 1 and 2. Assimilating the GRACE TWS without considering surface water storage within the area (case 3) causes larger errors especially in April and September. The largest error, however, is obtained for the model-free run. Hereafter, we use the result of data assimilation for case 1 since it performed slightly better than case 2 and significantly better than case 3 in terms of the RMSE (see Fig. 12.5).

The model's water storage variations computed by assimilating GRACE TWS data into W3RA are presented in Figs. 12.6 and 12.7. Temporal averages of soil moisture and groundwater storage variations for each grid point from data assimilation, WGHM, and W3 in the study area are displayed in Figs. 12.6 and 12.7, respectively. We find large correlations between assimilation results and WGHM (0.76 on average for soil moisture and 0.82 on average for groundwater) and W3 (0.71 on average for soil moisture) outputs. The results show more negative groundwater variations within

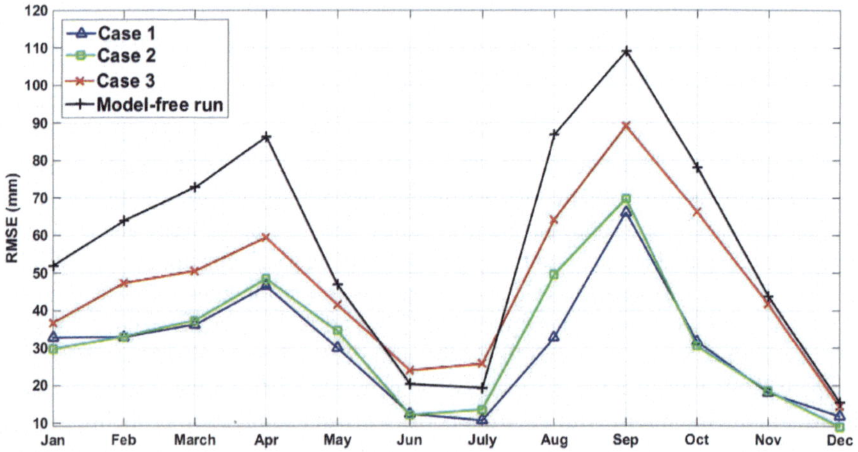

Fig. 12.5 Comparison between RMSE achieved from implementing each data assimilation scenario as well as model-free run during 2007. In case 1, surface storages is removed from GRACE TWS, in case 2, surface storages is added to W3RA surface water, and case 3 refers to the data assimilation with no surface storage correction

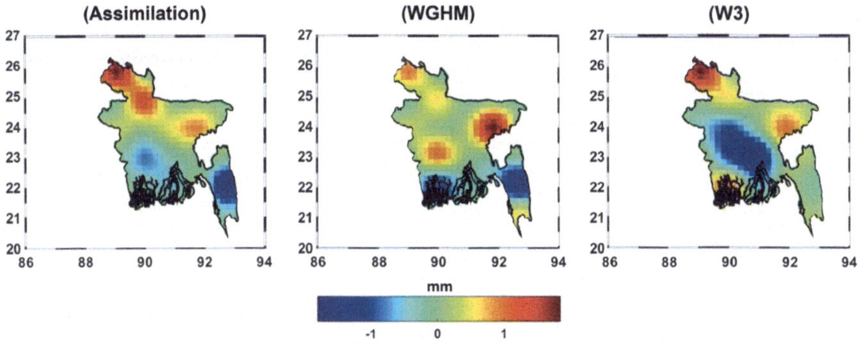

Fig. 12.6 Spatial distribution of average soil water storage variations from data assimilation, WGHM, and W3

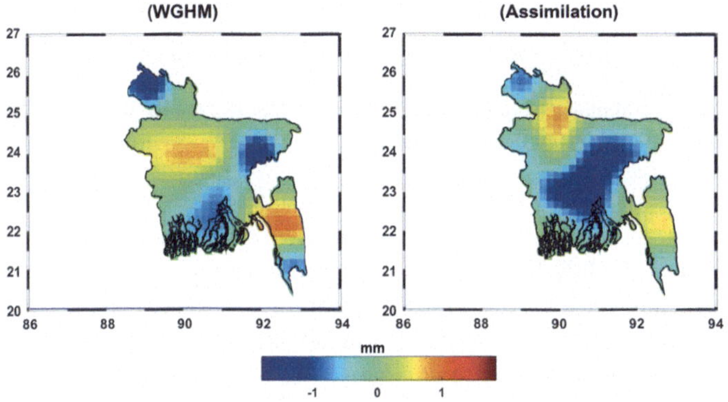

Fig. 12.7 Spatial distribution of average groundwater storage variations from data assimilation and WGHM

different parts of Bangladesh than soil water storage variations (see Fig. 12.6). Both water compartments indicate larger signals (in terms of amplitude) in the central and northwestern parts of Bangladesh. A positive soil moisture variations are found in the centre toward east and north within the study period, especially for the assimilation and WGHM maps. Larger groundwater variations are also captured in the same area. While assimilation results show negative groundwater changes over the central, eastern, and to a lesser degree southern parts, WGHM only indicates negative variations in the southern and eastern parts. Figure 12.7 indicate that smaller water storage variations in the northwestern and northeastern parts of Bangladesh during 2003–2013.

The average time series of soil moisture and groundwater storages from data assimilation are shown in Figs. 12.8a, b, respectively. We estimate spatial averages for all the time series at grid points for this figure. Figure 12.8a shows slight declines in the soil water storage after 2007, which can be related to variations of surface water storage in the same period. The correlation between the surface water storage and soil

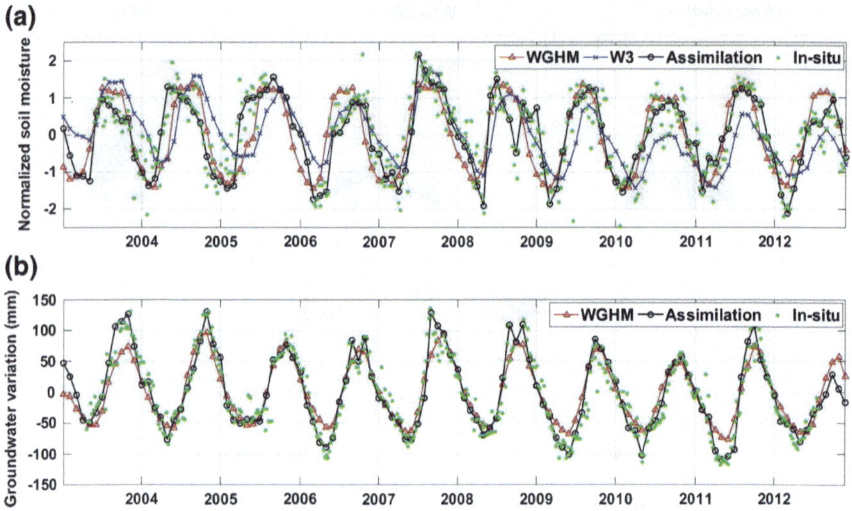

Fig. 12.8 Average soil moisture storage (**a**) and groundwater storage (**b**) time series from assimilation, WGHM, W3, and in-situ measurements

moisture time series (after removing seasonal effects) is found to be 0.92 (for a 95% confidence interval), 34% higher than the correlation between groundwater and soil moisture. This indicates that a stronger connection exists between the surface water storage and soil moisture over the area. Annual variations of groundwater storages, however, show a larger decline in comparison to soil moisture storage variations, especially between 2008 and 2012. A significant decrease in groundwater storage is seen in Fig. 12.8b with an average rate of 8.73 ± 2.45 mm/year, showing an overall $\sim$46% reduction. The decline in water availability can be due to over-extraction of groundwater resources since such a decrease is not seen in precipitation (see Sect. 12.5.2 for more details).

12.5.2 Statistical Analyses

First, the precipitation and TWS over Bangladesh is analyzed. To explore the climate variability and its relationship with water storages, precipitation will be compared to the data assimilation results. Principal component analysis (PCA Lorenz 1956) is applied on GRACE TWS and precipitation time series at each grid point to explore their spatio-temporal variations. The first three most dominant empirical orthogonal functions (EOF1, EOF2, EOF3) for each variable are presented in Fig. 12.9. The spatial distribution of precipitation within Bangladesh indicates larger rainfall in south-eastern parts. TWS distribution in EOF2 follows the same pattern. A large water storages are captured by EOF3 in the northwest. More information can be extracted

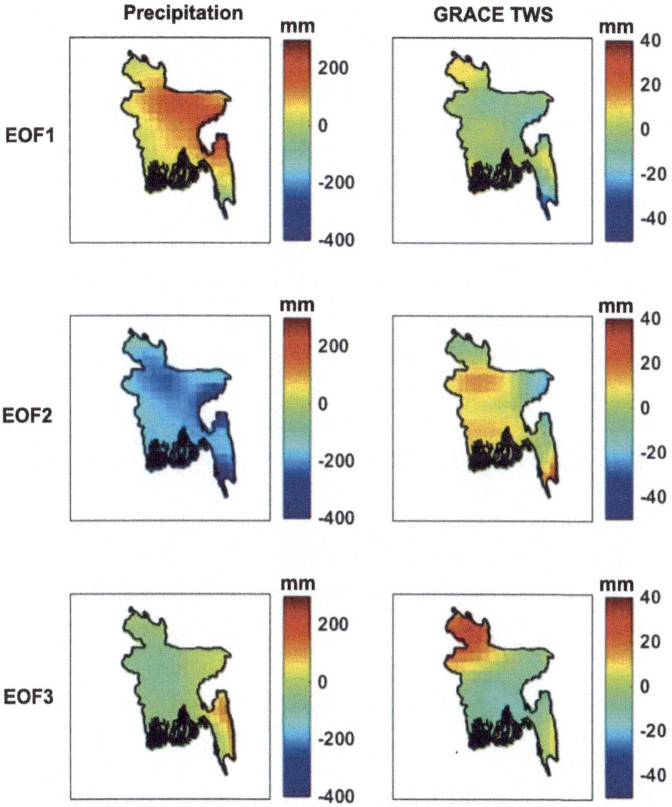

Fig. 12.9 Spatial distribution of EOF1, EOF2, and EOF3 from applying PCA on precipitation and GRACE TWS

from precipitation and TWS time series. The first three principal components (PC1, PC2, and PC3) of each data set are shown in Fig. 12.10. Large precipitation impacts are found in 2003 and 2007. A negative anomaly in rainfall is found in 2010 and 2012, as well as in the period between 2005 and 2007 (PC1). TWS time series demonstrate declines between 2005 and 2007, and particularly after 2009 following a drop event (Mondol et al. 2017). Results, however, show an increase in 2007 in agreement with ENSO rainfall. The overall average of TWS variations during the study period is negative ($\sim$11.48 $\pm$ 3.19 mm/year) for the entire country. A similar trend, however, is not observed in precipitation even though there is a shorter period negative decline in 2005 and after 2010. Figure 12.10 illustrates that although in some cases a variation in precipitation results in a changes in TWS, continuous TWS reduction possibly has different explanations that could become clear through the separation of the groundwater and surface water storages (cf. Section 12.4.4).

Details on surface and groundwater storage variations and their relationship to precipitation and rivers' level heights are presented in Table 12.4. For each grid point

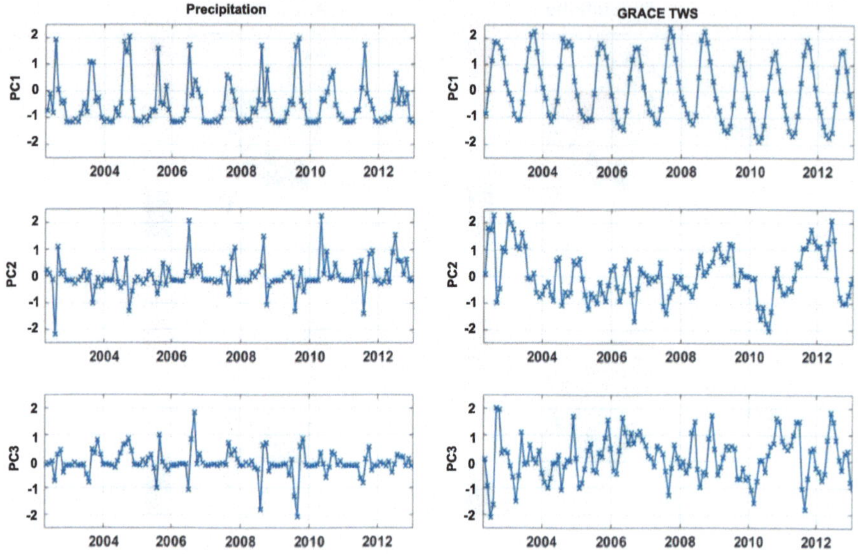

Fig. 12.10 The first three principal components from applying PCA on precipitation and GRACE TWS

Table 12.4 Statistics of water storage variations

Water storage	Variation rate (mm/year)	Depletion (%)			Correlation (95% confidence interval)	
		Min.	Max.	Mean	Precipitation	Water level height
Surface water	−1.54	0	38	11	0.74	0.81
Groundwater	−8.73	12	41	32	0.59	0.63

in the study area, we calculate water storage variation rates and depletions, and also a correlation coefficient between their time series and both precipitation and river height variations. Note that we use lag-correlation (cross correlation) to achieve the maximum correlation between each two time series. Table 12.4 illustrates that there is a water decline in both surface and groundwater storages at different rates. This can be inferred from the negative storage variation rates. An approximately 32% depletion in groundwater storage causes a significant decrease in TWS as shown in Fig. 12.10. This remarkable water reduction, unlike the rainfall pattern, is highly related to excessive groundwater usages, especially for irrigation. It can be concluded from Table 12.4, therefore, that groundwater storages are less correlated (16.5%) to river height variations and precipitation, respectively, in comparison to surface water storage. Consequently, variations in rainfalls and river heights are more reflected in surface storage variations.

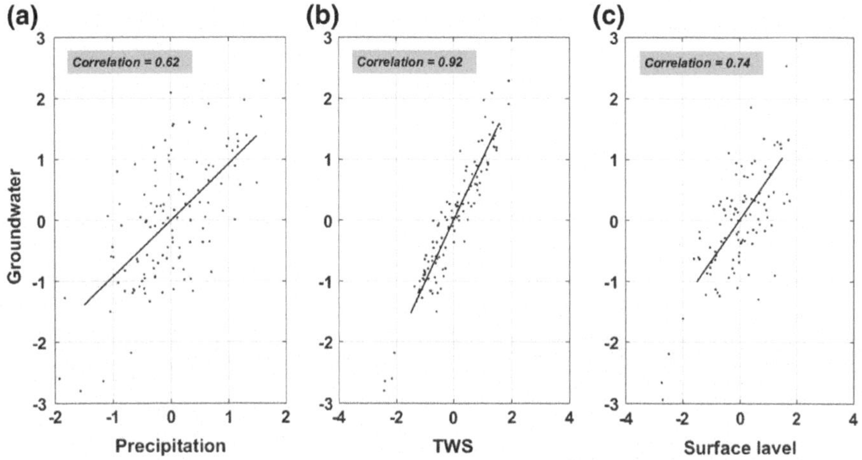

Fig. 12.11 Relationships between normalized Intrinsic Mode Functions (IMF) time series of groundwater and precipitation, TWS, and surface river height

To better analyze groundwater storage changes, we apply empirical mode decomposition (EMD) on time series in each grid point. EMD is used to extract Intrinsic Mode Functions (IMFs) of time series that are found to be most representative of the initial signals. The relationships between the groundwater IMFs and those of precipitation, TWS, and surface river fluctuations are shown in Fig. 12.11, which contains scatter bi-plots and the interpolated line representing the correspondence between two variables. The trend lines in the sub-figures show that the computed IMFs for the different variables are close to each other. The concentration of distributed points after applying EMD is more symmetric than for the initial time series especially for the groundwater and TWS, as well as the groundwater and water level variations. Table 12.5 contains the average correlation between the time series of groundwater and the variables of precipitation, TWS, and river height variation. The more symmetric distributed points in between the groundwater IMF and that of GRACE TWS shows the greater relationship between these two variables corresponding to a higher correlation presented in Table 12.5. The reason for this can be due to the use of GRACE TWS in data assimilation. A higher correlation is also obtained between the IMF of groundwater and those of river height. The least relationship is obtained for the groundwater IMF and precipitation, that implies the different pattern in variations of these two variables, which could be related to the non-climatic effects in the groundwater.

The extracted first two IMFs for the groundwater time series are illustrated in Fig. 12.12. In the both subfigures, a decline in groundwater storages is observed. Such a trend, however, is more significant for IMF 2. We also plot the first and second precipitation's IMFs for comparison. The precipitation's IMF 1 in Fig. 12.12, better indicates rainfall variation from Fig. 12.10. Two periods with larger rainfall can be seen for the years 2006 and 2009. A decrease in rainfall over Bangladesh is found

Table 12.5 Groundwater storage correlation to precipitation, TWS, and river level height variations

	Precipitation	GRACE TWS	River level height
Before EMD	0.57	0.73	0.63
After EMD	0.71	0.88	0.77
Improvement (%)	12	15	14

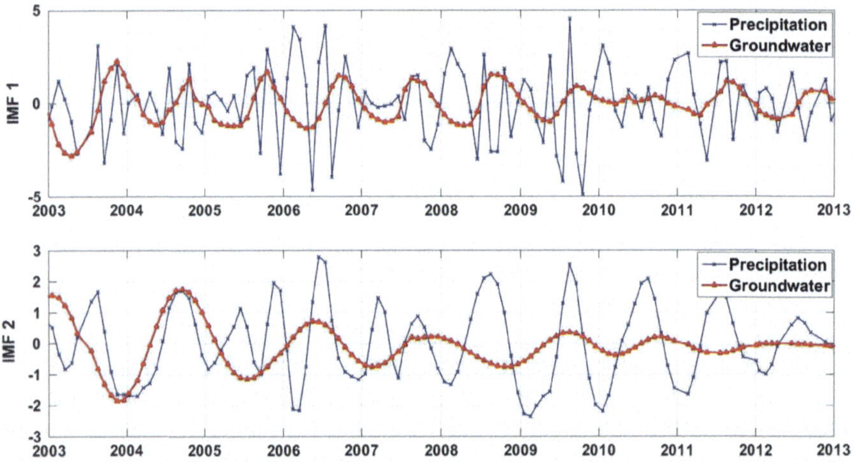

Fig. 12.12 The first and second extracted Intrinsic Mode Functions (IMF) time series of the groundwater storage (red) and precipitation (blue)

from 2010 onwards, with smaller amplitudes during 2010 and 2012. This may impact the groundwater levels during similar temporal periods. There are several similar patterns in both time series (groundwater and precipitation) especially for IMF 1. Both groundwater and precipitation IMFs increase during 2006 and mid-2008 to mid-2009. Figure 12.12b, presenting IMF 2 time series of assimilated groundwater, clearly shows the groundwater depletion despite having minimum changes in precipitations. This suggests that other factors (e.g., human impacts) affect groundwater storages in Bangladesh.

12.6 Conclusion

Terrestrial waters, as an essential factor for both human life and environment, can be affected by climate changes, especially over the south Asian areas. Bangladesh, in particular, is a highly vulnerable region in facing climate changes suffering from serious water issues, especially for irrigation. In this study, we analyze groundwater variations within Bangladesh using multi-mission satellite measurements, as well as by running a hydrological model during 2003 to 2013. The Gravity Recovery And Climate Experiment (GRACE) terrestrial water storage (TWS) data after removing surface water storages is assimilated into W3RA model using the ensemble-based sequential technique of the Square Root Analysis (SQRA) filter. This is done to improve the World-Wide Water Resources Assessment system (W3RA) simulations of groundwater, as well as soil water storages. We also apply the empirical mode decomposition (EMD) on water storages, precipitation, and altimetry-derived rivers level variations time series to explore the relationship between them in the area. The larger correlation is found between river level heights and rainfalls (78% average) in comparison to groundwater storage variations and rainfalls (57% average). The considerable difference between correlation coefficients indicates a different impact of rainfall on surface and groundwater variations, which could imply influences of groundwater depletion by population (especially for excessive irrigations) across the country. The results show an approximately 26%, groundwater depletion with a remarkable influence on a total water stored in the area. A significant decline in groundwater storage (~32% reduction over the study period) over the country is found by the assimilation results with an average rate of 8.73 mm/year. In the absence of any considerable decrease in precipitation over the region, a remarkable groundwater reduction is observed from the first and second Intrinsic Mode Functions (IMFs), which can be referred to human impacts. High spatio-temporal resolution remote sensing products along with the data assimilation methodology show a high capability for studying water storages in Bangladesh. Developing Earth observation missions dedicated to hydrology (GRACE follow-on and SWOT) can be very important to pursue and improve such modeling and assimilation studies.

Acknowledgements We would like to thank Dr. Ashraf Dewan and who provided us valuable groundwater and soil moisture in-situ measurements over Bangladesh. Authors are also grateful to BWDB (http://www.hydrology.bwdb.gov.bd/) for in-situ data. M. Khaki is grateful for the research grant of Curtin International Postgraduate Research Scholarships (CIPRS)/ORD Scholarship provided by Curtin University (Australia). This work is a TIGeR publication and supported by the Belmont Forum/G8 Project on coastal vulnerability, in the US via the National Science Foundation (Grant No. ICER-1342644).

Chapter 13
Multi-mission Satellite Data Assimilation over South America

Assimilation of GRACE TWS data allows users to consistently separate TWS (since both model and observation errors are considered) into different water compartments that include groundwater and soil moisture. This is due to the fact that the hydrologic model relies on the physical processes implemented in the model equations. Besides, GRACE-derived TWS observations are spatially downscaled using this approach, and therefore, higher spatial resolution estimations of water storages will be available within the study region

—M. Khaki (This chapter is presented following Khaki et al. (2018d), "The application of multi-mission satellite data assimilation for studying water storage changes over South America")

13.1 Summary

Constant monitoring of total water storage (TWS; surface, groundwater, and soil moisture) is essential for water management and policy decisions, especially due to the impacts of climate change and anthropogenic factors. Moreover, for most countries in Africa, Asia, and South America that depend on soil moisture and groundwater for agricultural productivity, monitoring of climate change and anthropogenic impacts on TWS becomes crucial. Hydrological models are widely being used to monitor water storage changes in various regions around the world. Such models, however, comes with uncertainties mainly due to data limitations that warrant enhancement from remotely sensed satellite products. In this study over South America, remotely sensed TWS from the Gravity Recovery And Climate Experiment (GRACE) satellite mission is used to constrain the World-Wide Water Resources Assessment (W3RA) model estimates in order to improve their reliabilities. To this end, GRACE-derived TWS and soil moisture observations from the Advanced Microwave Scanning Radiometer—Earth Observing System (AMSR-E) and Soil Moisture and Ocean Salinity (SMOS) are assimilated into W3RA using the Ensemble Square-Root Filter (EnSRF) in order to separately analyze groundwater and soil

M. Khaki, *Satellite Remote Sensing in Hydrological Data Assimilation*,
https://doi.org/10.1007/978-3-030-37375-7_13

moisture changes for the period 2002–2013. Following the assimilation analysis, Tropical Rainfall Measuring Mission (TRMM)'s rainfall data over 15 major basins of South America and El Niño/Southern Oscillation (ENSO) data are employed to demonstrate the advantages gained by the model from the assimilation of GRACE TWS and satellite soil moisture products in studying climatically induced TWS changes. From the results, it can be seen that assimilating these observations improves the performance of W3RA hydrological model. Significant improvements are also achieved as seen from increased correlations between TWS products and both precipitation and ENSO over a majority of basins. The improved knowledge of sub-surface water storages, especially groundwater and soil moisture variations, can be largely helpful for agricultural productivity over South America.

13.2 Introduction

South America, with unique ecosystems and a high biodiversity, has extreme geographic variations and diverse patterns of weather and climate that include tropical, subtropical and extratropical features (Garreaud et al. 2008). The region is largely under the influence of large-scale ocean-atmosphere phenomena including mainly El Niño Southern Oscillation (ENSO) and the North Atlantic Oscillation (NAO), which affects climate and its phases associated with droughts, floods, and extreme weather events within different parts of the continent (Magrin et al. 2007; Tedeschi and Collins 2016). Climate variability throughout South America can be categorized based on the distance from the equator and the altitude of the area. The Andes mountain ranges, running along South America's western side, plays an important role in tropical as well as subtropical latitudes by keeping dry conditions on the west and moist conditions on the east (Garreaud et al. 2008). These climate variabilities, e.g., due to the different climatic zones across the continent and/or large-scale ocean-atmosphere phenomena, have significant impacts on the continent's water storages (surface water, groundwater, soil moisture, and vegetation water). There are other important factors that largely threaten water resources such as excessive water use, especially for agricultural purposes (Grau and Aide 2008; Magrin et al. 2014). Therefore, the study of South America's water storage changes in light of the climate change and anthropogenic impacts is necessary for any future water use planning.

To study South America's water storage changes at high spatio-temporal resolutions, hydrological models have come in handy (e.g., Betts et al. 1996; Koster et al. 1999; Döll et al. 2003; van Dijk 2010; De Paiva et al. 2013; Getirana et al. 2014), particularly over the regions with a few ground-based observations such as Venezuela, Ecuador, Chile, and Peru. The applications of these models are especially important for agriculture and sustainable water managements (e.g., Bharati et al. 2008; Yu et al. 2015; Kourgialas and Karatzas 2015). However, in general, data limitations and other factors, e.g., imperfect modeling and uncertainties of model parameters can weaken performances of the models for simulation of hydrological processes (van Dijk et al. 2011; Vrugt et al. 2013). In this regards, data assimilation provides a unique opportunity to improve model reliabilities (Bertino et al. 2003). This approach integrates additional observations that have not been considered in those models into their dynamics to constrain its state estimates (Bertino et al. 2003; Hoteit et al. 2012).

Data assimilation has been used in different applications, e.g., atmospheric fields (Elbern and Schmidt 2001; Schunk et al. 2004; Altaf et al. 2014), oceanic (Bennett 2002; Lahoz 2007) and magnetospheric (Garner et al. 1999) studies. The method has also been applied in hydrological contexts to increase models' performances for estimating various water compartments (e.g., Reichle et al. 2002; Alsdorf et al. 2007; Goncalves et al. 2009; Renzullo et al. 2014; Dillon et al. 2016; Khaki et al. 2018b, c). The use of models to study hydrological variables over South America are reported, e.g., in the works of Yates et al. (1997), Chou et al. (2002), Grimson et al. (2013), and Erfanian et al. (2017), who investigate the application of the models on hydrological resources, droughts, and water storage changes. In the works above, the limitations have been that the models have not incorporated remotely sensed hydrological products such as the Gravity Recovery And Climate Experiment (GRACE) with a large capability of estimating terrestrial water storage (TWS) changes.

The main objective of the present study is, therefore, to use multimission satellite data products to improve hydrological model estimates of sub-surface water storages over South America. For this purpose, GRACE-derived TWS and soil moisture observations from the Advanced Microwave Scanning Radiometer - Earth Observing System (AMSR-E) and Soil Moisture and Ocean Salinity (SMOS) are assimilated into the World-Wide Water Resources Assessment (W3RA) hydrological model (van Dijk 2010). The model has been applied at different continental and global studies including South America (e.g., van Dijk et al. 2013, 2014; Beck et al. 2016; Schellekens et al. 2017). In terms of observations, several studies indicate that using GRACE TWS (e.g., Zaitchik et al. 2008; Houborg et al. 2012; Li et al. 2012; Eicker et al. 2014; Li et al. 2015; Reager et al. 2015; Li and Rodell 2015; Kumar et al. 2016; Girotto et al. 2016; Khaki et al. 2017a, b; Girotto et al. 2017; Khaki et al. 2018f) and satellite soil moisture (e.g., Tian et al. 2008; Renzullo et al. 2014; Dumedah et al. 2015; Tian et al. 2017; Kolassa et al. 2017) for data assimilation can successfully constrain the hydrological models simulations. The present study aims at investigating the effectiveness of multi-satellite data assimilation for studying sub-surface water storage changes using a non-regional hydrological model. It should be pointed out that although similar studies by the authors have been undertaken for other regions and using different products (e.g., Khaki et al. 2017c, 2018e), the main distinction and innovativeness between the current work over South America and those undertaken by the authors above, is that for the first time, both GRACE TWS and soil moisture products are employed in assimilation over the area. Furthermore, the contribution of climate variability on South America's water storage derived from assimilation using satellite precipitation products is also investigated.

Assimilation of GRACE TWS data allows users to consistently separate TWS (since both model and observation errors are considered) into different water compartments that include groundwater and soil moisture. This is due to the fact that the W3RA model relies on the physical processes implemented in the model equations.

Besides, GRACE-derived TWS observations are spatially downscaled using this approach, and therefore, higher spatial resolution estimations of water storages will be available within the study region (see, e.g., Schumacher et al. 2016). Moreover, the application of soil moisture observations in the assimilation can improve the performance of the process by separately updating model soil moisture estimates (e.g., Tian et al. 2017). For the purpose of data assimilation, here, we use the ensemble-based sequential technique of the Ensemble Square-Root Filter (EnSRF) filtering scheme (Whitaker and Hamill 2002) to integrate GRACE TWS into W3RA. EnSRF, as shown in Khaki et al. (2017b), is preferred over the traditional ensemble Kalman filter (e.g., Evensen 2003, 2007; Eicker et al. 2014) due to its higher computational speed, simplicity, and independence of perturbed observations.

Following the assimilation step, in-situ measurements are used to assess the performance of the approach. Furthermore, the study investigates the use of the model to study climate induced water storage changes by comparing correlations between assimilated and non-assimilated results with climate variability indicators of the Tropical Rainfall Measuring Mission (TRMM) rainfall and ENSO ocean-atmospheric couple indicator. For a better discussion, the study area is divided into 15 major basins selected (Fig. 13.1) based on their importance and large hydro-climatic effects, which also allow us to spatially have a better analysis. We also apply principal component analysis (PCA, Lorenz 1956) on the TRMM rainfall data, groundwater, soil moisture results from model over each basin to better understand the spatial and temporal variations of water storages and their interactions with precipitation. Frappart et al. (2013) found that PCA modes can better represent spatiotemporal variations in time series compared to the full signals by separating dominant water mass change signals, especially over South America (see also Abelen et al. 2015).

13.3 Materials and Methods

13.3.1 W3RA Hydrological Model

Vertical water compartments (e.g., soil moisture, groundwater, and surface water) of the globally distributed $1° \times 1°$ World-Wide Water Resources Assessment system (W3RA; http://www.wenfo.org/wald/data-software/) model are used to simulate water storage over South America. The model was developed in 2008 by the Commonwealth Scientific and Industrial Research Organisation (CSIRO; Australia) to simulate water storages (van Dijk 2010). In terms of forcing data, minimum and maximum temperature, downwelling short-wave radiation, and precipitation products provided by Princeton University (http://hydrology.princeton.edu) are used. Daily W3RA estimates of top, shallow, and deep root soil layers, groundwater storage, and surface water storage in a one-dimensional system (vertical variability) are used for data assimilation (see details in Sect. 13.3.6).

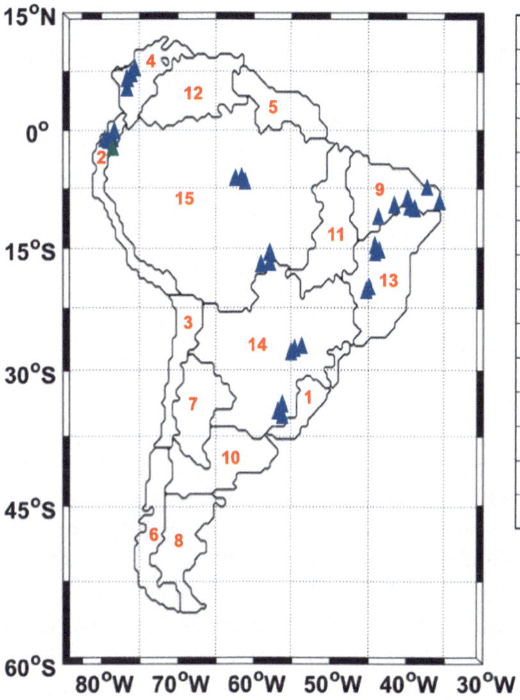

1	South-east Atlantic
2	Pacific Coast, Peru
3	Pacific Coast, North Chile
4	Magdalena
5	Atlantic North Coast
6	Pacific Coast, South Chile
7	Colorado Basin
8	Atlantic South Coast
9	North-east Atlantic
10	Negro Basin
11	Tocantins
12	Orinoco
13	Sao Francisco
14	La Plata
15	Amazon Basin

Fig. 13.1 Overview of the study area. The black polygons indicate the 15 river basins considered. These basins are selected according to HydroSHEDS (http://www.hydrosheds.org/) classification with small modifications to combine smaller basins and also for a better representation. The basins are sorted according to their areas. Data from in-situ groundwater stations (blue triangles) are used to provide independent validation of the assimilation results

13.3.2 *Remotely Sensed Observations (GRACE, Soil Moisture and TRMM Products)*

13.3.2.1 GRACE TWS

Monthly TWS observations at a $3° \times 3°$ spatial resolution (suggested by Khaki et al. 2017a, for data assimilation objectives) derived from the GRACE level 2 (L2) monthly Stokes's coefficients (following Wahr et al. 1998) up to degree and order 90 are used for the assimilation. L2 products along with their full error information are obtained from the ITSG-Grace2014 gravity field model (Mayer-Gürr et al. 2014) for the period between 2002 and 2013. Before converting L2 data into TWS, low degree coefficients of 1 and 2 (C20) are respectively replaced by those estimated by Swenson et al. (2008) and Satellite Laser Ranging solutions, respectively, to account for the change in the Earth's center of mass and large uncertainties (e.g., Cheng and Tapley 2004; Chen et al. 2007). The DDK2 smoothing filter by Kusche et al.

(2009) is applied to tackle colored/correlated noises in spherical harmonics. In order to reduce leakage effects, for every one of the 15 basins considered, an isotropic kernel using a Lagrange multiplier filter proposed by Swenson and Wahr (2002) is applied. This approach reduces short wavelength effects using Lagrange multiplier to minimize the leakage for a given value of satellite error. Here, the satellite error is selected based on the acquired GRACE full error covariance matrix. Khaki et al. (2018a) showed that this filtering technique can effectively reduce leakage errors, e.g., over Amazon basin. Finally, the mean TWS for the study period is taken from W3RA and added to the GRACE TWS change time series to obtain absolute values and make them comparable with model outputs (Zaitchik et al. 2008).

13.3.2.2 Satellite Soil Moisture

In addition, soil moisture observations from the Advanced Microwave Scanning Radiometer for EOS (AMSR-E), between 2002 and 2011, and ESA's Soil Moisture Ocean Salinity (SMOS) Earth Explorer mission, between 2011 and 2013, are used in the data assimilation to update model soil moisture variabilities. The AMSR-E measurements are correlated to the surface 0–2 cm soil moisture content (Njoku et al. 2003), while SMOS maps land soil moisture for the 0–5 cm depth. Level 3 CATDS (Centre Aval de Traitement des Donnees SMOS) products (Jacquette et al. 2010) are acquired. SMOS and AMSR-E are selected from ascending and descending passes, respectively, subject to their higher agreement to in-situ measurements (see, e.g., De Jeu and Owe 2003; Draper et al. 2009; Jackson and Bindlish 2012; Su et al. 2013). Both data products with a daily temporal resolution are spatially rescaled from $0.25^\circ \times 0.25^\circ$ to $1^\circ \times 1^\circ$ resolution using the nearest neighbor interpolation to match W3RA. Note that these soil moisture observations are used in different periods during the assimilation process, i.e., AMSR-E soil moisture is assimilated for the period 2002–2011 and SMOS soil moisture is assimilated for the period 2011–2013.

13.3.2.3 Precipitation

Furthermore, monthly precipitation data of the Tropical Rainfall Measuring Mission Project (TRMM-3B43 products; version 7, TRMM 2011; Huffman et al. 2012) are used to assess climate induced water storage changes. Due to the fact that ground validation over land is applied for TRMM-3B43 products, uncertainty in measured precipitation are smaller compared to those of the oceans. Several studies have implemented and validated these products over South America and proved their capabilities (see, e.g., Condom et al. 2011; Ceccherini et al. 2015; Cabrera et al. 2016). The rainfall data are provided on a $0.25^\circ \times 0.25^\circ$ gridded spatial resolution and to make them comparable to those of the model (cf. Sect. 13.3.1), they are converted to $1^\circ \times 1^\circ$ using the nearest neighbor interpolation for the period of 2002–2013.

13.3.3 Surface Storage Data

Although the focus of the present study is on sub-surface water storage compartments, in order to efficiently assimilate GRACE TWS data into W3RA, however, a special focus should be invested on surface water storage variations due to their large contribution in water storage changes over South America (Getirana et al. 2017). In particular, this is important because many surface water sources (in different forms, e.g., lakes and rivers, except for a few major ones) are not modeled in W3RA. To address this problem, the recently developed surface water storage data provided by Getirana et al. (2017) is used. The data is based on a coupled system compromising Noah land surface model (LSM) with multi-parameterization options (Noah-MP; Niu et al. 2011) and the Hydrological Modeling and Analysis Platform (HyMAP) river routing scheme (Getirana et al. 2012). Multiple meteorological forcings and precipitation datasets are used to generate an ensemble of 12 runs, and to establish reference product with associated uncertainties (see details in Getirana et al. 2017). The $1^\circ \times 1^\circ$ monthly gridded surface water data for the period of 2012 to 2013 are subtracted from GRACE TWS before data assimilation.

13.3.4 In-Situ Groundwater Measurements

In order to evaluate the obtained data assimilation results, independent in-situ groundwater measurements over 34 stations obtained from Global Groundwater Network (GGMN; https://ggmn.un-igrac.org/) and propagated within the study area (see Fig. 13.1) are compared with estimated groundwater storage changes obtained from data assimilation. Groundwater level measurements should be converted into groundwater (GW) storage, which requires specific yield values. In the absence of such information, following Tangdamrongsub et al. (2015), TWS variation from GRACE and Global Land Data Assimilation System (GLDAS, Rodell et al. 2004) soil moisture are used to calculate the specific yield and scale the observed head. The scaled in-situ groundwater level fluctuations are then used to assess the results. Afterwards, the assimilation results are spatially interpolated to the location of the in-situ measurements using the nearest neighbor (the closest four grid values). The Root-Mean-Squared Error (RMSE) and correlations between the in-situ and estimated groundwater storage measurements are then computed.

13.3.5 Data Assimilation Filtering Method

The filtering technique of Ensemble Square-Root Filter (EnSRF) proposed by Whitaker and Hamill (2002) is used to assimilate GRACE TWS and soil moisture data into the W3RA model. The method is based on a traditional Ensemble Kalman

Filter (EnKF) that poses a new sampling scheme. The filtering process starts with the forecast step, which includes integrating N (ensemble number) samples of model state X that contains top soil, shallow soil, and deep soil water, snow, vegetation, and groundwater by a dynamical model. The forecast state, thus, can be shown as,

$$X^f = [X_1{}^f \dots X_N{}^f], X_i{}^f \, i = 1 \dots N, \tag{13.1}$$

where 'f' stands for forecast ('a' in following represents analysis). The corresponding model state forecast error covariance of P^f and the mean state forecast $\bar{X}^f$ are defined by:

$$P^f = \frac{1}{N-1} \sum_{i=1}^{N} (X_i{}^f - \bar{X}^f)(X_i{}^f - \bar{X}^f)^T, \tag{13.2}$$

$$\bar{X}^f = \frac{1}{N} \sum_{i=1}^{N} (X_i). \tag{13.3}$$

The update stage in EnSRF contains two steps. First, it updates the ensemble-mean following,

$$\bar{X}^a = \bar{X}^f + K(y - H\bar{X}^f), \quad i = 1 \dots N, \tag{13.4}$$

$$K = P^f (H)^T (H P^f (H)^T + R)^{-1}, \tag{13.5}$$

where K is the Kalman gain, y is the observation vector and transition matrix is indicated by H. R represents the observation covariance matrix. Data assimilation methods are largely sensitive to the observations uncertainties. Therefore, it is important to assign accurate error values to each observation used in data assimilation. Here, for GRACE observations, TWS error covariance matrix is constructed from Full error information about the GRACE Stokes' coefficients. There is no covariance error information available for satellite soil moisture observations, thus, we assume their error covariances to be uncorrelated and consider various uncertainties to monitor their impacts on data assimilation by comparing the results with independent measurements. This allows us to obtain optimum error values for soil moisture part of observation error covariance. Accordingly, **R** is assumed to be diagonal with an error standard deviation of 0.04 (m^3 m^{-3}) for SMOS (suggested by Leroux et al. 2016) and 0.05 (m^3 m^{-3}) for AMSR-E (suggested by De Jeu et al. 2008). In addition, for the observation error covariance in simultaneous data assimilation case, GRACE data and both SMOS and AMSR-E observations are assumed to be uncorrelated. It is worth mentioning that more study is still required to efficiently estimate the spatially varying observations uncertainties, which also account for error correlations. This can lead to different results and potentially improved data assimilation. $\bar{X}^a$ in Eq. 13.4 is the analysis ensemble-mean. In the next step, i.e., the analysis step, the filter updates the forecast ensemble of anomalies,

$$A^f = [A_1{}^f \dots A_N{}^f], \tag{13.6}$$
$$A_i{}^f = X_i{}^f - \bar{X}^f, \tag{13.7}$$

into the analysis ensemble deviation A^a in Eq. 13.8. EnSRF exploits the serial formulation of the Kalman filter analysis step in which the observations are assimilated each at a time to compute the analysis perturbations that exactly match the Kalman filter covariance (Hoteit et al. 2008) using the modified gain ($\tilde{K} = \alpha K$) with,

$$A^a = (I - \tilde{K}H)A_i{}^f, \tag{13.8}$$
$$\alpha = \left(1 + \sqrt{\frac{R}{HP^fH^T + R}}\right)^{-1}, \tag{13.9}$$

where I is an identity matrix. This definition requires the observation errors to be uncorrelated, which can always be satisfied by scaling the observations with the square-root inverse of the observational error covariance matrix (Hoteit et al. 2015). This, however, is not the case here because there is no rank deficiency on observation error covariance. We assume that soil moisture observations are uncorrelated. Furthermore, the correlation between GRACE TWS on the one hand and soil moisture observations, on the other hand, is also assumed to be zero. The rank deficiency issue raised from GRACE TWS block in the covariance matrix is mitigated by applying GRACE TWS observations in a $3^\circ \times 3^\circ$ spatial resolution along with the implementation of Local Analysis (LA) (Evensen 2003) scheme, which restricts the information used for the covariance matrix computation to a spatially limited area and uses only measurements located within a certain distance from a grid point (cf. Sect. 13.3.6, see also Khaki et al. 2017a). More details regarding the EnSRF algorithm and its performance in GRACE TWS data assimilation against other filters are described, e.g., in Whitaker and Hamill (2002) and Khaki et al. (2017b).

13.3.6 Experimental Setup

As already mentioned, the state vector includes different water storages, i.e., soil moisture, vegetation, snow, and groundwater, simulated by W3RA. Previous studies have investigated the surface water variations over South America (e.g., De Paiva et al. 2013; Getirana et al. 2017), thus, we only focus on the estimation of sub-surface compartments; groundwater and soil moisture. The modified GRACE TWS data (cf. Sect. 13.3.3) is then used to update the above water compartments excluding surface storage. The observation operator aggregates different water storages at each grid point (1688 points in total) to update with GRACE TWS and scales top-layer soil storage by the field capacity value to provide a relative wetness for updating with soil moisture products of AMSR-E and SMOS (Renzullo et al. 2014).

Considering the different temporal resolution of assimilation observations, e.g., monthly GRACE TWS and daily soil moisture measurements, both observation sets are temporally rescaled into a 5-day resolution for data assimilation. This is done to allow for a simultaneous data assimilation of GRACE TWS and satellite soil moisture measurements. Khaki et al. (2017a) showed that assimilating GRACE TWS in a 5-day temporal scale leads to a better improvement in state variables compared to daily and monthly scales. Therefore, in the analysis steps during the assimilation, the 5-day temporal average update increment (cf. Eq. 13.4) is applied. In order to produce ensemble for EnSRF filtering, we use Monte Carlo sampling of multivariate normal distribution, with the errors representing the standard deviations to perturb three main forcing parameters; precipitation, temperature, and radiation (see details in Renzullo et al. 2014). Afterwards, by integrating perturbed meteorological forcing forward in time with the model from 2000 to 2002, 72 sets of state vectors (ensemble; as suggested by Oke et al. 2008) is created at the beginning of the study period.

While implementing data assimilation with a large number of ensemble members results in a heavy computational burden, using a small ensemble size can also be problematic, as it can lead to filter divergence or inaccurate estimation (Tippett et al. 2003). To address this issue, two filter tuning is applied including ensemble inflation and LA. Ensemble inflation helps ensemble members to adequately span the model sub-space by inflating prior ensemble deviation from the ensemble-mean and increases their variations (Anderson 2001; Anderson et al. 2007). Various inflation factors ($[1 - 1.8]$) are tested and their impacts ensemble spreads are monitored to determine the best value (i.e., 1.12). Furthermore, LA (Evensen 2003; Ott et al. 2004) is applied to both account for a limited ensemble number and also GRACE limited spatial resolution. Applying GRACE TWS data on a high spatial resolution (e.g., $1° \times 1°$) causes correlation errors, which degrades the performance of data assimilation (Khaki et al. 2017a, b). Khaki et al. (2017a) showed that LA can successfully mitigate this problem by restricting the impact of the measurements in the update step to variables located within a certain distance only, e.,g, 5°, which is applied in the present study.

13.3.7 Climate Variability Impacts

In order to investigate the model's enhancement for studying climate induced impacts, TRMM rainfall and ENSO data are employed. At each grid point, correlations between TWS with and without data assimilation for both rainfall (at the same point) and ENSO are calculated. Afterwards, improvements achieved by data assimilation with respect to no assimilation of TWS are explored. Furthermore, principal component analysis (PCA; Lorenz 1956) is applied on the estimated groundwater and soil moisture storages (from assimilation), as well as on TRMM rainfall to better analyze the spatio-temporal changes of water storages and precipitation. This is done to examine the precipitation patterns within the area between 2002 and 2013 and to investigate their connections to water storage changes. Since precipitation is the

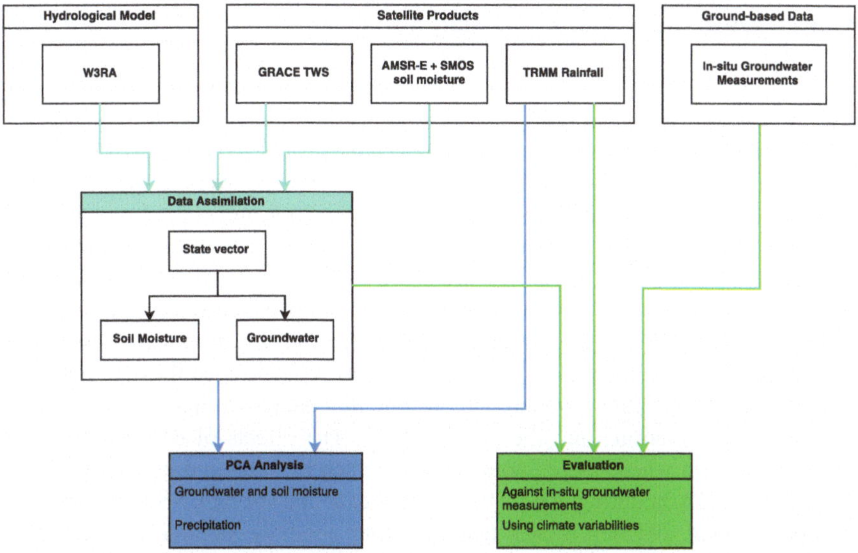

Fig. 13.2 Schematic illustration of the methodology steps including data used, assimilation scheme, and evaluation processes. In data assimilation process, W3RA is used for forecasting and GRACE TWS and satellite soil moisture measurements are used to update forecasts from the model. Once the groundwater and soil moisture are estimated from the assimilation process, their relationship with rainfall data is investigated using PCA analysis. The in-situ groundwater measurements as well as rainfall data are further used to assess the data assimilation estimates

major effective parameter on water storage recharge, the process helps to study the role of climate variability on water storage variations. A schematic illustration of the methodology steps is provided in Fig. 13.2.

13.4 Results and Discussions

13.4.1 Data Assimilation

In what follows, data assimilations results and their comparison with in-situ measurements are first discussed. First, we investigate the impacts of assimilated observations, e.g., GRACE TWS and satellite soil moisture on water storage estimates (cf. Sect. 13.4.1.1). Note that the results presented in this section are not used for validation and only show how the assimilation results differ from the open-loop (no data assimilation) results. Evaluation against independent measurements will also be discussed (cf. Sect. 13.4.1.2).

13.4.1.1 Observation Impacts on State Variables

The spatially averaged time series of TWS variations estimated by EnSRF over South America is presented in Fig. 13.3a, which shows that the application of data assimilation reduces misfits (Fig. 13.3b) between the results and GRACE TWS compared to the open-loop. Furthermore, Fig. 13.3c shows the average time series of soil moisture variations from the model top layer open-loop and assimilation, as well as satellite remote sensing. Similar to Fig. 13.3b, d indicate that the data assimilation successfully decreases the differences between soil moisture estimates and the observations. The average discrepancy between the estimated (assimilated) TWS and those by GRACE is approximately 46%, and between soil moisture estimates and satellite (AMSR-E and SMOS) observations is 34% less than those of between open-loop and observations, which demonstrate that data assimilation successfully incorporates observations into the system states. The effects of data assimilation can better be seen where large anomalies exist such as 2005 and 2011. The large anomaly in Fig. 13.3a during 2011 could be related to the strong ENSO impact (see, e.g., Boening et al. 2012). It is clear that this strong anomaly captured by GRACE is successfully reflected into assimilation TWS contrary to that of open-loop.

For a better discussion, we also calculate the correlation between the soil moisture variations from satellites and estimations for each of the 15 basins (Fig. 13.3e). This is done on a basin scale due to the fact that basin averaged time series can be more representative of water storage changes in the area. In all the cases, regardless of the method, data assimilation resulted in higher correlations with the observations compared to the open-loop (un-assimilated model products). Assimilation of only one satellite products, e.g., GRACE TWS or soil moisture, increases the correlation values in Fig. 13.3e. As expected, GRACE TWS data assimilation has more effects on enhancing TWS correlations, however, it can be seen that even soil moisture only data assimilation in most of the cases causes higher TWS correlation than the open-loop results. It can also be seen that the correlation between joint data assimilation (GRACE TWS and satellite soil moisture) results in Fig. 13.3e are largely in agreement with the observed variables compared to GRACE-only data assimilation. This indicates that assimilation of soil moisture products along with GRACE TWS leads to more improvements. One main reason for this is that while GRACE TWS data assimilation is generally an effective approach for updating models TWS (e.g., Zaitchik et al. 2008; Reager et al. 2015; Khaki et al. 2018b, c), it can also introduce artifact effects to different storage such as by assigning wrong increments to either groundwater or soil moisture (Li et al. 2015; Girotto et al. 2017; Khaki et al. 2018f). Assimilation of soil moisture products can account for this problem by independently constraining soil moisture estimates. Fig. 13.3e shows that this joint assimilation leads to better estimations of soil moisture.

In addition, the average correlation improvement from jointly assimilating GRACE TWS and soil moisture in comparison to the open-loop is presented in Fig. 13.3f. Note that only the results of joint data assimilation are presented in the figure due to its better performance (cf. Fig. 13.3e). Figure 13.3f demonstrates that higher correlations are achieved after data assimilation. For GRACE TWS, higher

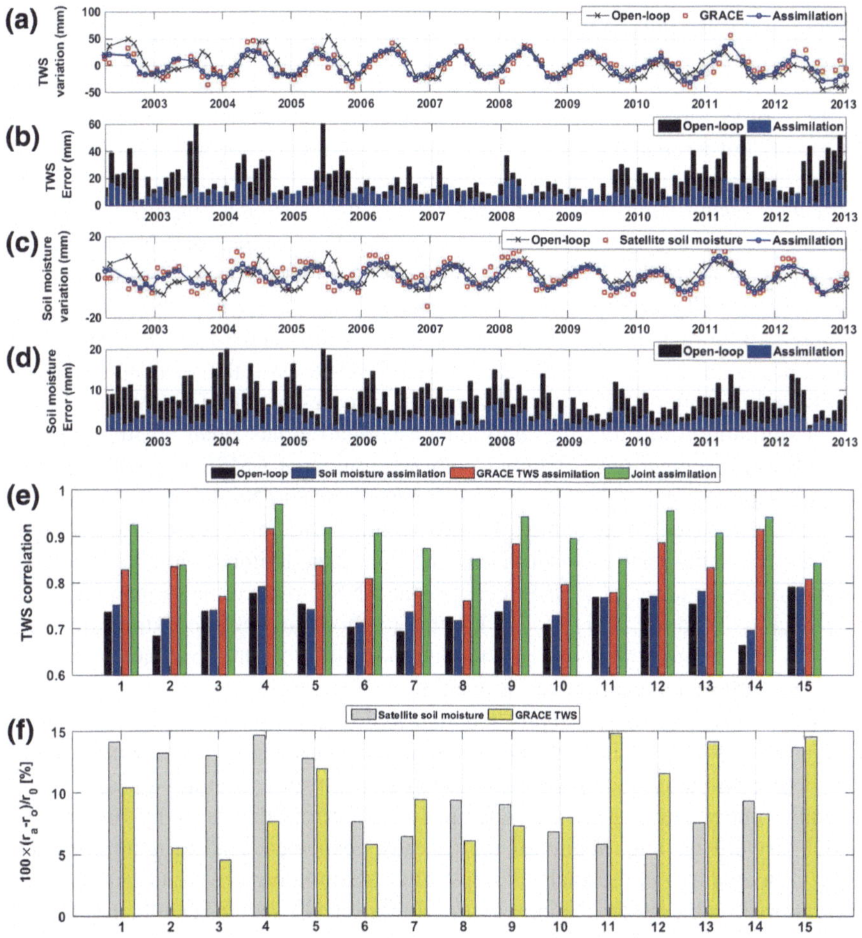

Fig. 13.3 **a** Comparison between the TWS time series from the assimilation process (blue), GRACE TWS (red), with the open-loop referring to the model estimation without applying data assimilation (black). **b** Absolute error bars before (black) and after (blue) data assimilation process in comparison to the GRACE TWS observations. **c** and **d** are similar to (**a**) and (**b**), respectively, but for soil moisture observations. **e** Average correlations between GRACE-derived TWS and TWS simulated by W3RA before assimilation, GRACE TWS only assimilation, soil moisture only assimilation, and joint data assimilation for each basin (cf. Fig. 13.1). **f** Correlation improvement between joint data assimilation results and GRACE TWS and soil moisture observation with respect to the open-loop results

correlations are achieved within larger basins such as Amazon (number 15) and Tocantins (number 11). This suggests that GRACE TWS data assimilation has larger impacts on these basins. It can be seen that the minimum improvement happens for the Pacific Coast, North Chile basin (number 3), where GRACE TWS data are expected to have larger errors in comparison to other basins like the Amazon basin with small leakage errors (Wiese 2015). Nevertheless, in general, the assimilation process increases the correlation between outputs and GRACE TWS.

13.4.1.2 Evaluation Results

In order to examine the validity of the data assimilation, groundwater in-situ measurements from various stations are spatially averaged to the location of nearest model grid points and compared with their estimates. As mentioned (cf. Sect. 13.3.4), we calculate RMSE and correlation for three tests including the open-loop, GRACE-only TWS data assimilation, and joint GRACE-soil moisture assimilation. Table 13.1 presents the average RMSE, corresponding RMSE reduction, and Nash-Sutcliffe coefficient (NSE) of the results before and after assimilation. In order to statistically assess the significance of the results, the student t-test is applied after considering the autocorrelation in time series. The estimated t-values and the distribution at 0.05 significant level are used to calculate p-values. Data assimilation results indicate significantly smaller RMSE and higher NSE in cases of GRACE TWS and joint data assimilation. Soil moisture only data assimilation has small positive impacts on NSE improvement (e.g., 3%) but with no significant RMSE improvement. An average improvement of 23.43% in RMSE and 14.08% in NSE (for all assimilation experiments) proves the capability of data assimilation to improve model simulations with respect to the reality. Nevertheless, the joint data assimilation indicates larger improvements in terms of RMSE reduction and NSE improvements than GRACE-only data assimilation. This shows that this method can better constrain different water storage compartments. It can be seen that soil moisture observations help in better controlling the distribution of increments between storages.

Furthermore, it is found that this joint data assimilation better reduces the forecast uncertainties. We calculate the average standard deviation (STD) of ensemble members before and after each data assimilation step for all assimilation cases. These ensemble uncertainties generally refer to forcing errors that grow by running the model forward in time. While all the cases, as expected, lead to a smaller STD (5.31% on average) in the analysis steps (after assimilating observations), the least uncertainty is achieved for the joint data assimilation (11.78% STD reduction). Note that the smaller ensemble STD can also lead to a weaker ensemble spread, however, this is not the case here. The achieved STD reduction means that the method can better propagate ensemble members by improving the spread of forecast ensemble

Table 13.1 Statistics of groundwater errors. For each case, the RMSE average and its range (±XX) at the 95% confidence interval is presented. Improvements in data assimilation results are calculated with respect to the groundwater storages from the model without implementing data assimilation

			Improvement (%)	
Experiment scenario	NSE	RMSE (mm)	NSE	RMSE
Open-loop	0.63	69.26 ± 7.38	–	–
GRACE-only data assimilation	0.75	54.19 ± 5.79	16.01	21.76
Soil moisture data assimilation	0.65	66.48 ± 7.12	3.07	–
Joint GRACE-soil moisture assimilation	0.82	51.87 ± 5.16	23.17	25.11

members based on the observations and their associated uncertainties. These results show that data assimilation can improve our understanding of water storage changes. More importantly, monitoring groundwater using reliable information is crucial over South America, where only a few studies have focused on it (e.g., International Groundwater Resources Assessment Centre 2004; Villar 2016). Groundwater is a major water resource along with surface storages within the area providing almost 60% of freshwater use (Villar 2016). This number is even higher for some countries such as Chile, Peru, Venezuela, Suriname and The Guyanas (Morris et al. 2003). The application of the proposed approach for studying groundwater variations can benefit many of these countries to better monitor groundwater using the enhanced estimates.

As previously mentioned, data assimilation, especially when using GRACE TWS data, can introduce artifacts to other variables. This can be the case not only for different variables in the state vector (e.g., groundwater and soil moisture) but also for non-assimilated variables such as water discharge. To monitor this effect, we compare model water discharge to streamflow in-situ measurements obtained from Hydrology and Geochemistry of the Amazon basin (HYBAM). The monthly in-situ discharge measurements, computed as the sum of the daily discharge, are spatially interpolated to the closest grid points and compared with the estimates at those grid points. Figure 13.4a shows the average discharge time series over the Amazon basin before and after data assimilation, as well as from in-situ measurements. It can be seen that assimilation of GRACE TWS and soil moisture reduces the misfits between model and in-situ water discharge time series. Furthermore, Fig. 13.4b, c present the average scatter plots of the discharge estimates from the open-loop and assimilation compared to the in-situ values. The average correlations between time series are also indicated in the plots, which show the larger agreement between the assimilation results and in-situ streamflow measurements. Data assimilation decreases the RMSE values from 6.47 to 2.88 cm and increases NSE from 0.47 to 0.71. Assimilation of GRACE TWS and soil moisture, thus, effectively reduces discharge error. This confirms the findings of Syed et al. (2005), who used GRACE TWS and additional model-derived fluxes observations to study water discharge over the Amazon basin. Undermining groundwater and moisture storage changes in their experiment, however, led to some degree of discrepancy between the estimated and observed discharge. In contrast, in this study, updating different water compartments including groundwater during the assimilation analysis results in a better agreement between the results and in-situ measurements. In general, Fig. 13.4 indicates that the joint assimilation process not only causes any artifact errors but also improve the discharge estimates (cf. Fig. 13.4).

To further investigate the effect of data assimilation, we compare the TWS estimates from the joint data assimilation and the open-loop run with precipitation over the Amazon Basin. The rationale behind this choice is due to the fact that various studies have reported different droughts (see, e.g., Chen et al. 2009; Frappart et al. 2013) over the basin and a successful data assimilation should be able to capture these phenomena. Figure 13.5 shows the TWS variations over the basin from the above approaches, as well as precipitation. The average correlation between TWS

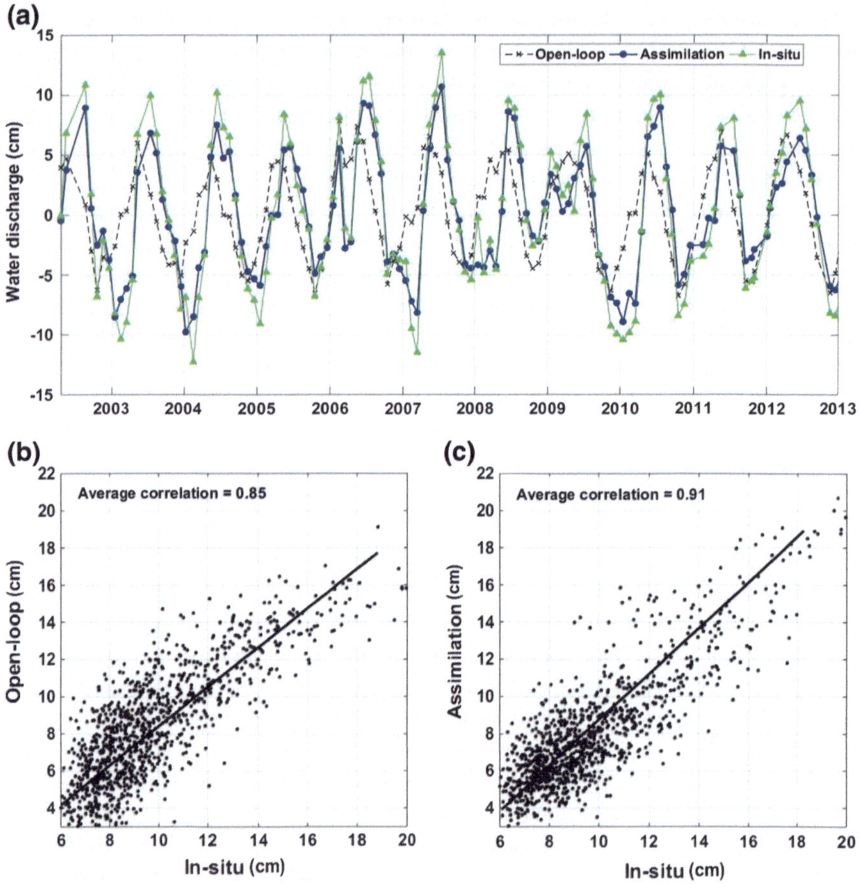

Fig. 13.4 **a** Average discharge time series before and after data assimilation, as well as from in-situ measurements. Scatter plots of average discharge from the open-loop and joint data assimilation with respect to in-situ measurements are presented in (**b**) and (**c**), respectively

estimates and precipitation is 0.89, ~17% larger than that of the open-loop run. It can be seen that the data assimilation time series better capture large anomalies such as in 2004 and 2009 reflected also in the precipitation time series. La Niña impact during 2011 (see also Boening et al. 2012) is better captured by the assimilation results. Furthermore, 2005 drought over the Amazon Basin (see, e.g., Chen et al. 2009) is reflected in both open-loop and assimilation results, while the later show a larger amplitude.

Table 13.2 contains the correlations between the open-loop run and joint data assimilation TWS results and precipitation. The table also reports the correlation improvements in the assimilation results with respect to those of the open-loop against both precipitation as well as ENSO (using Niño 3.4 indicator), as the dominant climate variability index over South America (Tourre et al. 2008; Xavier et al. 2010;

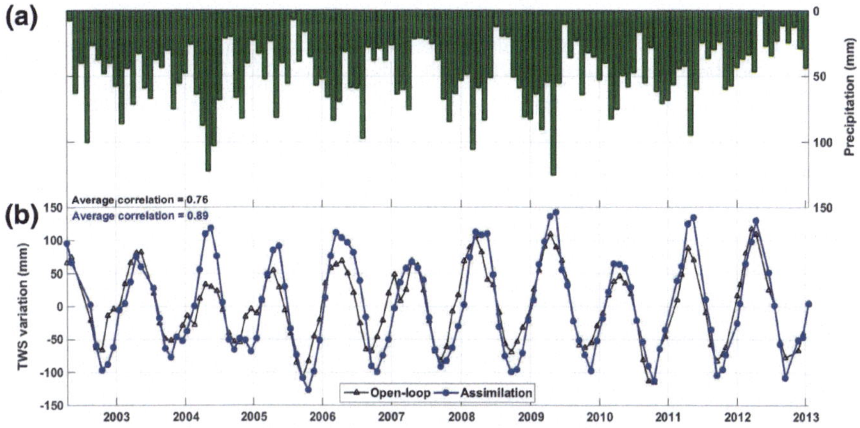

Fig. 13.5 Average precipitation (**a**) and TWS time series from the open-loop and joint data assimilation (**b**) over Amazon basin

Flantua et al. 2016) for each basin. It can be seen that significant improvements are achieved by assimilating remotely sensed TWS and soil moisture observations into W3RA hydrological model. Correlation with both precipitation and ENSO over a majority of the basins showcase these improvements. Note that only precipitation correlation improvements are statistically significant. An average correlation between rainfall and TWS anomalies within South America is found to be 0.89, ~11% larger than the open-loop results. This indicates that there is a larger agreement between the assimilation results and rainfall over the area than the case of the model simulations without data assimilation. The improvements in terms of correlations with ENSO are different for various basins. For example, larger correlations and corresponding improvements are estimated for Atlantic North Coast (basin 5), Pacific Coast, North Chile (basin 3), Negro (basin 10), Magdalena (basin 4), and Orinoco (basin 12) basins. The reason for this is due to the fact that ENSO effects on precipitation are larger over basins located within the north towards the northeast and the southeast parts of South America and partially over the Amazon basin (Flantua et al. 2016). These larger effects lead to a similar impact on water storage changes that is successfully captured by data assimilation results. In general, larger correlations between the estimated TWS and precipitation over larger basins, e.g., Amazon (basin 15), La Plata (basin 14), and Sao Francisco (basin 13) are also found. This could be due to the ability of GRACE to solve larger basins that better constrain system states during data assimilation.

Table 13.2 Average correlation between the open-loop and assimilation TWS and precipitation. Correlation improvements are calculated using the increase of correlation between TWS from data assimilation and both precipitation and ENSO with respect to open-loop TWS

	Correlation to precipitation		Correlation improvements	
Basins	Open-loop	Assimilation	Precipitation	ENSO
(1) South-east Atlantic	0.88	0.90	2.27	3.33
(2) Pacific Coast, Peru	0.84	0.89	5.95	7.78
(3) Pacific Coast, North Chile	0.79	0.91	15.19	8.17
(4) Magdalena	0.87	0.92	5.75	7.23
(5) Atlantic North Coast	0.91	0.95	4.39	8.45
(6) Pacific Coast, South Chile	0.84	0.89	5.95	3.12
(7) Colorado Basin	0.78	0.91	16.67	1.72
(8) Atlantic South Coast	0.80	0.87	8.75	2.60
(9) North-east Atlantic	0.85	0.88	3.53	–
(10) Negro Basin	0.67	0.83	23.88	11.35
(11) Tocantins	0.69	0.89	28.98	4.54
(12) Orinoco	0.73	0.86	17.81	9.85
(13) Sao Francisco	0.92	0.92	–	3.06
(14) La Plata	0.75	0.94	25.33	5.17
(15) Amazon Basin	0.92	0.94	2.17	5.35

13.4.2 Water Storage Changes and Climatic Impacts

Average monthly TWS variations over South America from joint data assimilation is shown in Fig. 13.6. Different time spans are used for the averaging period including 2003–2012 (the entire study period) and 2005, 2009, 2010, 2011, and 2012 with remarkable extreme climate event that could potentially affect TWS anomalies. Larger water storage changes can be seen generally for basins located in the northern (e.g, Amazon basin) and southern (e.g., Orinoco and Negro basins) parts of South America. Figure 13.6 suggests that more water content, and subsequently more TWS variations exist over these areas. This could be due to the abundance of precipitation over these regions (see, e.g., Sanso and Guenni 1999; Marengo 2009; Buytaert et al. 2013). On the other hand, basins located in the west and northwest parts, e.g., Magdalena, Pacific Coast-Peru, and Pacific Coast-north Chile basins experience smaller TWS anomalies. The negative water storage anomalies in the northern parts (e.g., Amazon basin) of South America are observed during 2005 and 2010, and also in the southern parts (e.g., Negro basin) during 2009, 2010, 2011, and 2012. These results are supported by the findings of Humphrey et al. (2016) who also demonstrated water storage deficits, e.g., over northern parts (2004–2005), majority parts of Amazon basin (2010), and the western parts of South America (2011). The impact of the 2012 drought, which can be attributed to the anomalous SST in the Atlantic Ocean (Pereira et al. 2014) can clearly be seen within the eastern and southern parts

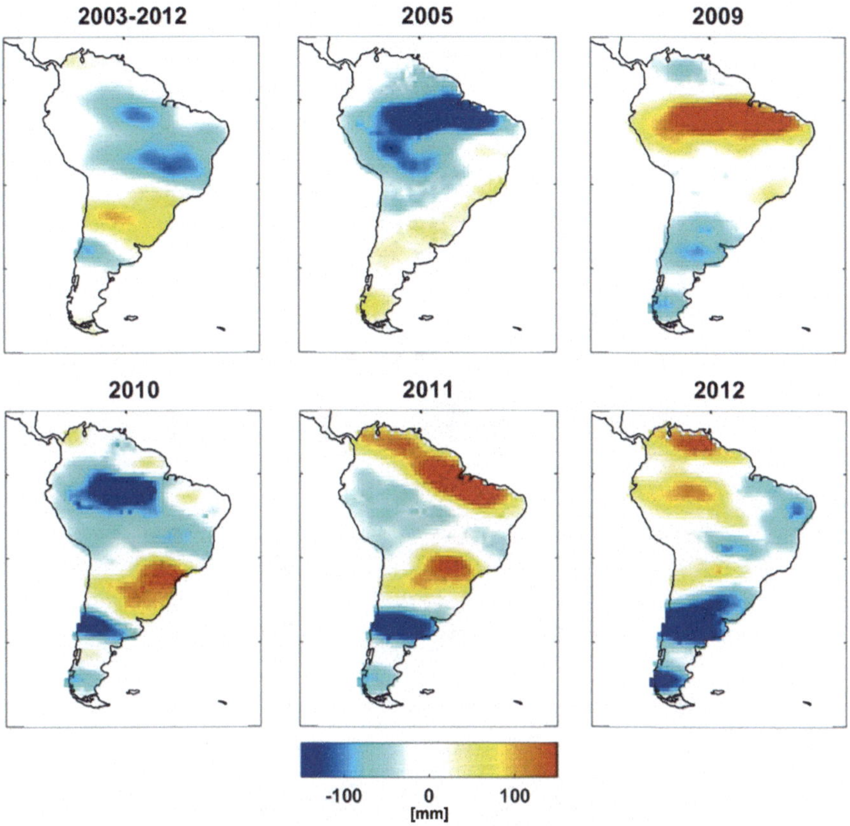

Fig. 13.6 Average monthly TWS variations from data assimilation for different time periods

of South America (see also Sun et al. 2016). Furthermore, El Niño effect in 2009 (Tedeschi et al. 2013) and La Niña effect in 2011 (Boening et al. 2012) can be seen through large anomalies, e.g., in the north, northeast, and southern parts.

To better analyze spatio-temporal variations of sub-surface water storages within South America, PCA is applied to groundwater and soil moisture results. Figure 13.7 shows the first three dominant modes. Furthermore, rainfall variations both spatially and temporally are investigated to explore their connections to water storage variations. Major water storages can be found from central to northern parts of South America, areas with rainfall patterns dominated by ENSO phenomena (Carrillo et al. 2010). This shows a larger amount of storages over the area mainly due to more rainfall. Considerable soil moisture content variations are found over North-east Atlantic (mode 2 and mode 3) and La Plata (mode 3). Larger groundwater variations can also be seen in Amazon and La Plata. To a lesser degree, the Orinoco and Atlantic North Coast basins contain large signal variations both for groundwater and soil moisture. It can also be seen that both groundwater and soil moisture variations modestly follow

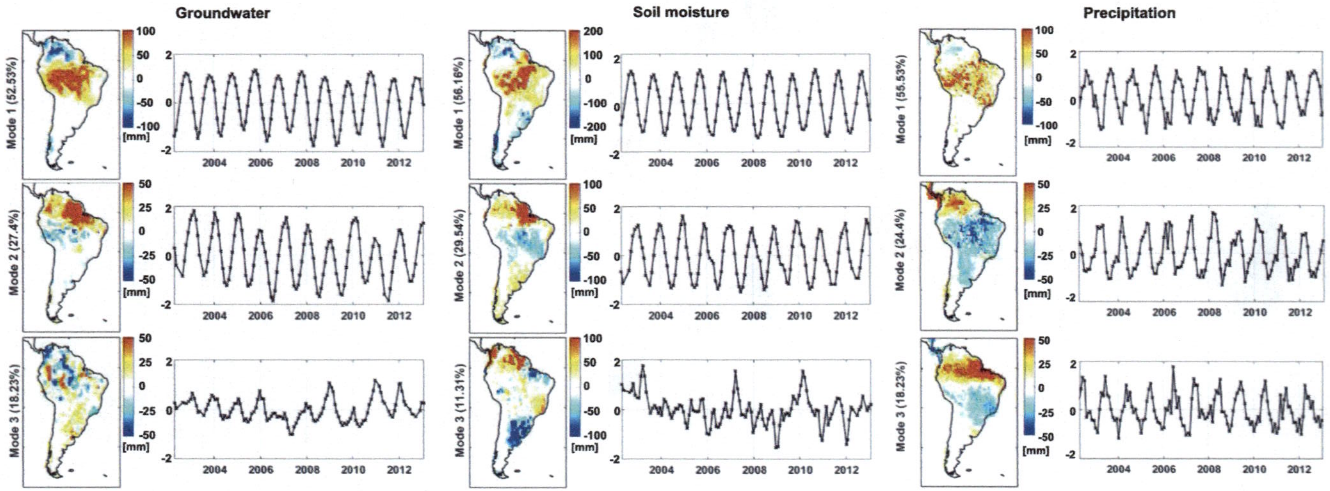

Fig. 13.7 Three first modes of spatial distribution and temporal variations from the application of PCA on groundwater, soil moisture, and precipitation

the same pattern except for mode 3, where negative variations exist in the soil moisture map over the south-eastern parts while the negative variations in groundwater map can be found over the central to northern parts. The positive anomalies over northern parts in soil moisture variations, as it will be shown, matches precipitation patterns in the same areas. This suggests that between the two water compartments, soil moisture variations follow precipitation more closely, whereas groundwater changes which can largely be affected by non-climatic factors, e.g., anthropogenic impacts. In general, based on these maps, more sub-surface water variabilities exist over the central towards northern and western parts of the continent compared to the south-western areas.

In terms of temporal variations, the first three extracted principal components (PC) of groundwater and soil moisture are also demonstrated in Fig. 13.7. The time series of both water storages largely depict annual effects dominant over the majority of South America's parts including its central to the north. These parts are also affected by the Strong influence of La Niña for 2010-2011, as well as El Niño effect in 2008–2009. Negative trends in groundwater are captured by PC2 over the northeast and central toward western parts before 2006, between 2007 and 2009, and also between 2010 and 2012. Such trends cannot be seen in soil moisture time series suggesting that non-climatic impacts such as the water used for power generation in Brazil (Sun et al. 2016) could possibly be responsible for the groundwater depletion. The negative soil moisture variations (mode 2) in the central part can be attributed to the multiple drought conditions, e.g., in the La Plata basin (2008–2009, Abelen et al. 2015). This soil moisture reduction was also reported by Escobar (2015) over the Amazon basin, which could be due to the anthropogenic impact on forest conservation. Dry events from 2012 to 2014 suggested by Humphrey et al. (2016) and Getirana (2016) can be seen in the northern and eastern parts of the South America, also reflected in groundwater and soil moisture time series (PC3 in Fig. 13.7). Considerable anomalies are found in 2006–2007 and 2010 from groundwater and soil moisture mostly over the northern and eastern parts, which could be attributed to extreme climatic events in the same periods. On the other hand, a negative anomaly is detected in 2004 by both water storages. The 2005 dry condition effects on soil moisture is captured by soil moisture's second mode, which confirms the same impacts presented in Fig. 13.6. The third mode of soil moisture time series depicts a negative anomaly for the period of 2002-2006 mostly over Negro basin, which, as will be shown, matches the third precipitation mode. El Niño effect in 2009 causes groundwater negative anomaly in both modes 2 and 3 (see also Fig. 13.6) affecting the central and eastern parts. Similar negative anomalies can also be seen in 2006 for groundwater, and in 2005 for soil moisture. A big part of these variations (e.g., over 2005, 2009–2012) can be related to climate variabilities while some of them, e.g., groundwater negative trends between 2003 and 2006 and also 2007 and 2009, can be due to non-climate factors such as human usage and irrigations.

A major rainfall pattern is found over the central toward northern parts of South America, especially Amazon basin, where most of groundwater and soil moisture variations are explored. While the spatiotemporal distributions of rainfall are highly matched to those of groundwater and soil moisture in the first mode, the precipitation second mode is more correlated to that of soil moisture. This, as expected, indicates that precipitation has larger influence on soil moisture variations while groundwater can be largely affected by other factors (e.g., water usages). Next, Atlantic North Coast, La Plata, and Negro basins indicate larger signals in contrast to south-western basins, e.g., Pacific Coast, South Chile and Atlantic South Coast. Similar to groundwater and soil moisture mode 1, rainfall time series in Fig. 13.7 also displays strong seasonal variations. In contrast to the negative anomalies in groundwater second mode time series, precipitation mode 2 does not show similar trends. However, both modes 2 and 3 indicate a rainfall decline after 2012 mostly affecting the eastern toward northern parts (see, e.g., Humphrey et al. 2016; Getirana 2016). Similar to water storage time series, the La Niña effect can be observed for 2010–2011 (Boening et al. 2012). Large anomalies are also captured in 2005, 2006, and 2010, which considerably impacts water storages. El Niño effect in 2009 can also be seen in the second mode, which also affected groundwater and soil moisture variations within the central and south-eastern parts. A negative anomaly can be seen before 2004 in the mode 3, which can be related to a weak El Niño causing negative anomalies of precipitation during the wet season (Juarez et al. 2009). Sun et al. (2016) suggested that this period exhibits months drier than the normal seasonal cycle of TWS due to the rainfall rates lower than the average. These prolonged reductions in rainfall can explain similar negative anomalies that occurred in groundwater and soil moisture time series seen.

Average trends for groundwater and soil moisture from the open-loop run and joint assimilation are presented in Table 13.3 for the basins. To this end, the modified Mann–Kendall trend test (Yue and Wang 2002) is applied on deseasonalized time series. Note that the autocorrelation analysis is also used to compute an effective sample size and to correct the Mann–Kendall statistic. The trends when the p-values fall below 0.05 are considered statistically significant. It is worth mentioning that in addition to groundwater and soil moisture, the basin averaged precipitation time series are also considered here to investigate the climatic impacts on water storage changes. It can be seen that the application of data assimilation in many cases causes changes in either amplitude of trends or their signs. For example, over the Orinoco, the signs of variations become negative after data assimilation while these remained the same for La Plata with different amplitudes. For some other basins like South-east Atlantic, the trend values are close before and after data assimilation, which could be due to the smaller impacts of data assimilation.

It is also evident that there are larger agreements between precipitation and soil moisture trends. This further indicates that soil moisture changes mostly rely on rainfall pattern within South America. The mismatch between precipitation and groundwater trend signs over most of the basins, e.g., Pacific Coast, South Chile, Orinoco, South-east Atlantic and Colorado, suggests that non-climatic factors mostly influence the groundwater changes. This, in all of the cases, leads to a groundwater depletion while precipitation shows neither negative nor statistically significant trends. Nevertheless, negative trends are found in precipitation, soil moisture, and also groundwater over La Plata, Atlantic North Coast, Atlantic South Coast, and North-East Atlantic. Even though one can conclude that a majority of groundwater depletion over these

Table 13.3 Statistics of groundwater and soil moisture variation rates (mm/year) from the open-loop run and joint data assimilation. The statistically significant values at 95% confidence limit are demonstrated in bold

	Open-loop		Assimilation		Precipitation
Basins	Groundwater	Soil moisture	Groundwater	Soil moisture	
(1) South-east Atlantic	**−0.37**	**−0.44**	**−0.57**	**+0.44**	**+0.09**
(2) Pacific Coast, Peru	**−0.10**	**−0.13**	**−1.09**	+0.12	+0.06
(3) Pacific Coast, North Chile	–	–	−0.03	−0.05	+0.13
(4) Magdalena	−0.02	**+0.16**	**+0.62**	+0.05	−0.01
(5) Atlantic North Coast	+0.02	+0.01	**−0.17**	**−0.23**	**−0.22**
(6) Pacific Coast, South Chile	–	−0.07	**−1.52**	**−0.76**	**+0.22**
(7) Colorado Basin	**−0.25**	**+0.20**	**−0.21**	**+0.23**	**+0.28**
(8) Atlantic South Coast	−0.06	–	**−0.57**	**−0.27**	**−0.12**
(9) North-east Atlantic	**+0.26**	+0.36	**−0.98**	**−0.47**	**−0.17**
(10) Negro Basin	−0.04	–	**−0.21**	–	–
(11) Tocantins	+0.02	+0.07	**+0.54**	+0.03	+0.02
(12) Orinoco	**+0.17**	**+0.35**	**−0.53**	+0.09	+0.01
(13) Sao Francisco	−0.07	−0.06	**−0.47**	**−0.24**	−0.03
(14) La Plata	**−0.89**	**+0.09**	**−1.09**	**−0.14**	**−0.10**
(15) Amazon Basin	+0.05	**+0.14**	+0.07	−0.02	+0.05

basins can be caused by the precipitation decline, human impacts can still be an effective factor whereas assessing their contributions require additional information.

From Table 13.3, negative groundwater trends can be seen in most of the basins. For example, Sao Francisco and North-east Atlantic basins show the largest groundwater depletion compared to the other basins. This can be attributed to the fact that these basins have been under an unprecedented water depletion as can be inferred from the studies of Getirana (2016) and Sun et al. (2016). Trend signs of soil moisture changes generally follow precipitation's. This, however, is different for some basins such as Pacific Coast, South Chile and Amazon. This mismatch over the Amazon basin can be explained by the fact that anthropogenic impacts on forest conservation results in soil moisture decline (see, e.g., Escobar 2015). Similar negative trends are observed for both groundwater and soil moisture over La Plata, and also groundwater over Orinoco (Ramirez et al. 2017), which can be attributed to deforestation and excessive water use that have also been reported, e.g., by Pereira et al. (2011) for La Plata and Ramirez et al. (2017) for Orinoco basins (see also Frappart et al. 2015). There are also discrepancies between soil moisture and groundwater trend signs, e.g., over La Plata and Pacific Coast, North Chile. While the rate of the changes are smaller over Pacific Coast North Chile, La Plata, which are located in the most populated areas of South America, they have larger negative groundwater trends that could possibly be due to increased agricultural and livestock water usage in the basin (see also Chen et al. 2010). For some of the basins (e.g., the Amazon basin), the trends are not significant, especially the soil moisture and precipitation changes (e.g., over the Tocantins). In general, the annual rate of groundwater anomaly is −0.24 (mm/year) in South America, suggesting its depletion between 2002 and 2013. This could be due to climatic impacts (e.g., droughts, see, e.g., Bates et al. 2008; Chen et al. 2010; Treidel et al. 2011; Getirana 2016; Sun et al. 2016) and/or exponential increase of agriculture and industrial activities (Bocanegra et al. 2010). This negative trend is very important due to its effects on South America's water and its use for agriculture. Groundwater is a major source of irrigation over most of the countries within South America such as major rice-growing regions of North Eastern Argentina, South Brazil and Uruguay (Herring 2012). Besides, groundwater depletion can largely increase water quality challenges (e.g., Arsenic growth) as a potential issue over South America (see, e.g., Munoz et al. 2002; Perez-Carrera and Cirelli 2009; Herring 2012).

13.5 Conclusion

Multimission satellite datasets including Terrestrial water storages (TWS) from the Gravity Recovery And Climate Experiment (GRACE) satellite mission and soil moisture products from the Advanced Microwave Scanning Radiometer—Earth Observing System (AMSR-E) and Soil Moisture and Ocean Salinity (SMOS) are assimilated into the World-Wide Water Resources Assessment (W3RA) model using the Ensemble Square-Root Filter (EnSRF) to increase the model performance for estimating groundwater and soil moisture over South America. The application of joint data assimilation causes improvements in W3RA estimates against groundwater in-situ measurements. This effect could clearly be seen for TWS estimates and importantly for groundwater simulations, which emphasize the potentials of assimilating remotely sensed products to increase the reliability of the W3RA hydrological model. We further investigate the correlation between assimilation results and precipitation from the Tropical Rainfall Measuring Mission (TRMM), as well as El Niño/Southern Oscillation (ENSO). The results indicate that assimilation TWS are more correlated to the TRMM rainfall and ENSO data compared to open-loop TWS estimates. Both of these assessments demonstrate the capability of data assimilation for improving model simulations of water resources over South America. Based on the results, the new information of groundwater and soil moisture are more reliable, which can be used for water management and agriculture objectives. From the application of principal component analysis (PCA) on water storage variations within South America and its 15 major basins, more soil moisture and groundwater anomalies are found over central toward northern and western parts of South America. Based on the results, a negative trend for groundwater is observed over most parts of South America. Negative trends are found for groundwater and to a lesser degree for soil moisture variations over the majority of the studied basins. This study shows that application of data assimilation can successfully improve our understanding of water storage changes. Nevertheless, more investigations are still needed to fully assess the approach's performance, e.g., by applying new observations such as GRACE follow-on and Surface Water and Ocean Topography (SWOT), sensitivity analysis regarding data uncertainties, and the impacts of GRACE data assimilation on non-assimilated variables.

Acknowledgements We would like to thank Dr. Augusto Getirana for providing the surface water storage data and also his review suggestions, which contributed to the improvement of this study. M. Khaki is grateful for the research grant of Curtin International Postgraduate Research Scholarships (CIPRS)/ORD Scholarship provided by Curtin University (Australia). J. Awange is grateful for the Brazilian Science Without Borders Program/CAPES Grant No. 88881.068057/2014-01. This work is a TIGeR publication.

Bibliography

Abdullah Aziz M, Majumder Abul Kashem M, Shahjahan Kabir Md, Ismail Hossain Md, Farhat Rahman NM, Rahman F, Hosen S (2015) Groundwater depletion with expansion of irrigation in Barind tract: a case study of Rajshahi District of Bangladesh. Int J Geol Agric Environ Sci 3:32–38

Abelen S, Seitz F, Abarca-del-Rio R, Güntner A (2015) Droughts and floods in the la plata basin in soil moisture data and GRACE. Remote Sens 7:7324–7349. https://doi.org/10.3390/rs70607324

Adhikary SK, Das SK, Saha GC, Chaki T (2013) Groundwater drought assessment for barind irrigation project in Northwestern Bangladesh. In: 20th international congress on modelling and simulation. Adelaide, Australia. www.mssanz.org.au/modsim2013

Adler RF, Susskind J, Huffman GJ, Bolvin D, Nelkin E, Chang A et al (2003) Global precipitation climatology project V2.1 monthly 2.5 deg global 1979-present (satellite only and gauge adjusted) 2003: the version-2 global precipitation climatology project (GPCP) monthly precipitation analysis (1979-present). J Hydrometeorol 4:1147–1167

Adnan S (1993) Living without floods: lessons from the drought of 1992. Research and Advisory Services, Dhaka

ADB, Technical Assistance Consultant's Report (2011) BANGLADESH: Khulna Water Supply Project (Financed by the Technical Assistance Special Fund), Institute of Water Modelling (IWM), Project Number: P42171-012. https://www.adb.org/sites/default/files/project-document/77210/42171-012-ban-tacr-01.pdf

Afshar AA, Joodaki GR, Sharifi MA (2016) Evaluation of groundwater resources in Iran using GRACE gravity satellite data. JGST 5(4):73–84. http://jgst.issge.ir/article-1-381-fa.html

Ahmed UA (2006) Bangladesh climate change impacts and vulnerability: a synthesis. Climate change cell. Department of Environment Component 4b, Comprehensive Disaster Management Programme, Bangladesh

Aires F (2014) Combining datasets of satellite retrieved products. Part I: methodology and water budget closure. J Hydrometeorol 15(4):1677–1691

Ait-El-Fquih B, Hoteit I (2015) Fast Kalman-like filtering in large-dimensional linear and gaussian state-space models. IEEE Trans Signal Process 63:5853–5867

Ait-El-Fquih B, Hoteit I (2016) A variational bayesian multiple particle filtering scheme for large-dimensional systems. IEEE Trans Signal Process 64(20):5409–5422. https://doi.org/10.1109/TSP.2016.2580524

Alder RF, Huffman GJ, Bolvin DT, Curtis S, Nelkin EJ (2000) Tropical rainfall distributions determined using TRMM combined with other satellite and rain gauge Information. J Appl Meteorol 39:2007–2023

Alimuzzaman UA (2017) Study of ground water recharge from rainfall in Dhaka City. Int J Sci Eng Investig 6(6). ISSN: 2251-8843

M. Khaki, *Satellite Remote Sensing in Hydrological Data Assimilation*,
https://doi.org/10.1007/978-3-030-37375-7

Allen R, Tasumi M, Trezza R (2007) Satellite-based energy balance for mapping evapotranspiration with internalized calibration (METRIC)-model. J Irrig Drain Eng 133(4):380–394

Alsdorf DE, Rodriguez E, Lettenmaier DP (2007) Measuring surface water from space. Rev Geophys 45:RG2002. https://doi.org/10.1029/2006RG000197

Altaf MU, Butler T, Mayo T, Luo X, Dawson C, Heemink AW, Hoteit I (2014) A comparison of ensemble Kalman filters for storm surge assimilation. Mon Weather Rev 142(8):2899–2914

Alvarez-Garreton C, Ryu D, Western AW, Su C-H, Crow WT, Robertson DE, Leahy C (2015) Improving operational flood ensemble prediction by the assimilation of satellite soil moisture: comparison between lumped and semi-distributed schemes. Hydrol Earth Syst Sci 19:1659–1676

Amery HA, Wolf AT (2000) Water in the middle east: a geography of peace. University of Texas Press, Austin, Tex

Anderson J (2001) An ensemble adjustment Kalman filter for data assimilation. Mon Weather Rev 129:2884–2903. https://doi.org/10.1175/1520-0493(2001)129<2884:AEAKFF>2.0.CO;2

Anderson JL (2010) A non-gaussian ensemble filter update for data assimilation. Mon Weather Rev 138:4186–4198. https://doi.org/10.1175/2010MWR3253.1

Anderson MC, Norman JM, Mecikalski JR, Otkin JA, Kustas WP (2007) A climatological study of evapotranspiration and moisture stress across the continental United States based on thermal remote sensing: 1. model formulation. J Geophys Res 112:D10117. https://doi.org/10.1029/2006JD007506

Anderson J, Anderson S (1999) A Monte Carlo implementation of the nonlinear filtering problem to produce ensemble assimilations and forecasts. Mon Weather Rev 127:2741–2758

Andreadis KM, Clark EA, Lettenmaier DP, Alsdorf DE (2007) Prospects for river discharge and depth estimation through assimilation of swathaltimetry into a raster-based hydrodynamics model. Geophys Res Lett 34: https://doi.org/10.1029/2007GL02972

Ardakani R (2009) Overview of water management in Iran. In: Proceeding of regional center on urban water management. Tehran, Iran

Arkin PA, Meisner BN (1987) The relationship between large-scale convective rainfall and cold cloud over the Western Hemisphere during 1982–1984. Mon Weather Rev 115:51–74

Arnold H, Moroz I, Palmer T (2013) Stochastic parametrizations and model uncertainty in the Lorenz 1996 system. Philso Trans R Soc A 371:20110479

Arulampalam MS, Maskell S, Gordon N, Clapp T (2002) A tutorial on particle filters for online nonlinear/non-Gaussian Baysian tracking. IEEE Trans Signal Process 50(2):174–188

Asadzadeh Jarihani A, Callow JN, Johansen K, Gouweleeuw B (2013) Evaluation of multiple satellite altimetry data for studying inland water bodies and river floods. J Hydrol 505:78–90. https://doi.org/10.1016/j.jhydrol.2013.09.010. ISSN 0022-1694

Awange JL, Gebremichael M, Forootan E, Wakbulcho G, Anyah R, Ferreira VG, Alemayehu T (2014) Characterization of Ethiopian mega hydrogeological regimes using GRACE, TRMM and GLDAS datasets. Adv Water Resour 74:64–78. https://doi.org/10.1016/j.advwatres.2014.07.012

Awange JL, Mpelesoka F, Goncalves RM (2016) When every drop counts: analysis of droughts in Brazil for the 1901–2013 period. Sci Total Environ 566–567:1472–1488

Awange J, Sharifi M, Baur O, Keller W, Featherstone W, Kuhn M (2009) GRACE hydrological monitoring of Australia: current limitations and future prospects. J Spat Sci 54(1):23–35

Awwad HM, Valdés JB, Restrepo PJ (1994) Streamflow forecasting for Han River basin. Korea J Water Resour Plann Manage 120:651–673

Baguis P, Roulin E (2017) Soil moisture data assimilation in a hydrological model: a case study in Belgium using large-scale satellite data. Remote Sens 9:820

Balmaseda MA, Dee D, Vidard A, Anderson DLT (2007) A multivariate treatment of bias for sequential data assimilation: application to the tropical oceans. Q J R Meteorol Soc 133:167–179. https://doi.org/10.1002/qj.12

Bannari A, Morin D, Bonn F, Huete AR (1995) A review of vegetation indices. Int J Remote Sens 13(1–2):95–120

Barrett EC (1989) Satellite remote sensing of rainfall. In: Toselli F (eds) Applications of remote sensing to agrometeorology. Ispra Courses (On Remote Sensing). Springer, Dordrecht

Bartalis Z, Wagner W, Naeimi V, Hasenauer S, Scipal K, Bonekamp H et al (2018) Initial soil moisture retrievals from the METOP-A Advanced Scatterometer (ASCAT). Geophys Res Lett 34:L20401. https://doi.org/10.1029/2007GL091088
Bates BC, Kundzewicz ZW, Wu S, Palutikof JP (2008) Climate change and water, Chapter 3 Linking climate change and water resources: impacts and responses. IPCC Secretariat, Geneva, p 210
Bauer E, Hasselmann K, Young IR, Hasselman S (1996) Assimilation of wave data into the wave model WAM using an impulse response function method. J Geophys Res 101(C2):3801–16
Beck HE, van Dijk AIJM, de Roo A, Miralles DG, McVicar TR, Schellekens J, Bruijnzeel LA (2016) Global-scale regionalization of hydrologic model parameters. Water Resour Res 52:3599–3622. https://doi.org/10.1002/2015WR018247
Beller-Simms N, Brown E, Fillmore L, Metchis K, Ozekin K, Ternieden C, Lackey K (2014) Water/Wastewater utilities and extreme climate and weather events: case studies on community response, lessons learned, adaptation, and planning needs for the future. Project No. CC7C11 by the Water Environment Research Foundation. Alexandria, VA
Benada JR (1997) PODAAC Merged GDR (TOPEX/POSEIDON) Generation B User's Handbook, Version 2.0. JPL D-11007. Pasadena: Jet Propulsion Laboratory, California Institute of Technology
Bennett AF (2002) Inverse modeling of the ocean and atmosphere. Cambridge University Press, New York, p 234. ISBN 0-521-81373-5
Bergemann K, Reich S (2010) A mollified ensemble Kalman filter. Q.J.R. Meteorol. Soc. 136:1636–1643. https://doi.org/10.1002/qj.672
Berger H, Forsythe M (2004) Satellite wind superobbing. Met Office Forecasting Research Technical Report, 451, Doc ID :NWPSAF-MO-VS-016
Berry PAM, Garlick JD, Freeman JA, Mathers EL (2005) Global Inland water monitoring from multi-mission altimetery. Geophys. Res. Lett. 32:L16401. https://doi.org/10.1029/2005GL022814
Berry T, Harlim J (2016) Variable bandwidth diffusion Kernels. Appl Comput Harmon Anal 40:68
Berry T, Sauer T (2013) Adaptive ensemble Kalman filtering of non-linear. Tellus A: Dyn Meteorol Ocean 65(1). Article: 20331. https://doi.org/10.3402/tellusa.v65i0.20331
Bertino L, Evensen G, Wackernagel H (2003) Sequential data assimilation techniques in oceanography. Int Stat Rev 71(2):223–241. https://doi.org/10.1111/j.1751-5823.2003.tb00194.x
Betts AK, Ball JH, Beljaars ACM, Miller MJ, Viterbo PA (1996) The land surface-atmosphere interaction: a review based on observational and global modeling perspectives. J Geophys Res 101(D3):7209–7225
Bharati L, Rodgers C, Erdenberger T, Plotnikova M, Shumilov S, Vlek P, Martin N (2008) Integration of economic and hydrologic models: exploring conjunctive irrigation water use strategies in the Volta Basin. Agric Water Manag 95(8):925–936. https://doi.org/10.1016/j.agwat.2008.03.009. ISSN 0378-3774
Birkett CM (1995) The Contribution of Topex/Poseidon to the global monitoring of climatically sensitive lakes. J Geophys Res 25:25–179
Birkett CM (1998) Contribution of the TOPEX NASA radar altimeter to the global monitoring of large rivers and wetlands. Water Resour Res 34(5):1223–1239. https://doi.org/10.1029/98WR00124
Birkett CM, Mertes LAK, Dunne T, Costa MH, Jasinski MJ (2002) Surface water dynamics in the Amazon basin: application of satellite radar altimetry. J Geophys Res 107(D20):LBA 26-1–LBA 26-21. https://doi.org/10.1029/2001JD000609
Bishop CH, Etherton B, Majumdar SJ (2001) Adaptive sampling with the ensemble transform Kalman filter, Part I: theoretical aspects. Mon Weather Rev 129:420–436
Blei David M, Alp Kucukelbir, McAuliffe Jon D (2017) Variational inference: a review for statisticians. J Am Stat Assoc 112(518):859–877. https://doi.org/10.1080/01621459.2017.1285773
Bocanegra E, Silva GC, Custodio E, Manzano M, Montenegro S (2010) State of knowledge of coastal aquifer management in South America. Hydrogeol J 18(1):261–267. https://doi.org/10.1007/s10040-009-0520-5

Boening C, Willis JK, Landerer FW, Nerem RS, Fasullo J (2012) The 2011 La Niña: so strong, the oceans fell. Geophys Res Lett 39:L19602. https://doi.org/10.1029/2012GL053055

Boergens E, Dettmering D, Schwatke C, Seitz F (2016) Treating the hooking effect in satellite altimetry data: a case study along the mekong river and its tributaries. Remote Sens 8:91. https://doi.org/10.3390/rs8020091

Bormann N, Collard A, Bauer P (2010) Estimates of spatial and interchannel observation-error characteristics for current sounder radiances for numerical weather prediction. II: application to AIRS and IASI data. Q J R Meteorol Soc 136:1051–1063

Botterill LC, Fisher M (2012) Beyond Drought: People, Policy and Perspectives, CSIRO Publishing, Melbourne, DAFF (Department of Agriculture, Fisheries and Forestry), Drought and Natural Disaster Declaration: 12 September. Queensland Government, Accessed at https://www.longpaddock.qld.gov.au/queenslanddroughtmonitor/queenslanddroughtreport/

Bouttier F, Mahfouf J-F, Noilhan J (1993) Sequential assimilation of soil moisture from atmospheric low-level parameters. Part I: sensitivity and calibration studies. J Appl Meteorol 32(8):1335–1351

Boyer JF, Dieulin C, Rouchie N, Cres A, Servat E, Paturel JE, Mahe G (2006) SIEREM an environmental information system for water resources. In: 5th World FRIEND conference. La Havana-Cuba, vol 308. November 2006 in Climate Variability and Change—Hydrological Impacts IAHS Publ., pp 19–25

Bras RL, Restrepo-Posada P (1980) Real time automatic parameter calibration in conceptual runoff forecasting models. In: Proceedings of third international symposium on stochastic hydraulics, pp 61–70

Brekke LD, Kiang JE, Olsen JR, Pulwarty RS, Raff DA, Turnipseed DP, Webb RS, White KD (2009) Climate change and water resources management—a federal perspective: U.S. Geological Survey Circular, vol 1331, p 65

Brocca L, Melone F, Moramarco T, Wagner W, Naeimi V, Bartalis Z, Hasenauer S (2010) Improving runoff prediction through the assimilation of the ASCAT soil moisture product. Hydrol Earth Syst Sci 14:1881–1893. https://doi.org/10.5194/hess-14-1881-2010

Brocca L, Moramarco T, Melone F, Wagner W, Hasenauer S, Hahn S (2012) Assimilation of surface- and Root-Zone ASCAT soil moisture products into rainfall-runoff modeling. IEEE Trans Geosci Remote Sens 50:2542–2555

Brown GS (1977) The average impulse response of a rough surface and its applications. IEEE Trans Antennas Propag 25:67–74

Brown NJ, Tregoning P (2010) Quantifying GRACE data contamination effects on hydrological analysis in the Murray-Darling Basin. Southeast Australia. Aust J Earth Sci 57(3):329–335. https://doi.org/10.1080/08120091003619241

Burdanowitz J, Nuijens L, Stevens B, Klepp C (2015) Evaluating light rain from satellite- and ground-based remote sensing data over the subtropical North Atlantic. J Appl Meteor Climatol 54:556–572. https://doi.org/10.1175/JAMC-D-14-0146.1

Burgers G, van Leeuven PJ, Evensen G (1998) Analysis scheme in the ensemble Kalman filter. Mon Weather Rev 126:1719–1724

Buytaert W, Breuer L, Buytaert W, Breuer L et al (2013) Water resources in South America: sources and supply, pollutants and perspectives. In: Symposium on Understanding Freshwater Quality Problems in a Changing World/Joint Assembly of IAHS, IAPSO and IASPEI, Publisher: INT ASSOC HYDROLOGICAL SCIENCES, pp 106–113. ISSN: 0144-7815

BWDB (Bangladesh Water Development Board) (1994) Report on the compilation of aquifer test analysis results, BWDB Water Supply Paper 534. Ground Water Circle II, BWDB, Dhaka

Cabrera J, Yupanqui RT, Rau P (2016) Validation of TRMM daily precipitation data for extreme events analysis. the case of piura watershed in peru. Procedia Eng 154:154–157. https://doi.org/10.1016/j.proeng.2016.07.436. ISSN 1877-7058

Calmant S, Seyler F, Crétaux JF (2008) Monitoring continental surface waters by satellite altimetry. Surv Geophys 29(4–5):247–269. https://doi.org/10.1007/s10712-008-9051-1

Calvet J-C, Noilhan J, Bessemoulin P (1998) Retrieving root-zone soil moisture from surface soil moisture of temperature estimates: a feasibility study based on field measurements. J Appl Meteor 37:371–386

Carrillo CM (2010) The rainfall over tropical South America generated by multiple scale processes. Graduate theses and dissertations, pp 11536. http://lib.dr.iastate.edu/etd/11536

Ceccherini G, Ameztoy I, Hernández C, Moreno C (2015) Highresolution precipitation datasets in South America and West Africa based on satellite-derived rainfall. Enhanc Veg Index Digit Elev Model, Remote Sens 7:6454–6488. https://doi.org/10.3390/rs70506454

Chang B, Kruger U, Kustra R, Zhang J (2013) Canonical correlation analysis based on hilbert-schmidt independence criterion and centered kernel target alignment. In: Volume 28: Proceedings of the 30th international conference on machine learning, vol 2, 28, pp 316–324. http://jmlr.csail.mit.edu/proceedings/papers/v28/chang13.pdf

Chen JL, Wilson CR, Famiglietti JS, Rodell M (2007) Attenuation effect on seasonal basin-scale water storage changes from GRACE time-variable gravity. J Geod 81(4):237–245. https://doi.org/10.1007/s00190-006-0104-2

Chen JL, Wilson CR, Tapley BD (2013) Contribution of ice sheet and mountain glacier melt to recent sea level rise. Nat Geosci 6:549–552. https://doi.org/10.1038/ngeo1829

Chen JL, Wilson CR, Tapley BD, Longuevergne L, Yang ZL, Scanlon BR (2010) Recent La Plata basin drought conditions observed by satellite gravimetry. J Geophys Res 115:D22108. https://doi.org/10.1029/2010JD014689

Chen JL, Wilson CR, Tapley BD, Yang ZL, Niu GY (2009) 2005 drought event in the Amazon River basin as measured by GRACE and estimated by climate models. J Geophys Res 114:B05404. https://doi.org/10.1029/2008JB006056

Cheng MK, Tapley BD (2004) Variations in the Earth's oblateness during the past 28 years. J Geophys Res, Solid Earth 109:B09402. https://doi.org/10.1029/2004JB003028

Chen M, Xie P, Janowiak JE, Arkin PA (2002) Global land precipitation: a 50-yr monthly analysis based on gauge observations. J Hydrometeorol 3:249–266

Chiew FHS, Stewardson MJ, McMahon TA (1993) Comparison of six rainfall-runoff modelling approaches. J Hydrol 147:1–36. https://doi.org/10.1016/0022-1694(93)90073-I

Chou SC, Tanajura CAS, Xue Y, Nobre CA (2002) Validation of the coupled Eta/SSiB model over South America. J Geophys Res 107(D20):8088. https://doi.org/10.1029/2000JD000270

Christiansen L, Krogh PE, Bauer-Gottwein P, Andersen OB, Leiriäo S, Binning PJ, Rosbjerg D (2007) Local to regional hydrological model calibration for the Okavango River Basin from In-situ and space borne gravity observations. In: Proceedings of 2nd Space for Hydrology Workshop. Geneva, Switzerland, pp 12–14

Cisneros BE, Oki T, Arnell NW, Benito G, Cogley JG, Döll P, Jiang T, Mwakalila SS (2014) Freshwater resources. In: Climate change 2014: impacts, adaptation, and vulnerability. Part A: global and sectoral aspects. contribution of working group II to the fifth assessment report of the intergovernmental panel on climate change [Field CB, VR Barros, DJ Dokken, KJ Mach, MD Mastrandrea, TE Bilir, M Chatterjee, KL Ebi, YO Estrada, RC Genova, B Girma, ES Kissel, AN Levy, S MacCracken, PR Mastrandrea, LL White (eds)]. Cambridge University Press, Cambridge, pp 229–269

Clark MP, Rupp DE, Woods RA, Zheng X, Ibbitt RP, Slater AG (2008) Data assimilative modeling investigation of Gulf Stream Warm Core Ring interaction with continental shelf and slope circulation, Hydrological data assimilation with the ensemble Kalman filter: use of streamflow observations to update states in a distributed hydrological model. Adv Water Res 31(10):1309–1324. https://doi.org/10.1016/j.advwatres.2008.06.005. ISSN 0309-1708

Claverie M, Matthews JL, Vermote EF (2016) Justice CO (2016) A 30+ year AVHRR LAI and FAPAR climate data record: algorithm description and validation. Remote Sens 8:263

Claverie M, Vermote E, NOAA CDR Program (2014) NOAA Climate Data Record (CDR) of Leaf Area Index (LAI) and Fraction of Absorbed Photosynthetically Active Radiation (FAPAR), Version 4. [indicate subset used]. NOAA National Centers for Environmental Information. https://doi.org/10.7289/V5M043BX[access date]

Condom T, Rau P, Espinoza JC (2011) Correction of TRMM 3B43 monthly precipitation data over the mountainous areas of Peru during the period 1998–2007. Hydrol Process 25:1924–1933. https://doi.org/10.1002/hyp.7949

Coumou D, Rahmstorf S (2012) A decade of weather extremes. Nat Clim Chang 2(7):1–6. https://doi.org/10.1038/nclimate1452

Courtier P, Thépaut JN, Hollingsworth A (1994) A strategy for operational implementation of 4DVAR, using an incremental approach. Q J R Meteorol Soc 120:1367–1387. https://doi.org/10.1002/qj.49712051912

Crow W, Wood E (2003) The assimilation of remotely sensed soil brightness temperature imagery into a land surface model using ensemble filtering: a case study based on ESTAR measurements during SGP97. Adv Water Res 26:137–149

Dando ML, Thorpe AJ, Eyre JR (2007) The optimal density of atmospheric sounder observations in the Met Office NWP system. Q J R Meteorol Soc 133:1933–1943. https://doi.org/10.1002/qj.175

Dattalo P (2014) A demonstration of canonical correlation analysis with orthogonal rotation to facilitate interpretation. Unpublished manuscript, School of Social Work, Virginia Commonwealth University, Richmond, Virginia

Davis CH (1997) A robust threshold retracking algorithm for measuring ice-sheet surface elevation change from satellite radar altimeters. IEEE Trans Geosci Remote Sens 35(4). https://doi.org/10.1109/36.602540

de Goncalves LG, Shuttleworth WJ, Vila D, Larroza E, Bottino MJ, Herdies DL, Aravequia JA, De Mattos JG, Toll DL, Rodell M, Houser P (2009) The South American Land Data Assimilation System (SALDAS) 5-Yr Retrospective Atmospheric Forcing Datasets. J Hydrometeor 10:999–1010. https://doi.org/10.1175/2009JHM1049.1

De Jeu RAM, Owe M (2003) Further validation of a new methodology for surface moisture and vegetation optical depth retrieval. Int J Remote Sens 24:4559–4578. https://doi.org/10.1080/0143116031000095934

De Jeu RAM, Wagner W, Holmes TRH, Dolman AJ, van de Giesen NC, Friesen J (2008) Global soil moisture patterns observed by space borne microwave radiometers and scatterometers. Surv Geophys 29(4–5):399–420. https://doi.org/10.1007/s10712-008-9044-0

De Lannoy GJM, de Rosnay P, Reichle RH (2015) Soil moisture data assimilation. In: Duan Q, Pappenberger F, Thielen J, Wood A, Cloke H, Schaake J (eds) Handbook of hydrometeorological ensemble forecasting. Springer, Heidelberg

De Lannoy GJM, Houser PR, Pauwels VRN, Verhoest NE (2009) Assessment of model uncertainty for soil moisture through ensemble verification. J Geophys Res 111:D10101. https://doi.org/10.1029/2005JD006367

De Lannoy G, Pauwels VRN, Houser PR, Verhoest NEC, Gish T (2007) Representativeness of point soil moisture observations, upscaling and assimilation. IUGG general assembly. Perugia Italy, Session HS2004

De Paiva RCDRCD, Buarque DCDC, Collischonn W, Bonnet M-PMP, Frappart F, Calmant S, Bulhões Mendes CACA (2013) Large-scale hydrologic and hydrodynamic modeling of the Amazon River basin. Water Resour Res 49:1226–1243. https://doi.org/10.1002/wrcr.20067

Devi GK, Ganasri BP, Dwarakish GS (2015) A review on hydrological models. Aquatic Procedia 4:1001–1007. https://doi.org/10.1016/j.aqpro.2015.02.126

Dey NC, Alam MS, Sajjan AK, Bhuiyan MA, Ghose L, Ibaraki Y, Karim F (2011) Assessing environmental and health impact of drought in the Northwest Bangladesh. J Environ Sci Nat Resour 4(2):89–97. https://doi.org/10.3329/jesnr.v4i2.10141

Dillon ME, Skabar YG, Ruiz J, Kalnay E, Collini EA, Echevarría P, Saucedo M, Miyoshi T, Kunii M (2016) Application of the WRF-LETKF data assimilation system over southern South America: sensitivity to model physics. Weather Forecast 31:217–236. https://doi.org/10.1175/WAF-D-14-00157.1

Dingman SL (2002) Physical hydrology, 2nd edn. Prentice Hall, Upper Saddle River, p 646

Döll P, Kaspar F, Lehner B (2003) A global hydrological model for deriving water availability indicators: model tuning and validation. J Hydrol 270:105–134. https://doi.org/10.1016/S0022-1694(02)00283-4

Dong Q (2015) Lu F (2015) Performance assessment of hydrological models considering acceptable forecast error threshold. Water 7:6173–6189

Doucet A (1998) On sequential simulation-based methods for Bayesian filtering. Technical report CUED/F-INFENG/TR 310, Department of Engineering, Cambridge University, Cambridge, UK

Doucet A, Cappe O, Moulines E (2005) Comparison of resampling schemes for particle filtering. In: 4th International Symposium on Image and Signal Processing and Analysis (ISPA)

Doucet A, Freitas N, Gordon N (2001) Sequential Monte Carlo methods in practice. Springer, New York

Draper CS, Mahfouf J-F, Walker JP (2009) An EKF assimilation of AMSR-E soil moisture into the ISBA land surface scheme. J Geophys Res 114:D20104. https://doi.org/10.1029/2008JD011650

Dreano D, Mallick B, Hoteit I (2015) Filtering remotely sensed chlorophyll concentrations in the Red Sea using a space-time covariance model and a Kalman filter. Spat Stat 13:1–20. https://doi.org/10.1016/j.spasta.2015.04.002. ISSN 2211-6753

Droogers P, Kite G (2002) Remotely sensed data used for modelling at different hydrological scales. Hydrol Process 16:1543–1556. https://doi.org/10.1002/hyp.1019

Drusch M, Wood EF, Gao H (2005) Observation operators for the direct assimilation of TRMM microwave imager retrieved soil moisture. Geophys Res Lett 32:L15403

Dumedah G, Walker JP, Merlin O (2015) Root-zone soil moisture estimation from assimilation of downscaled soil moisture and ocean salinity data. Adv Water Resour 84:14–22. https://doi.org/10.1016/j.advwatres.2015.07.021. ISSN 0309-1708

Edwards CA, Moore AM, Hoteit I, Cornuelle BD (2015) Regional ocean data assimilation. Ann Rev Mar Sci 7:6.1–6.22. https://doi.org/10.1146/annurev-marine-010814-015821.

Eicker A, Forootan E, Springer A, Longuevergne L, Kusche J (2016) Does GRACE see the terrestrial water cycle "intensifying"? J Geophys Res: Atmos 121:733–745. https://doi.org/10.1002/2015JD023808

Eicker A, Schumacher M, Kusche J, Döll P, Müller-Schmied H (2014) Calibration/data assimilation approach for integrating GRACE data into the WaterGAP global hydrology model (WGHM) using an ensemble Kalman filter: first results. Surv Geophys 35(6):1285–1309. https://doi.org/10.1007/s10712-014-9309-8

Elbern H, Schmidt H (2001) Ozone episode analysis by fourdimensional variational chemistry data assimilation. J Geophys Res 106:3569–3590

Entekhabi D, Nakamura H, Njoku EG (1994) Solving the inverse problem for soil moisture and temperature profiles by sequential assimilation of multifrequency remotely senses observations. IEEE Trans Geosci Remote Sens 32:438–448

Ercolani G, Castelli F (2017) Variational assimilation of streamflow data in distributed flood forecasting. Water Resour Res 53:158–183. https://doi.org/10.1002/2016WR019208

Erfanian A, Wang G, Fomenko L (2017) Unprecedented drought over tropical South America in 2016: significantly under-predicted by tropical SST. Sci Rep 7. Article number 5811: https://doi.org/10.1038/s41598-017-05373-2

Escobar H (2015) Drought triggers alarms in Brazil's biggest metropolis. Science 347:812–812

Evensen G (1994) Sequential data assimilation with a nonlinear quasi-geostrophic model using Monte Carlo methods to forecast error statistics. J Geophys Res: Ocean 99(C5):10143–10162. https://doi.org/10.1029/94JC00572

Evensen G (2003) The ensemble Kalman filter: theoretical formulation and practical inplementation. Ocean Dyn 53:343–367. https://doi.org/10.1007/s10236-003-0036-9

Evensen G (2004) Sampling strategies and square root analysis schemes for the EnKF. Ocean Dyn 54(6):539–560

Evensen G (2007) Data assimilation: the ensemble Kalman filter. Springer, Berlin, p 279

Evensen G (2009) Data assimilation. the ensemble Kalman filter, 2nd edn. Springer, Berlin. ISBN 978-3-642-03711-5

Famiglietti JS, Ryu D, Berg AA, Rodell M, Jackson TJ (2008) Field observations of soil moisture variability across scales. Water Resour Res 44:W01423. https://doi.org/10.1029/2006WR005804

FAO (Food and Agriculture Organization of the United Nations) (2009) FAO water report, 34

Farg E, Arafat SM, Abd El-Wahed MS, El-Gindy AM (2012) Estimation of evapotranspiration ETc and crop coefficient Kc of wheat, in South Nile Delta of Egypt using integrated FAO-56 approach and remote sensing data. Egypt J Remote Sens Space Sci 15(1):83–89

Fatolazadeh F, Voosoghi B, Naeeni MR (2016) Wavelet and Gaussian approaches for estimation of groundwater variations using GRACE data. Groundwater 54:74–81. https://doi.org/10.1111/gwat.12325

Flandrin P, Gonçalvès P, Rilling G (2004) Detrending and denoising with empirical mode decompositions, 12th European signal processing conference. Vienna, pp 1581–1584

Flantua SGA, Hooghiemstra H, Vuille M, Behling H, Carson JF, Gosling WD, Hoyos I, Ledru MP, Montoya E, Mayle F, Maldonado A, Rull V, Tonello MS, Whitney BS, González-Arango C (2016) Climate variability and human impact in South America during the last 2000 years: synthesis and perspectives from pollen records. Clim Past 12:483–523. https://doi.org/10.5194/cp-12-483-2016

Fleming K, Awange JL (2013) Comparing the version 7 TRMM 3B43 monthly precipitation product with the TRMM 3B43 version 6/6A and Bureau of Meteorology datasets for Australia. Aust Meteorol Ocean J 63(3):421–426

Forootan E, Awange J, Kusche J, Heck B, Eicker A (2012) Independent patterns of water mass anomalies over Australia from satellite data and models. J Remote Sens Environ 124:427–443. https://doi.org/10.1016/j.rse.2012.05.023

Forootan E, Rietbroek R, Kusche J, Sharifi MA, Awange J, Schmidt M, Omondi P, Famiglietti J (2014a) Separation of large scale water storage patterns over Iran using GRACE, altimetry and hydrological data. Remote Sens Environ 140:580–595. https://doi.org/10.1016/j.rse.2013.09.025

Forootan E, Didova O, Schumacher M, Kusche J, Elsaka B (2014b) Comparisons of atmospheric mass variations derived from ECMWF reanalysis and operational fields, over 2003 to 2011. J Geod 88:503–514. https://doi.org/10.1007/s00190-014-0696-x

Forootan E, Khandu Awange J, Schumacher M, Anyah R, van Dijk A, Kusche J (2016) Quantifying the impacts of ENSO and IOD on rain gauge and remotely sensed precipitation products over Australia. Remote Sens Environ 172:50–66. https://doi.org/10.1016/j.rse.2015.10.027

Forootan E, Safari A, Mostafaie A, Schumacher M, Delavar M, Awange J (2017) Large-scale total water storage and water flux changes over the arid and semi-arid parts of the Middle East from GRACE and reanalysis products. Surv Geophys. https://doi.org/10.1007/s10712-016-9403-1

Frappart F, Calmant S, Cauhope M, Seyler F, Cazenave A (2006) Preliminary results of envisat ra-2-derived water levels validation over the amazon basin. Remote Sens Environ 100(2):252–264

Frappart F, Papa F, Malbeteau Y, León JG, Ramillien G, Prigent C, Seoane L, Seyler F, Calmant S (2015) Surface freshwater storage variations in the orinoco floodplains using multi-satellite observations. Remote Sens 7:89–110

Frappart F, Papa F, Santos da Silva J, Ramillien G, Prigent C, Seyler F, Calmant S (2012) Surface freshwater storage and dynamics in the Amazon basin during the 2005 exceptional drought. Environ Res Lett 7:044010(7pp). https://doi.org/10.1088/1748-9326/7/4/044010

Frappart F, Seoane L, Ramillien G (2013) Validation of GRACE-derived terrestrial water storage from a regional approach over South America. Remote Sens Environ 2013(137):69–83. Elsevier

Frappart F, Seyler F, Martinez JM, Leon JG, Cazenave A (2005) Floodplain water storage in the Negro River basin estimated from microwave remote sensing of inundation area and water levels. Remote Sens Environ 99(4):387–399

Fu LL, Cazenave A (2001) Satellite altimetry and Earth sciences. a handbook of techniques and applications. In: International Geophysics Series. Academic, San Diego, p 69

Gaiha R, Hill K, Thapa G (2010) Natural disasters in South Asia, ASARC working papers, The Australian National University, Australia South Asia Research Centre. http://EconPapers.repec.org/RePEc:pas:asarcc:2010-06

Gain AK, Giupponi C (2012) Renaud FG (2012) Climate change adaptation and vulnerability assessment of water resources systems in developing countries: a generalized framework and a feasibility study in Bangladesh. Water 4:345–366

Garatuza-Payan J, Tamayo A, Watts C, Rodriguez J (2003) Estimating large area wheat evapotranspiration from remote sensing data. In: Proceedings of the IGARSS (2003) 2003 IEEE international geoscience and remote sensing symposium. Toulouse, France, pp 380–382

Garner TW, Wolf RA, Spiro RW, Thomsen MF (1999) First attempt at assimilating data to constrain a magnetospheric model. J Geophys Res 104(A11):25145–25152. https://doi.org/10.1029/1999JA900274

Garreaud RD, Vuille M, Compagnucci R, Marengo J (2008) Present-day South American climate. Paleogeogr Palaeoclimatol Palaeoecol. https://doi.org/10.1016/j.palaeo.2007.10.032

Geruo A, Wahr J, Zhong S (2013) Computations of the viscoelastic response of a 3-D compressible Earth to surface loading: an application to Glacial Isostatic Adjustment in Antarctica and Canada. Geophys J Int 192(2):557–572. https://doi.org/10.1093/gji/ggs030

Getirana A (2016) Extreme water deficit in Brazil detected from space. J Hydrometeorol 17:591–599. https://doi.org/10.1175/JHM-D-15-0096.1

Getirana AB, Yamazaki D, Decharme B, Papa F, Mognard N (2012) The Hydrological Modeling and Analysis Platform (HyMAP): evaluation in the Amazon basin. J Hydrometeorol 13:1641–1665. https://doi.org/10.1175/JHM-D-12-021.1

Getirana A, Dutra E, Guimberteau M, Kam J, Li H, Decharme B, Zhang Z, Ducharne A, Boone A, Balsamo G, Rodell M, Toure AM, Xue Y, Drapeau G, Arsenault K, Kumar SV, Leung LR, Peters-Lidard C, Ronchail J, Sheffield J (2014) Water balance in the Amazon basin from a land surface model ensemble. J Hydrometeorol 15:2586–2614. https://doi.org/10.1175/JHM-D-14-0068.1

Getirana A, Kumar S, Girotto M, Rodell M (2017) Rivers and floodplains as key components of global terrestrial water storage variability. Geophys Res Lett 44: https://doi.org/10.1002/2017GL074684

Gharamti ME, Valstar J, Janssen G, Marsman A, Hoteit I (2016) On the efficiency of the hybrid and the exact second-order sampling formulations of the EnKF: a reality-inspired 3-D test case for estimating biodegradation rates of chlorinated hydrocarbons at the port of Rotterdam. Hydrol Earth Syst Sci 20:4561–4583. https://doi.org/10.5194/hess-20-4561-2016

Girotto M, De Lannoy GJ, Reichle RH, Rodell M (2016) Assimilation of gridded terrestrial water storage observations from GRACE into a land surface model. Water Resour Res 52(5):4164–4183. https://doi.org/10.1002/2015WR018417

Girotto M, De Lannoy GJ, Reichle RH, Rodell M, Draper C, Bhanja SN, Mukherjee A (2017) Benefits and pitfalls of GRACE data assimilation: a case study of terrestrial water storage depletion in India. Geophys Res Lett 44(9):4107–4115. https://doi.org/10.1002/2017GL072994

Girotto M, Reichle RH, Rodell M, Liu Q, Mahanama S, De Lannoy GJ (2019) Multi-sensor assimilation of SMOS brightness temperature and GRACE terrestrial water storage observations for soil moisture and shallow groundwater estimation. Remote Sens Environ 227:12–27. https://doi.org/10.1016/j.rse.2019.04.001

Giustarini L, Matgen P, Hostache R, Montanari M, Plaza D, Pauwels VRN, De Lannoy GJM, De Keyser R, Pfister L, Hoffmann L, Savenije HHG (2011) Assimilating SAR-derived water level data into a hydraulic model: a case study. Hydrol Earth Syst Sci 15:2349–2365. https://doi.org/10.5194/hess-15-2349-2011

Gleason CJ, Wada Y, Wang J (2018) A hybrid of optical remote sensing and hydrological modelling improves water balance estimation. J Adv Model Earth Syst 10:2–17. https://doi.org/10.1002/2017MS000986

Glenn EP, Neale CMU, Hunsaker DJ, Nagler PL (2011) Vegetation index-based crop coefficients to estimate evapotranspiration by remote sensing in agricultural and natural ecosystems". Hydrol Process 25(26):4050–4062

Golian S, Mazdiyasni O (2015) AghaKouchak A (2015) Trends in meteorological and agricultural droughts in Iran. Theor Appl Climatol 119:679–688. https://doi.org/10.1007/s00704-014-1139-6

Gordon NJ, Salmond DJ, Smith AFM (1993) Novel approach to nonlinear/non-Gaussian Bayesian state estimation. IEE Proc F 140:107–113

Gosling SN, Arnell NW (2016) A global assessment of the impact of climate change on water scarcity. Clim Change 134:371. https://doi.org/10.1007/s10584-013-0853-x

Grau HR, Aide M (2008) Globalization and land-use transitions in Latin America. Ecol Soc 13(2):16. www.ecologyandsociety.org/vol13/iss2/art16/

Grimson R, Montroull N, Saurral R, Vasquez P, Camilloni I (2013) Hydrological modelling of the Iberá Wetlands in southeastern South America. J Hydrol 503:47–54. https://doi.org/10.1016/j.jhydrol.2013.08.042. ISSN 0022-1694

Guo N (2003) Vegetation index and its advances. Arid Meteorol 21(4):71–75

Gutentag ED, Heimes FJ, Krothe NC, Luckey RR, Weeks JB (1984) Geohydrology of the High Plains aquifer in parts of Colorado, Kansas, Nebraska, New Mexico, Oklahoma, South Dakota, Texas, and Wyoming, U.S. Geol. Surv. Prof. Pap., 1400-B, 66 pp

Haile AT, Rientjes TH, Gieske A, Gebremichael M (2010) Rainfall estimation at the source of the Blue Nile: A multispectral remote sensing approach. Int J Appl Earth Obs Geoinformation 12S:76–83

Hairer M, Stuart A, Voss J (2005) A Bayesian approach to data assimilation. Physica D

Hamill TM, Snyder C (2002) Using improved background-error covariances from an ensemble Kalman filter for adaptive observations. Mon Weather Rev 130:1552–1572. https://doi.org/10.1175/1520-0493(2002)130<1552:UIBECF>2.0.CO;2

Hamilton F, Berry T, Sauer T (2015) Pred icting chaotic time series with a partial model. Phys Rev E 92:010902

Hamilton F, Berry T, Sauer T (2016) Ensemble Kalman filtering without a model. Phys Rev X 6, 6(1):011021. https://doi.org/10.1103/PhysRevX.6.011021

Hamilton F, Berry T, Sauer T (2017) Kalman-Takens filtering in the presence of dynamical noise. To appear Eur Phys J ST

Han SC, Shum CK, Jekeli C, Kuo CY, Wilson CR, Seo KW (2005) Non-isotropic filtering of GRACE temporal gravity for geophysical signal enhancement. Geophys J Int 16:18–25. https://doi.org/10.1111/j.1365-246X.2005.02756.x

Han X, Li X (2008) An evaluation of the nonlinear/non-Gaussian filters for the sequential data assimilation. Remote Sens Environ 112(4):1434–1449. https://doi.org/10.1016/j.rse.2007.07.008

Han R, Tian XJ, Fu Y, Cai ZN (2015) Real-data assimilation experiment with a joint data assimilation system: assimilating carbon dioxide mole fraction measurements from the Greenhouse gases Observing Satellite. Atmos Ocean Sci Lett 9(2):107–113. https://doi.org/10.1080/16742834.2016.1133070

Harries JE, Russell JE, Hanafin JA et al (2005) The geostationary earth radiation budget project. Am Meteorol Soc 86:945–960

Harris IC (2008) Climatic Research Unit (CRU) time-series datasets of variations in climate with variations in other phenomena. NCAS British Atmospheric Data Centre, date of citation, University of East Anglia Climatic Research Unit; Jones, P.D. http://catalogue.ceda.ac.uk/uuid/3f8944800cc48e1cbc29a5ee12d8542d

Hasan MR, Mostafa MG, Matin I (2013) Effect of rainfall on groundwater level fluctuations in Chapai Nawabgonj District. Int J Eng Res Technol (IJERT) 2(4). ISSN: 2278-0181

Helmert J, Şorman A, Montero R, De Michele C, De Rosnay P, Dumont M, Finger DC, Lange M, Picard G, Potopová V, Pullen S, Vikhamar-Schuler D (2018) Arslan AN (2018) Review of snow data assimilation methods for hydrological, land surface, meteorological and climate models: results from a COST harmosnow survey. Geosciences 8:489

Henck AH, Montgomery DR, Huntington KW, Liang C (2010) Monsoon control of effective discharge. Yunnan Tibet, Geol 38:975–978

Herring M (2012) Agriculture and Biodiversity: opportunities for a better marriage. In: Rice growers association annual conference, One Tree, NSW

Hersbach H, Stoffelen A, De Haan S (2007) An Improved C-band scatterometer ocean geophysical model function: CMOD5. J Geophys Res Oceans 112:C03006

Hoffman Y, Ribak E (1991) Constrained realizations of Gaussian fields—a simple algorithm. Astrophys J Part 2—Lett 380:L5–L8. (ISSN 0004-637X)

Hol JD, Schon TB, Gustafsson F (2006) On resampling algorithms for particle filters. Nonlinear statistical signal processing workshop, 2006 IEEE 79–82. https://doi.org/10.1109/NSSPW.2006.4378824

Hossain M (1990) Natural calamities, instability in production and food policy in Bangladesh, The Bangladesh Development Studies (BDS). Bangladesh Inst Dev Stud 18(1):

Hoteit I, Luo X, Pham DT (2012) Particle Kalman filtering: a nonlinear bayesian framework for ensemble kalman filters. Mon Weather Rev 140(2):528–542. https://doi.org/10.1175/2011MWR3640.1

Hoteit I, Pham DT, Blum J (2002) A simplified reducedorder kalman filtering and application to altimetric data assimilation in tropical Pacific. J Mar Syst 36:101–127. https://doi.org/10.1016/S0924-7963(02)00129-X

Hoteit I, Pham DT, Gharamti ME, Luo X (2015) Mitigating observation perturbation sampling errors in the stochastic EnKF. Mon Weather Rev 143(7):2918–2936. https://doi.org/10.1175/MWR-D-14-00088.1

Hoteit I, Pham DT, Triantafyllou G, Korres G (2008) A new approximate solution of the optimal nonlinear filter for data assimilation in meteorology and oceanography. Mon Weather Rev 136:317–334. https://doi.org/10.1175/2007MWR1927.1

Hoteit I, Triantafyllou G, Petihakis G (2005) Efficient data assimilation into a complex, 3-D physical-biogeochemical model using partially-local Kalman filters. Ann Geophys, Eur Geosci Union 23(10):3171–3185

Houborg R, Rodell M, Li B, Reichle RH, Zaitchik BF (2012) Drought indicators based on model-assimilated Gravity Recovery and Climate Experiment (GRACE) terrestrial water storage observations. Water Resour Res 48:W07525. https://doi.org/10.1029/2011WR011291

Houser PR, De Lannoy G, Walker J (2012) Hydrologic data assimilation. In: Tiefenbacher J (ed) Approaches to managing disaster - assessing hazards, emergencies and disaster impacts. In-Tech, Rijeka Croatia, pp 41–64

Houser PR et al (1998) Integration of soil moisture remote sensing and hydrologic modeling using data assimilation. Water Resour Res 34(12):3405–3420

Houtekamer PL (1995) The construction of optimal perturbations. Mon Weather Rev 123:2888–2898

Houtekamer PL, Mitchell HL (1998) Data assimilation using an ensemble Kalman filter technique. Mon Weather Rev 126:796–811

Houtekamer PL, Mitchell HL (2001) A sequential ensemble Kalman filter for atmospheric data assimilation. Mon Weather Rev 129(1):123–137

Huang NE, Shen Z, Long SR, Wu MC, Shih HH et al (1998) The empirical mode decomposition and the Hilbert spectrum for nonlinear and non-stationary time series analysis. Proc R Soc Lond A 454:903–995

Huang S, Kumar R, Flörke M, Yang T, Hundecha Y, Kraft P, Gao C, Gelfan A, Liersch S, Lobanova A, Strauch M, Ogtrop FV, Reinhardt J, Haberlandt U, Krysanova V (2016) Evaluation of an ensemble of regional hydrological models in 12 large-scale river basins worldwide. Clim Change 141(3):381–397. https://doi.org/10.1007/s10584-016-1841-8

Huffman GJ, Adler RF, Bolvin DT, Gu G, Nelkin EJ, Bowman KP, Hong Y, Stocker EF, Wolff DB (2007) The TRMM Multi-satellite precipitation analysis: quasi- global, multi-year, combined-sensor precipitation estimates at fine scale. J Hydrometeor 8(1):38–55

Huffman G, Adler R, Arkin P, Chang A, Ferraro R, Gruber A, Janowiak J, McNab A, Rudolf B, Schneider U (1997) The Global Precipitation Climatology Project (GPCP) Combined Precipitation Dataset. Bull Am Meteor Soc 78:5–20. https://doi.org/10.1175/1520-0477(1997)078<0005:TGPCPG>2.0.CO;2

Huffman G, Bolvin D (2012) TRMM and other data precipitation data set documentation. NASA Goddard Space Flight Center and Science Systems and Applications Inc, Mesoscale Atmospheric Processes Laboratory

Humphrey V, Gudmundsson L, Seneviratne SI (2016) Assessing global water storage variability from GRACE: trends, seasonal cycle, subseasonal anomalies and extremes. Surv Geophys 37:357. https://doi.org/10.1007/s10712-016-9367-1

Huntington TG (2006) Evidence for intensification of the global water cycle: review and synthesis. J Hydrol 319(1–4):83–95. https://doi.org/10.1016/j.jhydrol.2005.07.003

Hutton CJ, Kapelan Z, Vamvakeridou-Lyroudia L, Savić D (2014) Real-time data assimilation in urban rainfall-runoff models. Procedia Eng 70:843–852. https://doi.org/10.1016/j.proeng.2014.02.092. ISSN 1877-7058

International Groundwater Resources Assessment Centre (2004) Graphic groundwater and climate change, International Groundwater Resources Assessment Centre, UNESCO, International Hydrological Programme, Division of water sciences. www.graphicnetwork.net

Irmak A, Kamble B (2009) Evapotranspiration data assimilation with genetic algorithms and SWAP model for on-demand irrigation. Irrig Sci 28(1):101–112. https://doi.org/10.1007/s00271-009-0193-9

Islam MS, Alam MK, Rahman MA (2013) Groundwater dynamics in Pabna District, Bangladesh. IOSR J Environ Sci Toxicol Food Technol (IOSR-JESTFT) 4(2):65–72. e-ISSN: 2319-2402, p-ISSN: 2319-2399

Islam MN, Chowdhury A, Islam KM, Rahaman MZ (2014) Development of rainfall recharge model for natural groundwater recharge estimation in Godagari Upazila of Rajshahi district. Bangladesh. Am J Civ Eng 2(2):48–52. https://doi.org/10.11648/j.ajce.20140202.16

Jaakkola TS, Jordan MI (2000) Bayesian parameter estimation via variational methods. Stat Comput 10:25–37

Jackson T, Bindlish R (2012) Validation of Soil Moisture and Ocean Salinity (SMOS) soil moisture over watershed networks in the US. IEEE Trans Geosci Remote Sens 50:1530–1543

Jacobowitz H, Tighe R (1984) The earth radiation budget derived from the NIMBUS 7 ERB experiment. J Geophys Res 89:4997–5010

Jacquette E, Al Bitar A, Mialon A, Kerr Y, Quesney A, Cabot F et al (2010) SMOS CATDS level 3 global products over land. In: CMU Neale, A Maltese (eds), Remote sensing for agriculture, ecosystems, and hydrology XII. volume 7824 of Proceedings of SPIE-the international society for optical engineering. conference on remote sensing for agriculture, ecosystems, and hydrology XII, Toulouse, France

Jajarmizadeh M, Harun S, Salarpour M (2012) A review on theoretical consideration and types of models in hydrology. J Environ Sci Technol 5:249–261

Jardak M, Navon IM, Zupanski M (2007) Comparison of sequential data assimilation methods for the Kuramoto-Sivashinsky equation. Int J Numer Methods Fluids 62(4):374–402. https://doi.org/10.1002/fld.2020

Jazwinski AH (1970) Stochastic processes and filtering theory, vol 64. Academic Press, p 376. ISBN: 9780080960906

Jekeli C (1981) Alternative methods to smooth the Earth's gravity field. Technical report 327, Department of Geodesy and Science and Surveying, Ohio State University

Jing W, Song J, Zhao X (2018) A Comparison of ECV and SMOS soil moisture products based on OzNet monitoring network. Remote Sens 10(5):703. https://doi.org/10.3390/rs10050703

Joanes AH, Gill CA (1998) Comparing Measures of Sample Skewness and Kurtosis. The Statistician 47(1):183–189

Jonard F, Weihermüller L, Jadoon KZ, Schwank M, Vereecken H, Lambot S (2011) Mapping field scale soil moisture with L-band radiometer and ground-penetrating radar over bare soil. IEEE Trans Geosci Remote Sens 49:2863–2875. https://doi.org/10.1109/TGRS.2011.2114890

Jones DA, Wang W, Fawcett R, Grant I (2007) Climate data for the Australian water availability project. In: Australian Water Availability Project Milestone Report. Bur. Met., Australia, p 37

Joodaki G, Wahr J, Swenson S (2014) Estimating the human contribution to groundwater depletion in the Middle East, from GRACE data, land surface models, and well observations. Water Resour Res 50:2679–2692. https://doi.org/10.1002/2013WR014633

Joyce RJ, Janowiak JE, Arkin PA, Xie PP (2004) CMORPH: a method that produces global precipitation estimates from passive microwave and infrared data at high spatial and temporal resolution. J Hydrometeorol 5:487–503

Juarez RI, Li W, Fu R, Fernandes K, de Oliveira Cardoso A (2009) Comparison of precipitation datasets over the Tropical South American and African Continents. J. Hydrometeor 10:289–299. https://doi.org/10.1175/2008JHM1023.1

Julier SJ, Uhlmann JK (1997) A new extension of the kalman filter to nonlinear systems. In: Proceedings of AeroSense: The 11th international symposium on aerospace/defence sensing, simulation and controls

Julier SJ, Uhlmann JK (2004) Unscented filtering and nonlinear estimation. Proc IEEE 92:401–422

Julier S, Uhlmann J, Durrant-Whyte HF (2000) A new method for the nonlinear transformation of means and covariances in filters and estimators. IEEE Trans Automat Control 45:477–482

Kalman RE (1960) A new approach to linear filtering and prediction problems. Trans ASME—J Basic Eng

Kalnay E (2003) Atmospheric modelling, data assimilation and predictability. Cambridge University Press. ISBN: 9780521796293

Kamble B, Kilic A, Hubbard K (2013) Estimating crop coefficients using remote sensing-based vegetation index. Remote Sens 5(4):1588–1602

Kanoua W, Merkel BJ (2015) Groundwater recharge in Titas Upazila in Bangladesh. Arab J Geosci 8:1361. https://doi.org/10.1007/s12517-014-1305-2

Karamouzian M, Haghdoost AK (2015) Population control policies in Iran. The Lancet 385(9973):21–27. https://doi.org/10.1016/S0140-6736(15)60596-7. ISSN 0140-6736

Keppenne CL (2000) Data assimilation into a primitive-equation model with a parallel ensemble Kalman filter. Mon Weather Rev 128:1971–1981

Keppenne CL, Rienecker M (2002) Initial testing of a massively parallel ensemble Kalman filter with the poseidon isopycnal ocean general circulation model. Mon Weather Rev 130:2951–2965

Khaki M, Forootan E, Sharifi MA (2014) Satellite radar altimetry waveform retracking over the Caspian Sea. Int J Remote Sens 35(17):6329–6356. https://doi.org/10.1080/01431161.2014.951741

Khaki M, Forootan E, Sharifi MA, Awange J, Kuhn M (2015) Improved gravity anomaly fields from retracked multimission satellite radar altimetry observations over the Persian Gulf and the Caspian Sea. Geophys J Int 202(3):1522–1534. https://doi.org/10.1093/gji/ggv240

Khaki M, Schumacher MJ, Forootan Kuhn M, Awange E, van Dijk AIJM (2017a) Accounting for spatial correlation errors in the assimilation of GRACE into hydrological models through localization. Adv Water Resour 108:99–112. https://doi.org/10.1016/j.advwatres.2017.07.024

Khaki M, Hoteit I, Kuhn M, Awange J, Forootan E, van Dijk AIJM, Schumacher M, Pattiaratchi C (2017b) Assessing sequential data assimilation techniques for integrating GRACE data into a hydrological model. Adv Water Resour 107:301–316. https://doi.org/10.1016/j.advwatres.2017.07.001

Khaki M, Ait-El-Fquih B, Hoteit I, Forootan E, Awange J, Kuhn M (2017c) A two-update ensemble Kalman filter for land hydrological data assimilation with an uncertain constraint. J Hydrol 555:447–462. https://doi.org/10.1016/j.jhydrol.2017.10.032

Khaki M, Forootan E, Kuhn M, Awange J, Longuevergne L, Wada Y (2018a) Efficient basin scale filtering of GRACE satellite products. Remote Sens Environ 204:76–93. https://doi.org/10.1016/j.rse.2017.10.040

Khaki M, Forootan E, Kuhn M, Awange J, Papa F, Shum CK (2018b) A study of Bangladesh's subsurface water storages using satellite products and data assimilation scheme. Sci Total Environ 625:963–977. https://doi.org/10.1016/j.scitotenv.2017.12.289

Khaki M, Forootan E, Kuhn M, Awange J, van Dijk AIJM, Schumacher M, Sharifi MA (2018c) Determining water storage depletion within Iran by assimilating GRACE data into the W3RA hydrological model. Adv Water Resour 114:1–18. https://doi.org/10.1016/j.advwatres.2018.02.008

Khaki M, Awange J (2018d) The application of remotely sensed products to enhance model-derived water storage changes over South America. Sci Total Environ 647:1557–1572. https://doi.org/10.1016/j.scitotenv.2018.08.079

Khaki M, Ait-El-Fquih B, Hoteit I, Forootan E, Awange J, Kuhn M (2018e) Unsupervised ensemble Kalman filtering with an uncertain constraint for land hydrological data assimilation. J Hydrol 564:175–190. https://doi.org/10.1016/j.jhydrol.2018.06.080

Khaki M, Hamilton F, Forootan E, Hoteit I, Awange J, Kuhn M (2018f) Non-parametric data integration scheme for land hydrological applications. Water Resour Res 54:4946–4964. https://doi.org/10.1029/2018WR022854

Khaki M, Hoteit I, Kuhn M, Forootan E, Awange J (2018g) Assessing data assimilation frameworks for using multi-mission satellite products in a hydrological context. Sci Total Environ 647:1031–1043. https://doi.org/10.1016/j.scitotenv.2018.08.032

Khandu Awange JL, Kuhn M, Anyah R, Forootan E (2017) Changes and variability of precipitation and temperature in the Ganges-Brahmaputra-Meghna River Basin based on global high-resolution reanalyses. Int J Climatol 37:2141–2159. https://doi.org/10.1002/joc.4842

Kilpatrick KA, Podestá GP, Evans R (2001) Overview of the NOAA/NASA advanced very high resolution radiometer Pathfinder algorithm for sea surface temperature and associated matchup database. J Geophys Res 106(C5):9179–9197. https://doi.org/10.1029/1999JC000065

Klees R, Revtova EA, Gunter BC, Ditmar P, Oudman E, Winsemius HC (2008) The design of an optimal filter for monthly GRACE gravity models. Geophys J Int 175(2):417–432. https://doi.org/10.1111/j.1365-246X.2008.03922.x

Knappett PSK, Mailloux BJ, Choudhury I, Khan MR, Michael HA, Barua S, Mondal DR, Steckler MS, Akhter SH, Ahmed KM, Bostick B, Harvey CF, Shamsudduha M, Shuai P, Mihajlov I, Mozumder R, van Geen A (2016) Vulnerability of low-arsenic aquifers to municipal pumping in Bangladesh. J Hydrol 539:674–686

Koch KR (2007) Introduction to bayesian statistics, 2nd. Springer, Berlin. https://doi.org/10.1007/978-3-540-72726-2. ISBN:978-3-540-72726-2

Kolassa J, Reichle RH, Liu Q, Cosh M, Bosch DD, Caldwell TG, Colliander A, Holifield Collins C, Jackson TJ, Livingston SJ, Moghaddam M, Starks PJ (2017) Data assimilation to extract soil moisture information from SMAP observations. Remote Sens 9:1179

Koster RD, Suarez MJ (1999) A simple framework for examining the interannual variability of land surface moisture fluxes. J Clim 12:1911–1917

Kourgialas NN, Karatzas GP (2015) A modeling approach for agricultural water management in citrus orchards: cost-effective irrigation scheduling and agrochemical transport simulation. Environ Monit Assess 187(7):462. https://doi.org/10.1007/s10661-015-4655-7

Kumar SV, Peters-Lidard CD, Mocko D, Reichle RH, Liu Y, Arsenault KR, Xia Y, Ek M, Riggs G, Livneh B, Cosh M (2014) Assimilation of remotely sensed soil moisture and snow depth retrievals for drought estimation. J Hydrometeorol 15:2446–2469. https://doi.org/10.1175/JHM-D-13-0132.1

Kumar SV, Peters-Lidard CD, Santanello JA, Reichle RH, Draper CS, Koster RD, Nearing G, Jasinski MF (2015) Evaluating the utility of satellite soil moisture retrievals over irrigated areas and the ability of land data assimilation methods to correct for unmodeled processes. Hydrol Earth Syst Sci 19:4463–4478. https://doi.org/10.5194/hess-19-4463-2015

Kumar SV, Reichle RH, Harrison KW, Peters-Lidard CD, Yatheendradas S, Santanello JA (2012) A comparison of methods for a priori bias correction in soil moisture data assimilation. Water Resour Res 48:W03515. https://doi.org/10.1029/2010WR010261

Kumar SV, Reichle RH, Koster RD, Crow WT, Peters-Lidard CD (2009) Role of subsurface physics in the assimilation of surface soil moisture observations. J Hydromet 10(6):1534–1547. https://doi.org/10.1175/2009JHM1134.1

Kumar S, Zaitchik B, Peters-Lidard C, Rodell M, Reichle R, Li B, Jasinski M, Mocko D (2016) Assimilation of gridded GRACE terrestrial water storage estimates in the North American land data assimilation system. J Hydrometeorol 17:1951–1972. https://doi.org/10.1175/JHM-D-15-0157.1

Kusche J (2007) AApproximate decorrelation and non-isotropic smoothing of time-variable GRACE-type gravity field models. J Geod 81:733–749. https://doi.org/10.1007/s00190-007-0143-3

Kusche J, Klemann V, Bosch W (2012) Mass distribution and mass transport in the Earth system. J Geodyn 59–60:1–8. https://doi.org/10.1016/j.jog.2012.03.003

Kusche J, Schmidt R, Petrovic S, Rietbroek R (2009) Decorrelated GRACE time-variable gravity solutions by GFZ and their validation using a hydrological model. J Geod. https://doi.org/10.1007/s00190-009-0308-3

Lagergren J, Reeder A, Hamilton F, Smith RC, Flores KB (2018) Forecasting and uncertainty quantification using a hybrid of mechanistic and non-mechanistic models for an age-structured population model. Bull Math Biol. https://doi.org/10.1007/s11538-018-0421-7

Lahoz WA, Geer AJ, Bekki S, Bormann N, Ceccherini S, Elbern H, Errera Q, Eskes HJ, Fonteyn D, Jackson DR, Khattatov B (2007) The Assimilation of Envisat data (ASSET) project. Atmos Chem Phys 7:1773–1796. https://doi.org/10.5194/acp-7-1773-2007

Landerer FW, Swenson SC (2012) Accuracy of scaled GRACE terrestrial water storage estimates. Water Resour Res 48(4):W04531. https://doi.org/10.1029/2011WR011453

Lawson WG, Hansen JA (2004) Implications of stochastic and deterministic filters as ensemble-based data assimilation methods in varying regimes of error growth. Mon Weather Rev 132:1966–1981

Lee H, Seo D-J, Koren V (2011) Assimilation of streamflow and in situ soil moisture data into operational distributed hydrologic models: effects of uncertainties in the data and initial model soil moisture states. Adv Water Resour 34(12):1597–1615. https://doi.org/10.1016/j.advwatres.2011.08.012

Leroux DJ, Pellarin T, Vischel T, Cohard J-M, Gascon T, Gibon F, Mialon A, Galle S, Peugeot C, Seguis L (2016) Assimilation of SMOS soil moisture into a distributed hydrological model and impacts on the water cycle variables over the Ouémé catchment in Benin. Hydrol Earth Syst Sci 20:2827–2840. https://doi.org/10.5194/hess-20-2827-2016

Lguensat R, Tandeo P, Ailliot P, Pulido M, Fablet R (2017) The analog data assimilation. Mon Weather Rev. https://doi.org/10.1175/MWR-D-16-0441.1

Liang SL, Zheng T, Liu RG, Fang HL, Tsay SC, Running S (2006) Estimation of incident photosynthetically active radiation from moderate resolution imaging spectrometer data. J Geophys Res —Atmos 111:13

Liang X, Lettenmiar DP, Wood EF, Burges SJ (1994) A simple hydrologically based model of land-surfacewater and energy fluxes for general-circulationmodels. J Geophys Res 99:14415–14428. https://doi.org/10.1029/94JD00483

Lievens H, Kumar S, Al Bitar A, De Lannoy GJM, Drusch M, Dumedah G, Hendricks Franssen HJ, Kerr YH, Martens B, Pan M, Roundy JK, Vereecken H, Walker JP, Wood EF, Verhoest NEC, Pauwels VRN (2015) SMOS soil moisture assimilation for improved hydrologic simulation in the Murray Darling Basin. Aust, Remote Sens Environ 168(10):146–162

Lievens H, Reichle RH, Liu Q, De Lannoy GJM, Dunbar RS, Kim SB, Das NN, Cosh M, Walker JP, Wagner W (2017) Joint Sentinel-1 and SMAP data assimilation to improve soil moisture estimates. Geophys Res Lett 44:6145–6153. https://doi.org/10.1002/2017GL073904

Lindskog M, Salonen K, Jarvinen H, Michelson DB (2004) Doppler radar wind data assimilation with HIRLAM 3DVAR. Mon Weather Rev 132:1081–1092

Li B, Rodell M (2013) Spatial variability and its scale dependency of observed and modeled soil moisture over different climate regions. Hydrol Earth Syst Sci 17:1177–1188. https://doi.org/10.5194/hess-17-1177-2013

Li B, Rodell M (2015) Evaluation of a model-based groundwater drought indicator in the conterminous U.S. J Hydrol 526:78–88. https://doi.org/10.1016/j.jhydrol.2014.09.02

Li B, Rodell M, Zaitchik BF, Reichle RH, Koster RD, van Dam TM (2012) Assimilation of GRACE terrestrial water storage into a land surface model: Evaluation and potential value for drought

monitoring in western and central Europe. J Hydrol 446–447:103–115. https://doi.org/10.1016/j.jhydrol.2012.04.035
Li Y, Ryu D, Western AW, Wang QJ (2015) Assimilation of stream discharge for flood forecasting: Updating a semidistributed model with an integrated data assimilation scheme. Water Resour Res 51:3238–3258. https://doi.org/10.1002/2014WR016667
Liu ZQ, Rabier F (2003) The potential of high-density observations for numerical weather prediction: a study with simulated observations. Q J R Meteorol Soc 129:3013–3035. https://doi.org/10.1256/qj.02.170
Liu Y, Peters-Lidard CD, Kumar S, Foster JL, Shaw M, Tian Y, Fall GM (2013) Assimilating satellitebased snow depth and snow cover products for improving snow predictions in Alaska. Adv Water Res 54:208–227. https://doi.org/10.1016/j.advwatres.2013.02.005
Liu Y, Weerts AH, Clark M, Hendricks Franssen H-J, Kumar S, Moradkhani H, Seo D-J, Schwanenberg D, Smith P, van Dijk AIJM, van Velzen N, He M, Lee H, Noh SJ, Rakovec O, Restrepo P (2012) Advancing data assimilation in operational hydrologic forecasting: progresses, challenges, and emerging opportunities. Hydrol Earth Syst Sci 16:3863–3887. https://doi.org/10.5194/hess-16-3863-2012
Liu C, Xue M (2016) Relationships among four-dimensional hybrid ensemble-variational data assimilation algorithms with full and approximate ensemble covariance localization. Mon Weather Rev 144:591–606. https://doi.org/10.1175/MWR-D-15-0203.1
Liyanage CP (2017) Yamada K (2017) Impact of population growth on the water quality of natural water bodies. Sustainability 9:1405
Longuevergne L, Wilson CR, Scanlon BR, Crétaux JF (2013) GRACE water storage estimates for the Middle East and other regions with significant reservoir and lake storage. Hydrol Earth Syst Sci 17:4817–4830. https://doi.org/10.5194/hess-17-4817-2013
Long D, Yang Y, Wada Y, Hong Y, Liang W, Chen Y, Yong B, Hou A, Wei J, Chen L (2015) Deriving scaling factors using a global hydrological model to restore GRACE total water storage changes for China's Yangtze River Basin. Remote Sens Environ 168:177–193. https://doi.org/10.1016/j.rse.2015.07.003
Lorenz E (1956) Empirical orthogonal function and statistical weather prediction. Technical Report Science Report No 1, Statistical Forecasting Project. MIT, Cambridge
Luo L, Wood EF, Pan M (2007) Bayesian merging of multiple climate model forecasts for seasonal hydrological predictions. J Geophys Res 112:D10102. https://doi.org/10.1029/2006JD007655
Madani K (2014) Water management in Iran: what is causing the looming crisis? J Environ Stud Sci. https://doi.org/10.1007/s13412-014-0182-z
Madsen H, Skotner C (2005) Adaptive state-updating in real-time river flow forecasting-a combined filtering and error forecasting procedure. J Hydrol 308:302–312
Maggioni V, Houser PR (2017) Soil Moisture Data Assimilation. In: Park S, Xu L (eds) Data Assimilation for atmospheric, oceanic and hydrologic applications (Vol. III). Springer, Cham
Magrin GO, Marengo JA, Boulanger J-P, Buckeridge MS, Castellanos E, Poveda G, Scarano FR, Vicuña S (2014) Central and South America. In: Climate change 2014: impacts, adaptation, and vulnerability. part b: regional aspects. contribution of working group II to the fifth assessment report of the intergovernmental panel on climate change [Barros VR, CB Field, DJ Dokken, MD Mastrandrea, KJ Mach, TE Bilir, M Chatterjee KL Ebi YO, Estrada RC Genova, B Girma, ES Kissel AN Levy, S MacCracken, PR Mastrandrea, LL White (eds)]. Cambridge University Press, Cambridge, United Kingdom and New York, NY, USA, pp 1499–1566
Magrin G, Gay García C, Cruz Choque D, Giménez JC, Moreno AR, Nagy GJ, Nobre C, Villamizar A (2007) Latin America. In: Climate change 2007: impacts, adaptation and vulnerability. contribution ofworking group II to the fourth assessment report of the intergovernmental panel on climate change [Parry ML, OF, Canziani JP, Palutikof PJ, van der Linden, CE Hanson (eds)]. Cambridge University Press, Cambridge, UK and NewYork, NY, USA, pp 581–615
Mainuddin M (2002) Groundwater irrigation in Bangladesh: tool for poverty alleviation or cause of mass poisoning? In: Proceedings of the symposium on intensive use of groundwater: challenges and opportunities, Valencia, Spain, pp 10–14

Marengo JA (2009) Long-term trends and cycles in the hydrometeorology of the Amazon basin since the late 1920s. Hydrol Process 23:3236–3244

Massari C, Camici S, Ciabatta L, Brocca L (2018) Exploiting satellite-based surface soil moisture for flood forecasting in the mediterranean area: state update versus rainfall correction. Remote Sens 10(2):292

Massoud EC, Huisman J, Benincà E, Dietze MC, Bouten W, Vrugt JA, Adler F (2018) Probing the limits of predictability: data assimilation of chaotic dynamics in complex food webs. Ecol Lett 21:93–103. https://doi.org/10.1111/ele.12876

Mayer-Gürr T, Zehentner N, Klinger B, Kvas A (2014) ITSG-Grace2014: a new GRACE gravity field release computed in Graz. In: GRACE Science Team Meeting (GSTM). Potsdam am: 29.09.2014

McBean E, deJong A, Gharabaghi B (2011) Groundwater in Bangladesh: implications in a climate-changing world. Water Res Manag 1(3):3–8

McCarthy JJ, Canziani OF, Leary NA, Dokken DJ, White KS et al (2001) Climate Change 2001: Impacts, Adaptation. and Vulnerability, Inter-Governmental Panel on Climate Change (IPCC), Work Group II Input to the Third Assessment Report, Cambridge University Press, Cambridge

McLaughlin D (2002) An integrate approach to hydrologic data assimilation: Interpolation, smoothing, and filtering. Adv Water Resour 25:1275–1286. https://doi.org/10.1016/S0309-1708(02)00055-6

McMillan HK, Hreinsson EÖ, Clark MP, Singh SK, Zammit C, Uddstrom MJ (2013) Operational hydrological data assimilation with the recursive ensemble Kalman filter. Hydrol Earth Syst Sci 17(1):21–38. https://doi.org/10.5194/hess-17-21-2013

Mehra R (1970) On the identification of variances and adaptive Kalman filtering. IEEE Trans Auto Cont 15:175–184

Mehra R (1972) Approaches to adaptive filtering. IEEE Trans Auto Cont 17:693–698

Mercer D, Christesen L, Buston M (2007) Squandering the future-Climate change, policy failure and the water crisis in Australia. Futures 39:272–287

Michaelides S, Levizzani V, Anagnostou E, Bauer P, Kasparis T, Lane JE (2009) Precipitation: measurement, remote sensing, climatology and modeling. Atmos Res 94:512–533

Michel D (2017) Iran's impending water crisis. In Water, Security, and US Policy, ed by David Reed. P 438. ISBN-10:1138051519

Miralles DG, Holmes TRH, de Jeu RAM, Gash JH, Meesters AGCA, Dolman AJ (2011) Global land-surface evaporation estimated from satellite-based observations. Hydrol Earth Syst Sci 15:453–469

Mitchell HL, Houtekamer PL (2000) An adaptive ensemble Kalman filter. Mon Weather Rev 128:416–433

Mitchell HL, Houtekamer PL, Pellerin G (2002) Ensemble size, balance, and model-error representation in an ensemble Kalman filter. Mon Weather Rev 130:2791–2808

Mladenova I, Lakshmi V, Jackson TJ, Walker JP, Merlin O (2011) de Jeu RAM (2011) Validation of AMSR-E soil moisture using L-band airborne radiometer data from national airborne field experiment 2006. Remote Sens Environ. https://doi.org/10.1016/j.rse.2011.04.011

Mohammadi-Ghaleni M, Ebrahimi K (2011) Assessing impact of irrigation and drainage network on surface and groundwater resources—Case study: Saveh Plain, Iran. In: ICID 21st international congress on irrigation and drainage, Tehran, Iran

Mondol AH, Ara I, Das SC (2017) Meteorological drought index mapping in Bangladesh using standardized precipitation index during 1981–2010. Adv Meteorol 2017. Article ID 4642060 https://doi.org/10.1155/2017/4642060

Montaldo N, Albertson JD, Mancini M, Kiely G (2001) Robust simulation of root zone soil moisture with assimilation of surface soil moisture data. Water Resour Res 37:2889–2900

Montzka C, Pauwels VRN, Franssen H-JH, Han X, Vereecken H (2012) Multivariate and multiscale data assimilation in terrestrial systems: a review. Sensors 12:16291–16333

Moradkhani H (2008) Hydrologic remote sensing and land surface data assimilation. Sensors 8:2986–3004

Moradkhani H, DeChant CM, Sorooshian S (2012) Evolution of ensemble data assimilation for uncertainty quantification using the particle filter-Markov chain Monte Carlo method. Water Resour Res 48:W12520. https://doi.org/10.1029/2012WR012144

Moradkhani H, Hsu KL, Gupta H, Sorooshian S (2005) Uncertainty assessment of hydrologic model states and parameters: sequential data assimilation using the particle filter. Water Resour Res 41:W05012. https://doi.org/10.1029/2004WR003604

Moradkhani H, Hsu K, Hong Y, Sorooshian S (2006) Investigating the impact of remotely sensed precipitation and hydrologic model uncertainties on the ensemble streamflow forecasting. Geophys Res Lett 33:L12107

Moradkhani H, Sorooshian S (2009) General review of rainfall-runoff modeling: model calibration, data assimilation, and uncertainty analysis. In: Sorooshian S, Hsu KL, Coppola E, Tomassetti B, Verdecchia M, Visconti G (eds) Hydrological modelling and the water cycle. Water science and technology library, vol 63. Springer, Berlin

Morris BL, Lawrence AR, Chilton PJ, Adams B, Calow R, Klinck BA (2003) Groundwater and its susceptibility to degradation: A global assessment of the problems and options for management. Early Warning and Assessment Report Series, RS, pp 03–3

Mostovoy GV, King RL, Reddy KR, Kakani VG, Filippova MG (2006) Statistical estimation of daily maximum and minimum air temperatures from MODIS LST data over the state of Mississippi. Gisci Remote Sens 43(1):78–110

Motagh M, Walter TR, Sharifi MA, Fielding E, Schenk A, Anderssohn J et al (2008) Land subsidence in Iran caused by widespread water reservoir overexploitation. Geophys Res Lett 35:L16403. https://doi.org/10.1029/2008GL033814

Mu Q, Heinsch FA, Zhao M, Running SW (2007) Development of a global evapotranspiration algorithm based on MODIS and global meteorology data. Remote Sens Environ 111:519–536. https://doi.org/10.1016/j.rse.2007.04.015

Müller Schmied H, Eisner S, Franz D, Wattenbach M, Portmann F, Flörke M, Döll P (2014) Sensitivity of simulated global-scale freshwater fluxes and storages to input data, hydrological model structure, human water use and calibration. Hydrol Earth Syst Sci 18:3511–3538. https://doi.org/10.5194/hess-18-3511-2014

Munier S, Aires F, Schlaffer S, Prigent C, Papa F, Maisongrande P et al (2014) Combining datasets of satellite retrieved products for basin-scale water balance study. Part II: Evaluation on the Mississippi Basin and closure correction model. J Geophys Res Atmos Am Geophys Union 119:100–116. https://doi.org/10.1002/2014JD021953

Munoz SE, Dee SG (2017) El Niño increases the risk of lower Mississippi River flooding. Sci Rep 7. Article number 1772. https://doi.org/10.1038/s41598-017-01919-6

Munoz O, Leyton I, Nuñez N et al (2002) Vegetables collected in the cultivated Andean area of northern Chile: total and inorganic arsenic contents in raw vegetables. J Agric Food Chemi 50(3):642–7

Murray RS, Nagler PL, Morino K, Glenn EP (2009) An empirical algorithm for estimating agricultural and riparian evapotranspiration using MODIS enhanced vegetation index and ground measurements of ET. II. application to the lower Colorado river, U.S. Remote Sens 1(4):1125–1138

Mu Q, Zhao M, Running SW (2011) Improvements to a MODIS Global terrestrial evapotranspiration algorithm. Remote Sens Environ 115:1781–1800

Nagy GJ, Filho WL, Azeiteiro UM, Heimfarth J, Verocai JE (2018) Li C (2018) An assessment of the relationships between extreme weather events, vulnerability, and the impacts on human wellbeing in Latin America. Int J Environ Res Public Health 15:1802

Naschen K, Diekkrüger B, Leemhuis C, Steinbach S, Seregina LS, Thonfeld F (2018) Van der Linden R (2018) Hydrological modeling in data-scarce catchments: the kilombero floodplain in Tanzania. Water 10:599

Neal J, Schumann G, Bates P, Buytaert W, Matgen P, Pappenberger F (2009) A data assimilation approach to discharge esti- mation from space. Hydrol Process 23:3641–3649. https://doi.org/10.1002/hyp.7518

Nelson PR (1983) A comparison of sample sizes for the analysis of means and the analysis of variance. J Qual Technol 15:33–39
Nerger L (2004) Parallel filter algorithms for data assimilation in oceanography, PhD Thesis, University of Bremen
Nicolai-Shaw N, Hirschi M, Mittelbach H, Seneviratne SI (2015) Spatial representativeness of soil moisture using in situ, remote sensing, and land reanalysis data. J Geophys Res Atmos 120:9955–9964. https://doi.org/10.1002/2015JD023305
Niu G-Y, Yang Z-L, Mitchell KE, Chen F, Ek MB, Barlage M, Xia Y (2011) The community Noah land surface model withmultiparameterization options (Noah-MP): 1. Model description and evaluation with local-scale measurements. J Geophys Res 116:D12109. https://doi.org/10.1029/2010JD015139
Njoku EG (2004) AMSR-E/Aqua daily L3 surface soil moisture, interpretive parameters, QC EASE-Grids. Version 2. [indicate subset used]. Boulder, Colorado USA: NASA national snow and ice data center distributed active archive center. https://doi.org/10.5067/AMSR-E/AE_LAND3.002
Njoku EG et al (2003) Soil moisture retreival from AMSR-e. IEEE Trans Geo-Sci Remote Sens 41(2):215–229
Njoku E, Entekhabi D (1996) Passive microwave remote sensing of soil moisture. J Hydrol 184:101–129. https://doi.org/10.1016/0022-1694(95)02970-2
Okamoto K, Iguchi T, Takahashi N, Iwanami K, Ushio T (2005) The global satellite mapping of precipitation (GSMaP) project. In: 25th IGARSS Proceedings, pp 3414–3416
Oke PR, Brassington GB, Griffin DA, Schiller A (2008) The Bluelink Ocean Data Assimilation System (BODAS). Ocean Model 21:46–70. https://doi.org/10.1016/j.ocemod.2007.11.002
Oke PR, Sakov P, Corney SP (2007) Impacts of localisation in the EnKF and EnOI: experiments with a small model. Ocean Dyn 57:32–45
Orlowsky B, Seneviratne SI (2014) On the spatial representativeness of temporal dynamics at European weather stations. Int. J. Climatol. 34(10):3154–3160. https://doi.org/10.1002/joc.3903
Ott E, Hunt BR, Szunyogh I, Zimin AV, Kostelich EJ, Corazza M, Kalnay E, Patil DJ, Yorke JA (2004) A local ensemble Kalman Filter for atmospheric data assimilation. Tellus 56A:415–428
Packard NH, Crutchfield JP, Farmer JD, Shaw RS (1980) Geometry from a time series. Phys Rev Lett 45:712
Palmer TN (2001) A nonlinear dynamical perspective on model error: a proposal for non-local stochastic-dynamic parametrization in weather and climate prediction models. Q J R Meteorol Soc 127:279
Pan M, Sahoo AK, Troy TJ, Vinukollu RK, Sheffield J, Wood EF (2012) Multisource estimation of long-term terrestrial water budget for major global River Basins. J Clim 25(9):3191–3206. https://doi.org/10.1175/JCLI-D-11-00300.1
Pan M, Wood EF (2006) Data assimilation for estimating the terrestrial water budget using a constrained ensemble Kalman filter. J Hydrometeoro 7(3):534–547. https://doi.org/10.1175/JHM495.1
Papa F, Durand F, Rossow WB, Rahman A, Bala SK (2010a) Seasonal and Interannual Variations of the Ganges-Brahmaputra River Discharge, 1993–2008 from satellite altimeters. J Geophys Res 115:C12013. https://doi.org/10.1029/2009JC006075
Papa F, Prigent C, Aires F, Jimenez C, Rossow WB, Matthews E (2010b) Interannual variability of surface water extent at global scale, 1993–2004. J Geophys Res 115:D12111. https://doi.org/10.1029/2009JD012674
Papa F, Frappart F, Malbeteau Y, Shamsudduha M, Vuruputur V et al (2015) Satellite-derived surface and sub-surface water storage in the Ganges-Brahmaputra River Basin. J Hydrol: Reg Stud, Part A 4:15–35. https://doi.org/10.1016/j.ejrh.2015.03.004. ISSN 2214-5818
Papa F, Prigent C, Durand F, Rossow WB (2006) Wetland dynamics using a suite of satellite observations: A case study of application and evaluation for the Indian Subcontinent. Geophys Res Lett 33: https://doi.org/10.1029/2006GL025767

Pauwels VRN, Hoeben R, Verhoest NEC, De Troch FP (2002) Troch PA (2002) Improvement of TOPLATS-based discharge predictions through assimilation of ERS-based remotely sensed soil moisture values. Hydrol Process 16:995–1013

Peischl S, Walker JP, Rüdiger C, Ye N, Kerr YH, Kim E, Bandara R, Allahmoradi M (2012) The AACES field experiments: SMOS calibration and validation across the Murrumbidgee River catchment. Hydrol Earth Syst Sci 16(1697–1708):2012. https://doi.org/10.5194/hess-16-1697-2012

Pereira MPS, Justino F, Malhado ACM, Barbosa H (2014) Marengo J (2014) The influence of oceanic basins on drought and ecosystem dynamics in Northeast Brazil. Environ Res Lett 9:124013

Pereira A, Miranda S, Pacino MC, Forsberg R (2011) Water storage changes from GRACE Data in the La Plata Basin. Geod Planet Earth, Int Assoc Geod Symp 136:613–618. https://doi.org/10.1007/978-3-642-20338-1_75

Perez-Carrera A, Cirelli AF (2009) Arsenic and water quality challenges in South America. Water Sustain Arid Reg 275–293: https://doi.org/10.1007/978-90-481-2776-4_17

Pérez-Sánchez J, Senent-Aparicio J, Segura-Méndez F, Pulido-Velazquez D (2019) Srinivasan R (2019) Evaluating hydrological models for deriving water resources in Peninsular Spain. Sustainability 11:2872

Pham DT (2001) Stochastic methods for sequential data assimilation in strongly nonlinear systems. Mon Weather Rev 129:1194–1207

Pham DT, Verron J, Roubaud MC (1998) A singular evolutive extended Kalman Filter for data assimilation in oceanography. J Marine Syst 16:323–340

Pietrafesa LJ, Shaowu B, Huang NE, Gayes PT, Yan T et al (2016) Great lakes water levels: decomposing time series for attribution. J Clim Weather Forecast 4:153. https://doi.org/10.4172/2332-2594.1000153

Pipunic C, Walker P, Western A (2008) Assimilation of remotely sensed data for improved latent and sensible heat flux prediction: a comparative synthetic study. Remote Sens Environ 112(4):1295–1305. https://doi.org/10.1016/j.rse.2007.02.038

Plaza DA, Keyser R, Lannoy GJM, Giustarini L, Matgen P, Pauwels VRN (2012) The importance of parameter resampling for soil moisture data assimilation into hydrologic models using the particle filter. Hydrol Earth Syst Sci 16(2):375–390

Prigent C, Papa F, Aires F, Jimenez C, Rossow WB, Matthews E (2012) Changes in land surface water dynamics since the 1990s and relation to population pressure. Geophys Res Lett 39:L08403. https://doi.org/10.1029/2012GL051276

Qureshi AS, Ahmad ZU, Krupnik TJ (2015) Moving from resource development to resource management: problems prospects and policy recommendations for sustainable groundwater management in Bangladesh. Water Resour Manag 29(12):4269–4283. https://doi.org/10.1007/s11269-015-1059-y

Rafiuddin M, Uyeda H, Islam MN (2010) Characteristics of monsoon precipitation systems in and around Bangladesh. Int J Climatol 30:1042–1055. https://doi.org/10.1002/joc.1949

Rafn EB, Contor B, Ames DP (2008) Evaluation of a method for estimating irrigated crop-evapotranspiration coefficients from remotely sensed data in Idaho. J Irrig Drain Eng 134(6):722–729

Rajib MA, Mortuza MR, Selmi S, Ankur Asif Khan AK et al (2011) Increase of heat index over Bangladesh: impact of climate change. World Acad Sci Eng Technol 5(10):340–343

Ramirez BH, van der Ploeg M, Teuling AJ, Ganzeveld L, Leemans R (2017) Tropical Montane Cloud Forests in the Orinoco river basin: the role of soil organic layers in water storage and release. Geoderma 298:14–26. https://doi.org/10.1016/j.geoderma.2017.03.007. ISSN 0016-7061

Rasmussen J, Madsen H, Jensen KH, Refsgaard JC (2015) Data assimilation in integrated hydrological modeling using ensemble Kalman filtering: evaluating the effect of ensemble size and localization on filter performance. Hydrol Earth Syst Sci 19:2999–3013. https://doi.org/10.5194/hess-19-2999-2015

Reager JT, Thomas AC, Sproles EA, Rodell M, Beaudoing HK, Li B, Famiglietti JS (2015) Assimilation of GRACE terrestrial water storage observations into a land surface model for the assessment of regional flood potential. Remote Sens 7(11):14663–14679. https://doi.org/10.3390/rs71114663

Reichle RH, Crow WT, Keppenne CL (2008) An adaptive ensemble Kalman filter for soil moisture data assimilation. Water Resour Res 44:W03423. https://doi.org/10.1029/2007WR006357

Reichle RH, De Lannoy GJ, Forman BA, Draper CS, Liu Q (2013) Connecting satellite observations with water cycle variables through land data assimilation: Examples using the NASA GEOS-5 LDAS. J Hydrometeorol 35(3):577–606. https://doi.org/10.1007/978-94-017-8789-5_6

Reichle RH, Koster RD (2004) Bias reduction in short records of satellite soil moisture. Geophys Res Lett 31:L19501. https://doi.org/10.1029/2004GL020938

Reichle RH, Koster RD (2005) Global assimilation of satellite surface soil moisture retrievals into the NASA catchment land surface model. Geophys Res Lett 32:L02404

Reichle RH, Kumar SV, Mahanama SPP, Koster RD, Liu Q (2010) Assimilation of satellite-derived skin temperature observations into land surface models. J Hydrometeorol 11:1103–1122. https://doi.org/10.1175/2010JHM1262.1

Reichle RH, McLaughlin DB (2001) Variational data assimilation of microwave radiobrightness observations for land surface hydrologic applications. IEEE Trans Geosci Remote Sens 39(8):1708–1718

Reichle RH, McLaughlin DB, Entekhabi D (2002) Hydrologic Data Assimilation with the Ensemble Kalman Filter. Mon Weather Rev 130:103–114. https://doi.org/10.1175/1520-0493(2002)130<0103:HDAWTE>2.0.CO;2

Renard B, Kavetski D, Kuczera G, Thyer M, Franks SW (2010) Understanding predictive uncertainty in hydrologic modeling: the challenge of identifying input and structural errors. Water Resour Res 46:W05521. https://doi.org/10.1029/2009WR008328

Renzullo LJ, Van Dijk AIJM, Perraud JM, Collins D, Henderson B, Jin H, Smith AB, McJannet DL (2014) Continental satellite soil moisture data assimilation improves root-zone moisture analysis for water resources assessment. J Hydrol 519:2747–2762. https://doi.org/10.1016/j.jhydrol.2014.08.008

Reyes-González A, Kjaersgaard J, Trooien T, Hay C, Ahiablame L (2018) Estimation of crop evapotranspiration using satellite remote sensing-based vegetation index. Adv Meteorol. 12. Article ID 4525021. https://doi.org/10.1155/2018/4525021

Rieser D, Kuhn M, Pail R, Anjasmara IM, Awange J (2010) Relation between GRACE-derived surface mass variations and precipitation over Australia. Aust J Earth Sci 57(7):

Rilling G, Flandrin P, Goncalves P (2003) On empirical mode decomposition and its algorithms. In: Proceedings IEEE-EURASIP workshop on nonlinear signal and image processing NSIP-03, Grado I, p 3

Roads J et al (2003) GCIP water and energy budget synthesis (WEBS). J Geophys Res 108:8609. https://doi.org/10.1029/2002JD002583

Robert C, Blayo E, Verron J (2006) Comparison of reduced-order, sequential and variational data assimilation methods in the tropical Pacific Ocean. Ocean Dyn 56:624. https://doi.org/10.1007/s10236-006-0079-9

Rodell M (2002) Use of MODIS-derived snow fields in the Global Land Data Assimilation System. In: Proceedings GAPP mississippi river climate and hydrology conference. New Orleans

Rodell M, Chen J, Kato H, Famiglietti JS, Nigro J, Wilson CR (2007) Estimating groundwater storage changes in the Mississippi River basin (USA) using GRACE. Hydrogeol J 15:159–166

Rodell M, Houser P, Jambor U, Gottschalck J, Mitchell K, Meng C, Arsenault K, Cosgrove B, Radakovich J, Bosilovich M, Entin J, Walker J, Lohmann D, Toll D (2004) The global land data assimilation system. Bull Am Meteorol Soc 85:381–394. https://doi.org/10.1175/BAMS-85-3-381

Rodgers CD (2000) Inverse methods for atmopsheric sounding: theory and practice. World Scientific, Singapore

Roy DP, Kovalskyy V, Zhang HK et al (2016) Characterization of Landsat-7 to Landsat-8 reflective wavelength and normalized difference vegetation index continuity. Remote Sens Environ 185:57–70

Sahoo AK, Pan M, Troy TJ, Vinukollu RK, Sheffield J, Wood EF (2011) Reconciling the global terrestrial water budget using satellite remote sensing. Remote Sens Environ 115(8):1850–1865. https://doi.org/10.1016/j.rse.2011.03.009

Sakov P, Oke PR (2008) A deterministic formulation of the ensemble Kalman filter: an alternative to ensemble square root filters. Tellus 60A:361–371

Samaniego L, Kumar R, Jackisch C (2011) Predictions in a data-sparse region using a regionalized grid-based hydrologic model driven by remotely sensed data. Hydrol Res 42(5):338–355. https://doi.org/10.2166/nh.2011.156

Sanso B, Guenni L (1999) Venezuelan rainfall data analysed by using a Bayesian space-time model. Appl Stat 15:594–612

Santos da Silva J et al (2010) Water levels in the Amazon basin derived from the ERS 2 and ENVISAT radar altimetry missions. Remote Sens Environ 114(10):2160–2181

Sarraf M, Owaygen M, Ruta G, Croitoru L (2005) Islamic Republic of Iran: Cost assessment of environmental degradation. Technical report 32043-IR. Washington, D.C, World Bank

Sato M (2001) Online model selection based on the variational Bayes. Neural Comput 13:1649–1681

Sauer T (2004) Reconstruction of shared nonlinear dynamics in a network. Phys Rev Lett 93:198701

Sauer T, Yorke J, Casdagli M (1991) Embedology. J Stat Phys 65:579

Schellekens J, Dutra E, Martínez-de la Torre A, Balsamo G, van Dijk A, Sperna Weiland F, Minvielle M, Calvet J-C, Decharme B, Eisner S, Fink G, Flörke M, Peßenteiner S, van Beek R, Polcher J, Beck H, Orth R, Calton B, Burke S, Dorigo W, Weedon GP (2017) A global water resources ensemble of hydrological models: the eartH2Observe Tier-1 dataset. Earth Syst Sci Data 9:389–413. https://doi.org/10.5194/essd-9-389-2017

Schmidt AH, Montgomery DR, Huntington KW, Liang C (2011) The question of communist land degradation: new evidence from local erosion and basin-wide sediment yield in Southwest China and Southeast Tibet. Ann Assoc Am Geograph. First published on: 28 March 2011 (iFirst)

Schmidt R, Petrovic S, Güntner A, Barthelmes F, Wünsch J, Kusche J (2008) Periodic components of water storage changes from GRACE and global hydrology models. J Geophys Res 113:B08419. https://doi.org/10.1029/2007JB005363

Schneider U, Fuchs T, Meyer-Christoffer A, Rudolf B (2008) In: GPC Centre (ed), Internet publication

Schnur MT, Xie H, Wang X (2010) Estimating root zone soil moisture at distant sites using MODIS NDVI and EVI in a semi-arid region of southwestern USA. Ecol Inf 5:400–409. https://doi.org/10.1016/j.ecoinf.2010.05.001

Schumacher M, Forootan E, van Dijk AIJM, Müller Schmied H, Crosbie RS, Kusche J, Döll P (2018) Improving drought simulations within the Murray-Darling Basin by combined calibration/assimilation of GRACE data into the WaterGAP Global Hydrology Model. Remote Sens Environ 204:212–228. https://doi.org/10.1016/j.rse.2017.10.029

Schumacher M, Kusche J, Döll P (2016) A systematic impact assessment of GRACE error correlation on data assimilation in hydrological models. J Geod 90(6):537–559. https://doi.org/10.1007/s00190-016-0892-y

Schunk RW, Scherliess L, Sojka JJ, Thompson DC (2004) USU global ionospheric data assimilation models, atmospheric and environmental remote sensing data processing and utilization: an end-to-end system perspective. Proc SPIE 5548:327–336. https://doi.org/10.1117/12.562448

Schuurmans JM, Troch PA, Veldhuizen A, Bastiaanssen WGM, Bierkensd MFP (2003) Assimilation of remotely sensed latent heat flux in a distributed hydrological model. Adv Water Resour 26(2):151–159. https://doi.org/10.1016/S0309-1708(02)00089-1

Sellers PJ, Mintz Y, Sud Y, Dalcher A (1986) A Simple Biosphere Model (SiB) for use within general circulation models. J Atmos Sci 43:505–531

Sengupta S, Kang A, Jacob N (2013) Water wealth a briefing paper on the state of groundwater management in Bangladesh. India Environ Portal. http://www.cseindia.org/userfiles/groundwater_management_bangladesh.pdf

Seo DJ, Koren V, Cajina N (2003) Real-time variational assimilation of hydrologic and hydrometeorological data into operational hydrologic forecasting. J Hydrometeor 4:627–641. https://doi.org/10.1175/1525-7541(2003)004<0627:RVAOHA>2.0.CO;2

Seoane L, Ramillien G, Frappart F, Leblanc M (2013) Regional GRACE-based estimates of water mass variations over Australia: validation and interpretation. Hydrol Earth Syst Sci 17:4925–4939. https://doi.org/10.5194/hess-17-4925-2013

Seyler F, Calmant S, Santos da Silva J, Filizola N, Roux E, Cochonneau G, Vauchel P, Bonnet M-P (2008) Monitoring water level in large trans-boundary ungauged basins with altimetry: the example of ENVISAT over the Amazon basin. J Appl Remote Sens 7150:715017. https://doi.org/10.1117/12.813258

Shahid S (2010) Rainfall variability and the trends of wet and dry periods in Bangladesh. Int J Climatol 30:2299–2313. https://doi.org/10.1002/joc.2053

Shamsudduha M, Chandler RE, Taylor RG, Ahmed KM (2009) Recent trends in groundwater levels in a highly seasonal hydrological system: the Ganges-Brahmaputra-Meghna Delta. Hydrol Earth Syst Sci 13:2373–2385

Shamsudduha M, Taylor RG, Ahmed KM, Zahid A (2011) The impact of intensive groundwater abstraction on recharge to a shallow regional aquifer system: evidence from Bangladesh. Hydrogeol J 19:901. https://doi.org/10.1007/s10040-011-0723-4

Shamsudduha M, Taylor RG, Jones D, Longuevergne L, Owor M, Tindimugaya C (2017) Recent changes in terrestrial water storage in the Upper Nile Basin: an evaluation of commonly used gridded GRACE products. Hydrol Earth Syst Sci 21:4533–4549. https://doi.org/10.5194/hess-21-4533-2017

Shamsudduha M, Taylor RG, Longuevergne L (2012) Monitoring groundwater storage changes in the highly seasonal humid tropics: Validation of GRACE measurements in the Bengal Basin. Water Resour Res 48:W02508. https://doi.org/10.1029/2011WR010993

Shannon CE, Weaver W (1949) The mathematical theory of communication. University of Illinois Press, Urbana

Sheffield J, Ferguson CR, Troy TJ, Wood EF, McCabe MF (2009) Closing the terrestrial water budget from satellite remote sensing. Geophys Res Lett 26:L07403. https://doi.org/10.1029/2009GL037338

Sheffield J, Goteti G, Wood EF (2006) Development of a 50-yearhigh-resolution global dataset of meteorological forcings for land surfacemodeling. J Clim 19(13):3088–3111

Shokri A, Walker JP, Dijk AIJM, Pauwels VRN (2019) On the use of adaptive ensemble Kalman filtering to mitigate error misspecifications in GRACE data assimilation. Water Resour Res 55: https://doi.org/10.1029/2018WR024670

Simmons AJ, Uppala S, Dee D, Kobayashi S (2007) ERA-interim: New ECMWF reanalysis products from 1989 onwards, ECMWF Newsletter No. 110 — Winter 2006/07

Simon D (2006) Optimal state estimation: Kalman, H_{∞}, and nonlinear approaches. Wiley, New York

Simon D, Chia TL (2002) Kalman filtering with state equality constraints. IEEE Trans Aerosp Electron Syst 38:128–136

Smidl V, Quinn A (2006) The variational Bayes method in signal processing. Springer, Berlin

Smidl V, Quinn A (2008) Variational Bayesian filtering. IEEE Trans Signal Process 56:5020–5030

Smith AB, Walker JP, Western AW, Young RI, Ellett KM, Pipunic RC, Richter H (2012) The Murrumbidgee soil moisture monitoring network data set. Water Resour Res 48(7):1–6. https://doi.org/10.1029/2012WR011976

Smith PJ, Beven KJ, Tawn JA (2008) Detection of structural inadequacy in process-based hydrological models: A particle-filtering approach. Water Resour Res 44:W01410. https://doi.org/10.1029/2006WR005205

Smith PJ, Dance SL, Nichols NK (2011) A hybrid data assimilation scheme for model parameter estimation: application to morphodynamic modelling. Comput Fluids 46(1):436–441. https://doi.org/10.1016/j.compfluid.2011.01.010. ISSN 0045-7930

Sokolov AA, Chapman TG (1974) Methods for water balance computation an international guide for research and practice. The Unesco Press, Paris

Sondergaard M, Jeppesen E (2007) Anthropogenic impacts on lake and stream ecosystems, and approaches to restoration. J Appl Ecol 44:1089–1094. https://doi.org/10.1111/j.1365-2664.2007.01426.x

Song Q, He Y (2009) Adaptive unscented Kalman filter for estimation of modelling errors for helicopter. In: IEEE international conference on robotics and biomimetics (ROBIO), Guilin, pp 2463–2467. https://doi.org/10.1109/ROBIO.2009.5420406

Sood A, Smakhtin V (2015) Global hydrological models: a review. Hydrol Sci J 60(4):549–565. https://doi.org/10.1080/02626667.2014.950580

Steiger JH, Browne MW (1984) The comparison of interdependent correlations between optimal linear composites. Psychometrika 49:11–21

Stewart LM, Dance SL, Nichols NK (2008) Correlated observation errors in data assimilation. Int J Numer Methods Fluids 56:1521–1527. https://doi.org/10.1002/fld.1636

Stone JV (2004) Independent component analysis: a tutorial introduction. MIT Press, London

Strassberg G, Scanlon BR, Rodell M (2007) Comparison of seasonal terrestrial water storage variations from GRACE with groundwater-level measurements from the High Plains Aquifer (USA). Geophys Res Lett 34:L14402. https://doi.org/10.1029/2007GL030139

Su C-H, Ryu D, Young RI, Western AW, Wagner W (2013) Inter-comparison of microwave satellite soil moisture retrievals over the Murrumbidgee Basin, southeast Australia. Remote Sens Environ 134:1–11

Subramanian AC, Hoteit I, Cornuelle B, Miller AJ, Song H (2012) Linear versus nonlinear filtering with scale-selective corrections for balanced dynamics in a simple atmospheric model. Atmos Sci 69(11):3405–3419. https://doi.org/10.1175/JAS-D-11-0332.1

Sumon FR, Abul Kalam AKM (2014) Rainwater harvesting and the scope of enhancing ground water table in Dhaka City. Dhaka metropolitan development area and its planning problems, issues and policies. Bangladesh Institute of Planners (BIP) Available through

Sun YA, Morris A, Mohanty S (2009) Comparison of deterministic ensemble Kalman filters for assimilating hydrogeological data. Adv Water Resour 32(2):280–292. https://doi.org/10.1016/j.advwatres.2008.11.006

Sun T, Ferreira V, He X, Andam-Akorful S (2016) Water availability of São francisco River Basin based on a space-borne geodetic sensor. Water 8:213. https://doi.org/10.3390/w8050213

Sun Q, Xie Z, Tian X (2015) GRACE terrestrial water storage data assimilation based on the ensemble four-dimensional variational method PODEn4DVar: method and validation. Sci China Earth Sci 58:371. https://doi.org/10.1007/s11430-014-4978-1

Swenson S, Chambers D, Wahr J (2008) Estimating geocentervariations from a combination of GRACE and ocean model output. J Geophys Res 113:B08410. https://doi.org/10.1029/2007JB005338

Swenson S, Wahr J (2002) Methods for inferring regional surface-mass anomalies from Gravity Recovery and Climate Experiment (GRACE) measurements of time-variable gravity. J Geophys Res 107(B9):2193. https://doi.org/10.1029/2001JB000576

Swenson S, Wahr J (2006) Post-processing removal of correlated errors in GRACE data. Geophys Res Lett 33:L08402. https://doi.org/10.1029/2005GL025285

Syed TH, Famiglietti JS, Chen J, Rodell M, Seneviratne SI, Viterbo P, Wilson CR (2005) Total basin discharge for the Amazon and Mississippi River basins from GRACE and a land-atmosphere water balance. Geophys Res Lett 32:L24404. https://doi.org/10.1029/2005GL024851

Takens F (1981) Dynamical systems and Turbulence, Warwick 1980. Lect Notes Math 898:366

Talagrand O, Courtier P (1987) Variational assimilation of meteorological observations with the adjoint vorticity equation. I: theory. Q J R Meteorol Soc 113:1311–1328. https://doi.org/10.1002/qj.49711347812

Tandeo P, Coauthors, (2015) Combining analog method and ensemble data assimilation: application to the lorenz-63 chaotic system. Machine learning and data mining approaches to climate science. Springer, Berlin, pp 3–12

Tangdamrongsub N, Han S-C, Tian S, Müller Schmied H, Sutanudjaja EH, Ran J, Feng W (2018) Evaluation of Groundwater Storage Variations estimated from GRACE data assimilation and state-of-the-art land surface models in Australia and the North China plain. Remote Sens 10:483

Tangdamrongsub N, Steele-Dunne SC, Gunter BC, Ditmar PG, Sutanudjaja EH, Xie T, Wang Z (2017) Improving estimates of water resources in a semi-arid region by assimilating GRACE data into the PCR-GLOBWB hydrological model. Hydrol Earth Syst Sci 21:2053–2074. https://doi.org/10.5194/hess-21-2053-2017

Tangdamrongsub N, Steele-Dunne SC, Gunter BC, Ditmar PG, Weerts AH (2015) Data assimilation of GRACE terrestrial water storage estimates into a regional hydrological model of the Rhine River basin. Hydrol Earth Syst Sci 19:2079–2100. https://doi.org/10.5194/hess-19-2079-2015

Tapley BD, Bettadpur S, Watkins M, Reigber C (2004) The gravity recovery and climate experiment: mission overview and early results. Geophys Res Lett 31:L09607. https://doi.org/10.1029/2004GL019920

Tardif R, Hakim GJ, Snyder C (2015) Coupled atmosphere-ocean data assimilation experiments with a low-order model and CMIP5 model data. Clim Dyn 45:1415. https://doi.org/10.1007/s00382-014-2390-3

Tedeschi RG, Cavalcanti IF, Grimm AM (2013) Influences of two types of ENSO on South American precipitation. Int J Climatol 33:1382–1400. https://doi.org/10.1002/joc.3519

Tedeschi RG, Collins M (2016) The influence of ENSO on South American precipitation during austral summer and autumn in observations and models. Int J Climatol 36:618–635

Terejanu GA (2009) Unscented Kalman filter tutorial, workshop on large-scale quantification of uncertainty. Sandia National Laboratories, pp 1–6

Teweldebrhan AT, Burkhart JF, Schuler TV (2018) Parameter uncertainty analysis for an operational hydrological model using residual-based and limits of acceptability approaches. Hydrol Earth Syst Sci 22:5021–5039. https://doi.org/10.5194/hess-22-5021-2018

Thomas AC, Reager JT, Famiglietti JS, Rodell M (2014) A GRACE-based water storage deficit approach for hydrological drought characterization. Geophys Res Lett 41:1537–1545. https://doi.org/10.1002/2014GL059323

Tian S, Tregoning P, Renzullo LJ, van Dijk AIJM, Walker JP, Pauwels VRN, Allgeyer S (2017) Improved water balance component estimates through joint assimilation of GRACE water storage and SMOS soil moisture retrievals. Water Resour Res 53(3):1820–1840. https://doi.org/10.1002/2016WR019641

Tian X, Xie Z, Dai A (2008) A land surface soil moisture data assimilation system based on the dual-UKF method and the community land model. J Geophys Res 113:D14127. https://doi.org/10.1029/2007JD009650

Tippett MK, Anderson JL, Bishop CH, Hamill TM, Whitaker JS (2003) Ensemble square root filters. Mon Weather Rev 131:1485–90

Tourian MJ, Elmi O, Chen Q, Devaraju B, Roohi Sh, Sneeuw N (2015) A spaceborne multisensor approach to monitor the desiccation of Lake Urmia in Iran. Remote Sens Environ 156:349–4257. https://doi.org/10.1016/j.rse.2014.10.006. ISSN 0034-4257

Tourre YM, Jarlan L, Lacaux JP, Rotela CH (2008) Lafaye M (2008) Spatio-temporal variability of NDVI-precipitation over southernmost South America: possible linkages between climate signals and epidemics. Environ Res Lett 3:044008

Tregoning P, McClusky S, van Dijk AIJM, Crosbie RS, Peña-Arancibia JL (2012) Assessment of GRACE satellites for groundwater estimation in australia. National Water Commission, Canberra, p 82

Treidel H, Martin-Bordes JL, Gurdak JJ (2011) Climate change effects on groundwater resources: a global synthesis of findings and recommendations. CRC Press. ISBN 0203120760:9780203120767

Trigo RM, Gouveia CM, Barriopedro D (2010) The intense 2007–2009 drought in the Fertile Crescent: Impact and associated atmospheric circulation. Agric For Meteorol 150:1245–1257

Tropical Rainfall Measuring Mission (TRMM) (2011) TRMM (TMPA/3B43) Rainfall Estimate L3 1 month 0.25 degree x 0.25 degree V7, Greenbelt, MD, Goddard Earth Sciences Data and

Information Services Center (GES DISC), Accessed [Data Access Date]. https://disc.gsfc.nasa.gov/datacollection/TRMM_3B43_7.html

Tseng KH, Shum CK, Yi Y, Fok HS, Kuo CY, Lee H, Cheng X, Wang X (2013) Envisat altimetry radar waveform retracking of quasi-specular echoes over the ice-covered Qinghai Lake. Terr Atmos Ocean Sci 24:615–627. https://doi.org/10.3319/TAO.2012.12.03.01(TibXS)

Ullman NR (1989) The Analysis of Means (ANOM) for signal and noise. J Qual Technol 21:111–127

Ummenhofer CC, England MH, McIntosh PC, Meyers GA, Pook MJ, Risbey JS, Sen Gupta A, Taschetto AS (2009) What causes southeast Australia's worst droughts? Geophys Res Lett 36:L04706. https://doi.org/10.1029/2008GL036801

Van Camp M, Radfar M, Martens K, Walraevens K (2012) Analysis of the groundwater resource decline in an intramountain aquifer system in Central Iran. Geologica Belgica 15(3):176–180 https://popups.uliege.be:443/1374-8505/index.php?id=3644

Van der Merwe R (2004) Sigma-point Kalman filters for probability inference in dynamic state-space models, PhD Thesis, Oregon Health and Science University

van Dijk AIJM (2010) The Australian water resources assessment system: Technical Report 3, Landscape model (version 0.5) Technical Description. CSIRO: Water for a Healthy Country National Research Flagship

van Dijk AIJM, Peña-Arancibia JL, Wood EF, Sheffield J, Beck HE (2013) Global analysis of seasonal streamflow predictability using an ensemble prediction system and observations from 6192 small catchments worldwide. Water Resour Res 49:2729–2746. https://doi.org/10.1002/wrcr.20251

van Dijk AIJM, Renzullo LJ, Rodell M (2011) Use of Gravity Recovery and Climate Experiment terrestrial water storage retrievals to evaluate model estimates by the Australian water resources assessment system. Water Resources Research 47:W11524. https://doi.org/10.1029/2011WR010714

van Dijk AIJM, Renzullo LJ, Wada Y, Tregoning P (2014) A global water cycle reanalysis (2003–2012) merging satellite gravimetry and altimetry observations with a hydrological multi-model ensemble. Hydrol Earth Syst Sci 18:2955–2973. https://doi.org/10.5194/hess-18-2955-2014

van Leeuwen PJ, Evensen G (1996) Data assimilation and inverse methods in terms of a probabilistic formulation. Mon Weather Rev 124:2898–2913. https://doi.org/10.1175/1520-0493(1996)124<2898:DAAIMI>2.0.CO;2

Vetra-Carvalho S, van Leeuwen PJ, Nerger L, Barth A, Altaf MU, Brasseur P, Kirchgessner P, Beckers JM (2018) State-of-theart stochastic data assimilation methods for high-dimensional non-Gaussian problems. Tellus A: Dyn Meteorol Ocean 70(1):1445364. https://doi.org/10.1080/16000870.2018.1445364

Villar PC (2016) International cooperation on transboundary aquifers in South America and the Guarani Aquifer case. Revista Brasileira de Política Internacional. Rev Bras Polít Int 59(1) Brasília. https://doi.org/10.1590/0034-7329201600107. ISSN 1983-3121

Voss KA, Famiglietti JS, Lo M-H, de Linage C, Rodell M, Swenson SC (2013) Groundwater depletion in the Middle East from GRACE with implications for transboundary water management in the Tigris-Euphrates-Western Iran region. Water Resour Res 49(2):904–914. https://doi.org/10.1002/wrcr.20078

Vrugt JA, Diks CG, Gupta HV, Bouten W, Verstraten JM (2005) Improved treatment of uncertainty in hydrologic modeling: combining the strengths of global optimization and data assimilation. Water Resour Res 41:W01017. https://doi.org/10.1029/2004WR003059

Vrugt JA, Gupta HV, Nuallain BO (2006) Real-time data assimilation for operational ensemble streamflow forecasting. J Hydrometeorol 7(3):548–565. https://doi.org/10.1175/JHM504.1

Vrugt JA, ter Braak CJF, Diks CGH, Schoups G (2013) Advancing hydrologic data assimilation using particle Markov chain Monte Carlo simulation: theory, concepts and applications. Adv Water Resour 51:457–478. https://doi.org/10.1016/j.advwatres.2012.04.002

Wahr J, Molenaar M, Bryan F (1998) Time variability of the Earth's gravity field' Hydrological and oceanic effects and their possible detection using GRACE. J Geophys Res 103(B12):30, 205-30, 229. https://doi.org/10.1029/98JB02844

Wallis JR (1965) Multivariate statistical methods in hydrology-A comparison using data of known functional relationship. Water Resour Res 1(4):447–461. https://doi.org/10.1029/WR001i004p00447

Wang J, Xu K, Zhou K, Lin S, Hu S, Guo B (2006) Spherical harmonics scaling. Visual Comput 22:713. https://doi.org/10.1007/s00371-006-0057-8

Wan E, van der Merwe R (2000) The unscented Kalman filter for nonlinear estimation. In: Proceedings of the IEEE 2000 adaptive systems for signal processing, communications, and control symposium (Cat. No.00EX373). Lake Louise, Alta., pp 153–158. https://doi.org/10.1109/ASSPCC.2000.882463

Wan E, van der Merwe R (2001) The unscented Kalman filter. Wiley Publishing

Watkins MM, Wiese DN, Yuan D-N, Boening C, Landerer FW (2015) Improved methods for observing Earth's time variable mass distribution with GRACE using spherical cap mascons. J Geophys Res: Solid Earth 120(4):2648–2671. https://doi.org/10.1002/2014JB011547

Weerts AH, El Serafy GYH (2006) Particle filtering and ensemble Kalman filtering for state updating with hydrological conceptual rainfall-runoff models. Water Resour Res 42:W09403. https://doi.org/10.1029/2005WR004093

Western AW, Zhou S-L, Grayson RB, McMahon TA, Blöschl G, Wilson DJ (2004) Spatial correlation of soil moisture in small catchments and its relationship to dominant spatial hydrological processes. J Hydrol 286:113–134. https://doi.org/10.1016/j.jhydrol.2003.09.014

Whitaker JS, Hamill TM (2002) Ensemble data assimilation without perturbed observations. Mon Weather Rev 130:1913–1924. https://doi.org/10.1175/1520-0493(2002)130<1913:EDAWPO>2.0.CO;2

Wiese DN (2015) GRACE monthly global water mass grids NETCDF RELEASE 5.0. Ver. 5.0. PO.DAAC, CA, USA. Dataset accessed [YYYY-MM-DD] at https://doi.org/10.5067/TEMSC-OCL05

Wolf AT, Newton JT (2007) Case study transboundary dispute resolution: The Tigris-Euphrates Basin, Transboundary Freshwater Dispute Database (TFDD). Oregon State University. http://www.transboundarywaters.orst.edu/

Wooldridge SA, Kalma JD (2001) Regional-scale hydrological modelling using multiple-parameter landscape zones and a quasi-distributed water balance model. Hydrol Earth Syst Sci 5:59–74

Wouters B, Bonin JA, Chambers DP, Riva REM, Sasgen I, Wahr J (2014) GRACE, time-varying gravity, Earth system dynamics and climate change. Rep Prog Phys 77(11):116801. https://doi.org/10.1088/0034-4885/77/11/116801

Xavier L, Becker M, Cazenave A, Longuevergne L, Llovel W, Filho OCR (2010) Interannual variability in water storage over 2003–2008 in the Amazon Basin from GRACE space gravimetry, in situ river level and precipitation data. Remote Sens Environ 114:1629–1637. https://doi.org/10.1016/J.RSE.2010.02.005

Xu T, Bateni SM, Liang S, Entekhabi D, Mao K (2014) Estimation of surface turbulent heat fluxes via variational assimilation of sequences of land surface temperatures from geostationary operational environmental satellites. J Geophys Res Atmos 119(18):10780–10798. https://doi.org/10.1002/2014jd021814

Yao Y, Liang S, Xu T (2013) Remote sensing data products for land surface data assimilation system application.World Scientific Publishing Co. Pte. Ltd. https://doi.org/10.1142/9789814472616_0001

Yao C, Luo Z, Wang H, Li Q, Zhou H (2016) GRACE-derived terrestrial water storage changes in the inter-basin region and its possible influencing factors: a case study of the sichuan basin. China Remote Sens 8(6):444. https://doi.org/10.3390/rs8060444

Yates DN (1997) Climate change impacts on the hydrologic resources of South America: an annual, continental scale assessment.Clim Res 9(1–2):147–155

Yin J, Zhan C, Gu H et al (2014) A case study of evapotranspiration data assimilation based bon hydrological model. Adv Earth Sci 29(9):1075–1084

Young PC (2002) Advances in real-time flood forecasting. Philos Trans R Soc Lond 360:1433–1450

Yu Y, Disse M, Yu R, Yu G, Sun L, Huttner P, Rumbaur C (2015) Large-Scale hydrological modeling and decision-making for agricultural water consumption and allocation in the main stem tarim river, China. Water 7:2821–2839

Yue S, Wang C (2004) The Mann-Kendall test modified by effective sample size to detect trend in serially correlated hydrological series. Water Resour Manag 18:201. https://doi.org/10.1023/B:WARM.0000043140.61082.60

Zaitchik BF, Rodell M, Reichle RH (2008) Assimilation of GRACE terrestrial water storage data into a land surface model: results for the Mississippi River Basin. J Hydrometeorol 9(3):535–548. https://doi.org/10.1175/2007JHM951.1

Zeng Y (2014) Efficient radar forward operator for operational data assimilation within the COSMO-model, KIT Scientific Publishing, pp 187. ISBN 3731501287:9783731501282

Zhang S, Al-Asadi K (2019) Evaluating the effect of numerical schemes on hydrological simulations: HYMOD as a case study. Water 11:329

Zhang Y, Bocquet M, Mallet V, Seigneur C, Baklanov A (2012) Real-time air quality forecasting, Part I: history, techniques, and current status. Atmos Environ 60:632–655. https://doi.org/10.1016/j.atmosenv.2012.06.031

Zhang J, Campbell JR, Hyer EJ, Reid JS, Westphal DL, Johnson RS (2014) Evaluating the impact of multisensor data assimilation on a global aerosol particle transport model. J Geophys Res Atmos 119:4674–4689. https://doi.org/10.1002/2013JD020975

Zhang Y, Pan M, Wood EF (2016) On creating global gridded terrestrial water budget estimates from satellite remote sensing. Surv Geophys 37(2):249–268. https://doi.org/10.1007/s10712-015-9354-y

Zhao Y, Deng X, Zhang S, Liu Z, Liu C, Vecchi G, Han G, Wu X (2017) Impact of an observational time window on coupled data assimilation: simulation with a simple climate model. Nonlinear Process Geophys 24:681–694. https://doi.org/10.5194/npg-24-681-2017

Zobitz JM, Moore DJP, Quaife T, Braswell BH, Bergeson A, Anthony JA, Monson RK (2014) Joint data assimilation of satellite reflectance and net ecosystem exchange data constrains ecosystem carbon fluxes at a high-elevation subalpine forest. Agric For Meteorol 195–196:73–88. https://doi.org/10.1016/j.agrformet.2014.04.011. ISSN 0168-1923

Zou L, Zhan C, Xia J, Wang T, Gippel CJ (2017) Implementation of evapotranspiration data assimilation with catchment scale distributed hydrological model via an ensemble Kalman filter. J Hydrol 549:685–702. https://doi.org/10.1016/j.jhydrol.2017.04.036. ISSN 0022-1694

The manufacturer's authorised representative in the EU is Springer Nature Customer Service Centre GmbH, Europaplatz 3, 69115 Heidelberg, Germany. If you have any concerns regarding our products, please contact ProductSafety@springernature.com

Printed and bound by CPI Group (UK) Ltd, Croydon, CR0 4YY
30/09/2026
02225462-0011